CONTEMPORARY MATHEMATICS

357

Variational Methods: Open Problems, Recent Progress, and Numerical Algorithms

Variational Methods: Open Problems,
Recent Progress, and Numerical Algorithms
June 5–8, 2002
Northern Arizona University
Flagstaff, Arizona

John M. Neuberger
Editor

American Mathematical Society
Providence, Rhode Island

2000 *Mathematics Subject Classification.* Primary 35Axx, 35Jxx, 35J20, 35J60, 65K10, 65Nxx, 65N25, 65N30, 65N38.

Library of Congress Cataloging-in-Publication Data

Variational methods: open problems, recent progress, and numerical algorithms, June 5–8, 2002, Northern Arizona University, Flagstaff, Arizona / John Neuberger, editor.
p. cm.– (Contemporary mathematics, ISSN 0271-4132; 357)
Includes bibliographical references.
ISBN 0-8218-3339-1 (alk. paper)
1. Differential equations, Elliptic–Congresses. 2. Variational inequalities (Mathematics)–Congresses. I. Neuberger, John, 1962– II. Contemporary mathematics (American Mathematical Society); v. 357.

QA377.V37 2004
515′.3533–dc22 2004049919

∞ The paper used in this book is acid-free and falls within the guidelines
established to ensure permanence and durability.
Visit the AMS home page at `http://www.ams.org/`

10 9 8 7 6 5 4 3 2 1 09 08 07 06 05 04

Contents

Introduction to the Proceedings

1. Preface

This article introduces the proceedings of the conference *Variational Methods: Open Problems, Recent Progress, and Numerical Algorithms*, which was held at Northern Arizona University in Flagstaff, Arizona, June 5th through June 8th, 2002. Alfonso Castro, Goong Chen, and Wei-Ming Ni each gave three one-hour keynote presentations. There were 17 contributed 30-minute talks, a numerical workshop, and lots of time for discussion.

This volume contains 15 mathematical papers; 4 articles authored and/or co-authored by the 3 keynote speakers, and 11 articles authored and/or co-authored by 13 of the 17 contributed speakers. Most of the material concerns variational methods as applied to nonlinear elliptic PDE, although several articles concern nonlinear PDE that are nonvariational and/or nonelliptic. There was a focus on open problems; hence papers with conjectures were encouraged. Survey papers were also solicited, as were works featuring applications. This conference successfully brought together those doing cutting-edge nonlinear functional analysis with researchers more computationally oriented; hence many of the papers in this volume feature numerical investigations. It was our intent to survey what is known in the general subject area, list the important open questions, and suggest analytical and numerical techniques that might be beneficial to those seeking to make progress towards solving those open problems.

Many of the contributers have included an extensive list of relevant references, thoughtfully cited throughout the narrative. We hope that the reader will find this volume an excellent starting point for considering new problems and relevant variational techniques in nonlinear elliptic and related PDE.

This conference would not have been possible without support from NSF DMS Grants 0074326 and 0124121. The organizer would also like to thank Northern Arizona University's Department of Mathematics and Statistics, Dean of the College of Arts and Sciences, and Provost for their generous contributions.

2. Reviews of the Articles by the Editor

The first article by keynote speaker Alfonso Castro [**1**] is an overview of the solvability of semilinear elliptic equations where the linear part has discrete spectrum. This survey paper is intended to be accessible to the non expert and provides an extensive bibliography referencing much of the relevant literature. The paper provides a brief introduction to many of the definitions, theorems, and concepts in critical point theory required by those interested in researching in this area.

For example, resonance, growth, boundary conditions, compactness, critical exponents, the Landesman-Lazer condition, and various classes of nonlinearities are discussed. The influence of the spectrum on nonlinear problems is demonstrated through examples such as Hammerstein integrals and simple hyperbolic problems. Nodal structure theorems, applications, several open problems, and possible extensions are also included. One of the key motivating factors behind organizing the conference was the desire to make progress towards "solving" the often conjectured "infinitely many solutions to the superlinear elliptic Dirichlet problem". This paper gets right to the heart of that issue, clearly explaining the role of "crossing eigenvalues" and providing some insight into N. Dancer's famous counterexample of the so-called Lazer-Mckenna conjecture.

The next three articles [**4, 5, 6**] are co-authored by keynote speaker Goong Chen, Jianxin Zhou, and others, including keynote speaker Wei-Ming Ni. Combined, the works cite an extensive bibliography. The authors provide a unique blend of analytical and numerical results, proving existence, multiplicity, and nodal structure theorems as well as technical convergence, convergence rate, and error estimate theorems. Their development of numerical algorithms together with this hard analysis is a combination rarely found in the literature.

The paper [**4**] presents theoretical numerical results for semilinear elliptic boundary value problems (BVP) with *nonlinear oblique* boundary conditions. This research paper introduces the subject and provides new results, including convergence, convergence rate, and error estimate theorems for their numerical scheme. Some computational results are also provided.

The next article [**5**] is also a research article, providing existence and multiplicity results for the elliptic sine-Gordon equation. The physical relevance to Josephson junctions and π-junctions is first reviewed and explained, and then the relevant variational structure is revealed. The effect of the choice of domain is discussed, and then computational results for positive solutions on a dumbbell-shaped domain are provided for visualization purposes.

The final research paper [**6**] by these authors concerns convergence analysis for an optimal scaling algorithm for semilinear elliptic BVP. A nice introduction discusses the strengths and applicability of many related algorithms, including variants of the Mountain Pass Algorithm (MPA), Monotone Iteration Algorithm (MIA), Scaling Iteration Algorithm (SIA), High-Linking Algorithm (HLA), and so-called variational derivative iteration type algorithms such as Sobolev steepest descent and Newton's method (see also [**12**] and [**13**]). After introducing the SIA in some detail, the Optimal Scaling Iteration Algorithm (OSIA) is presented. A convergence result is proven. Numerical computations which reveal the advantages of the scheme are included.

The survey article [**13**] is another paper very much in the spirit of the conference. Via simple examples, it introduces the concept of and need for Sobolev gradient steepest descent methods. The methodology behind enforcing boundary conditions and using projections is explained, and connections between continuous and discrete methods are made. A motivating theme in the authors' presented and cited work is described in their quote (destined to become a classic): "Analytical difficulties and numerical difficulties always come in pairs." A tantalizing list of applications is presented, including Ginzburg-Landau, transonic flow, Monge-Ampere, and minimal surfaces. For each subtopic, problems are stated that range from "exercises" to "open ended" and perhaps "impossible". Much of the material follows one of the author's book on the subject, definitely recommended reading; some of the applications are new. The applications are far from trivial and are physically significant. It is pointed out that the researcher who makes progress towards solving some of the open problems has the potential to make a real impact on mathematics and science.

The research article [**7**] gives a new semilinear elliptic result on all of $\mathbb{R}^N$. A brief introduction into resonant Schrödinger equations on bounded domains is given and many relevant references are provided; it is noted that resonant problems on the whole space for such equations are almost absent in the literature. The authors' new result is perhaps the first to deal with unbounded nonlinearities in this already novel setting.

The article [**3**] provides a historical survey citing many relevant references, new existence and multiplicity results, and numerical computations for quasilinear BVP. Specifically, the authors are interested in the p-Laplacian and the relationship between so-called eigenvalues of associated quasilinear-linear problems and the existence of solutions to quasilinear-nonlinear problems, much in the same way that semilinear researchers are concerned with linear problems. The provided background into the subject is quite complete and will be extremely useful to both new and experienced researchers in the area. The new results extend what is known concerning parameter regions and matters of existence and multiplicity. Numerical simulations provide complex yet effective visualization and confirmation of their results.

Two of the participants explain in [**8**] the issues understood and conjectured behind the so-called eigenvalue problem mentioned in the summary of the preceeding article. Specifically, they are interested in the variational characterization of eigenvalues of the quasilinear p-Laplacian problem. Enough background and known results are given so that interested new and experienced researchers may begin considering four significant open problems. For example, they ask if their variational characterization describes all possible eigenvalues.

The survey article [**14**] introduces the variational duality whereby nonlinear elliptic PDE can model underlying (and fascinating) geometric phenomena. The general concept is explained in depth, and specific examples are given. In particular, pattern formation is exploited to answer questions concerning triple-junction and minimal length / surface area solutions. In keeping with the theme of the conference, the article is far-reaching and contemplative in nature; seven open problems are loosely stated. The solution of any of these interesting problems would extend the author's successful related results. This article ties in with several other papers in this volume; see for example [**5**]. The historical section contains many references and provides an excellent starting point for those interested in considering open problems related to pattern formation in nonlinear elliptic PDE.

The research paper [**15**] further extends what is known about quasilinear problems. In particular, an existence theorem for periodic solutions of the perturbed p-Laplacian is given. Galerkin techniques are used, suggesting to this editor that a corresponding numerical algorithm might be obtained. After this technical new result is proven, three interesting open problems are presented. The problems are stated in sufficient detail that they might be immediately considered by the interested reader.

The survey paper [**11**] introduces an important concept in real-world applications, namely the application of variational optimization to solve ill-posed problems. Researchers consider such problems notoriously difficult when numerically seeking approximations; many applications in science lead to ill-posed systems that necessitate such efforts. After some historical background and numerical results are given, a detailed explanation of an application from electrical impedance tomography is presented. Several conjectures and open questions are stated, ready for the reader's immediate and serious consideration.

The organizer and editor's own article [**12**] is a complete survey of his efforts to use and develop numerical algorithms for investigating many of the problems found in the other articles in this volume. Specifically, he presents results from past and current experiments where variants of the MPA and Newton's method were used to explore semilinear elliptic BVP. Much time is spent describing the development, implementation, and application of the so-called GNGA: Gradient Newton Galerkin Algorithm. The experiments require a basis in function space. Often times this is a basis of eigenfunctions, a collection whose generation is a difficult and interesting process in its own right. Current and recent experiments are described, where general domains and symmetry are considered in the quest to explain the existence, multiplicity, and nodal structure of solutions via continuation methods. Numerous well-defined and less well-defined open problems are stated; some are suitable for undergraduate or master's level research, while others represent much longer term efforts to solve problems whose answers have long been sought by experienced researchers. The interplay between existence theory and analysis with numerical investigations was a theme of the conference and is a key component of this paper.

The research article [**2**] discusses critical nonlinearities and symmetry for semilinear elliptic PDE, chiefly on all of $\mathbb{R}^N$. New existence and nodal structure theorems are provided via variational methods. Six open problems are precisely stated, suitable for immediate consideration by the interested reader.

The research article [**10**] considers a class of semilinear elliptic PDE on all of $\mathbb{R}^N$. Specifically, ODE techniques are used to prove the existence of radial solutions when the nonlinearity is nonconvergent. The result is related to a previous result, and an open problem is suggested.

The final article [**9**] in this volume presents a new result considering traveling solitary wave solutions to several equations, namely the Burgers-KdV, combined dissipative double-dispersive, and sine-Gordon equations. Under some conditions, explicit solutions are given in these three instances. A brief statement of a general direction for future research along these lines is given. In particular, it is asked if variational techniques might be used to duplicate or extend the results found in this research article.

The organizer and editor wishes to thank the conference participants for their uniformly excellent talks, and thanks the majority of these participants for contributing survey and research articles. He also thanks the participants and contributors for their patience and understanding as regards his first attempt at managing a project of this magnitude. Any typographical errors or accidental omissions are the sole responsibility of this editor.

John M. Neuberger, Conference Organizer and Proceedings Editor

Bibliography

[1] Castro, Alfonso; *Semilinear Equations with Discrete Spectrum.*

[2] Catrina, Florin; *Critical Nonlinearities and Symmetric Solutions.*

[3] Čepička, Jan; Drábek, Pavel & Girg, Petr; *Quasilinear Boundary Value Problems: Existence and Multiplicity Results.*

[4] Chen, Goong; Deng, Yuanhua; Ni, Wei-Ming; & Zhou, Jianxin; *Semilinear Elliptic Boundary Value Problems with Nonlinear Oblique Boundary Conditions, a Boundary Element Monotone Iteration Approach.*

[5] Chen, Goong; Ding, Zhonghai; Hu, Chia-Ren; Ni, Wei-Ming & Zhou, Jianxin; *A Note on the Elliptic Sine-Gordon Equation.*

[6] Chen, Goong; Englert, Berthold G. & Zhou, Jianxin; *Convergence Analysis of an Optimal Scaling Algorithm for Semilinear Elliptic Boundary Value Problems.*

[7] Costa, David G. & Tehrani, Hossein; *Unbounded Perturbations of Resonant Schrodinger Equations.*

[8] Drábek, Pavel & Robinson, Stephen P.; *Eigenvalue Problems, Resonance Problems, and Open Problems.*

[9] Feng, Zhaosheng; *Traveling Wave Solutions to Nonlinear Evolution Equations.*

[10] Iaia, Joseph A.; *Non-convergent Radial Solutions of a Semilinear Elliptic Equation in* $\mathbb{R}^N$.

[11] Knowles, Ian; *Variational Methods for Ill-Posed Problems.*

[12] Neuberger, John M.; *GNGA: Recent Progress and Open Problems for Semilinear Elliptic PDE.*

[13] Neuberger, J. W. & Renka, R. J.; *Sobolev Gradients: Introduction, Applications, Problems.*

[14] Padilla, Pablo; *Variational, Dynamic and Geometric Aspects of Some Nonlinear Problems.*

[15] Shapiro, Victor L.; *The Perturbed* p*-Laplacian and Quadratic Growth.*

Contemporary Mathematics
Volume **357**, 2004

Semilinear equations with discrete spectrum

Alfonso Castro

ABSTRACT. This is an overview of the solvability of semilinear equations where the linear part has discrete spectrum. Semilinear elliptic and hyperbolic equations, as well as Hammerstein integral equations, are used as motivating examples. The presentation is intended to be accessible to non experts.

1. INTRODUCTION

These lectures deal with the study of equations of the type

$$L(u) = N(u), \tag{1}$$

where L is linear operator with discrete spectrum $\sigma(L) = \{\lambda_i; i = 1, 2, \dots\}$. We recall that if L is a linear operator defined on a dense subset of a Hilbert space then $\sigma(L)$ is the complement in the complex plane of the resolvent $\rho(L) = \{\lambda; L - \lambda I$ has a bounded inverse $\}$.

A main model for equation (1) is the elliptic problem

$$\begin{cases} -\Delta u = f(x,u) & x \in \Omega \\ u(x) = 0 & x \in \partial\Omega, \end{cases} \tag{2}$$

where Ω is a bounded region in $\mathbf{R}^n$ with smooth boundary, Δ is the Laplace operator, and $f : \Omega \times \mathbf{R} \to \mathbf{R}$ is a function of class C^1.

We will see how the interaction of the *growth* of N, generally determined by the *numerical range* of N', with the spectrum of L provides information on the existence, multiplicity, stability, and other properties of the solutions to (1). Such interaction introduces terms such as resonance, jumping nonlinearities, superlinearities, asymptotic nonlinearities, critical growth, and others.

1991 *Mathematics Subject Classification.* Primary 34B15; Secondary 35J65.

Key words and phrases. Neumann problem, sign-changing solution, nodal region, tilable region.

While these lectures emphasize the *variational* case, hence $\sigma(L) \subset \mathbf{R}$, digressions will be made to consider the case where L is a nonselfadjoint perturbation of a selfadjoint operator. This area is yet to be adequately explored.

2. THE SPECTRUM OF SOME LINEAR OPERATORS

Let us begin by considering the spectrum of the second order derivative operator, that is,

$$L(u) \equiv -u'' = \lambda u, \quad Bu = 0, \tag{3}$$

where $B(u)$ is a boundary condition such as $u(0) = u(\pi) = 0$ (Dirichlet boundary condition) or $u(t) = u(t + 2\pi)$ (periodic boundary condition). In the case of the Dirichlet boundary condition $\sigma(L) = \{j^2; j = 1, 2, \dots\}$, where as in the case of periodic boundary condition $\sigma(L) = \{j^2; j = 0, 1, \dots\}$, a minor difference. However when we consider a first order perturbation of the second derivative, for example

$$L(u) \equiv -u'' + au' = \lambda u, \quad Bu = 0, \tag{4}$$

the spectrum changes radically from one boundary condition to the other. For the Dirichlet boundary condition, σ is, as before, an increasing sequence of positive real numbers. Namely, $\sigma(L) = \{j^2 - (a^2/4); j = 1, 2, \dots\}$. The corresponding eigenfunctions are multiples of $u(t) = e^{at/2}\sin(jt)$. On the other hand for the periodic boundary condition, $\sigma(L)$ contains only one real number $\lambda = 0$. This has a semilinear analog. Namely, while the periodic problem

$$-u'' + au' = u^3, \quad u(t) = u(t + 2\pi) \tag{5}$$

is satisfied only by $u \equiv 0$, the equation

$$-u'' + au' = u^3, \quad u(0) = u(\pi) = 0 \tag{6}$$

has infinitely many solutions. More precisely, for each nonnegative integer j the equation (6) has a solution with exactly j interior zeroes.

If $L = -\Delta$ is applied to functions defined on the smooth bounded region satisfying one of the following boundary conditions

$$\begin{aligned} u &= 0 \text{ on } \partial\Omega \text{ (Dirichlet)}, \\ \frac{\partial u}{\partial \eta} &= 0 \text{ on } \partial\Omega \text{ (Neumann), or} \\ \frac{\partial u}{\partial \eta} + b(x)u &= 0 \text{ on } \partial\Omega \text{ (mixed)}, \end{aligned} \tag{7}$$

then

$$\sigma(L) = \{\lambda_1, \lambda_2, \dots, \lambda_k, \dots\}, \quad \lambda_k \to \infty. \tag{8}$$

Except for the case when Ω is an interval in $\mathbf{R}$, the eigenvalues need not be simple. However, all the eigenvalues have finite multiplicity. This implies that if $c \notin \sigma(L)$ then the inverse of $L - cI$ is compact. The reader is referred to [**29**] for a detailed analysis of this topic. In particular, the reader should consult [**29**] for the proof that the eigenvalues $\{\lambda_k\}$ and their corresponding eigenfunctions $\{\phi_k\}$ may be characterized by the following variational property

$$\lambda_1 = \min\left\{\frac{\int_\Omega \|\nabla u\|^2 dx}{\int_\Omega \|u\|^2 dx}; u \in H - \{0\}\right\}, \text{ and for } k > 1$$

$$(9) \quad \lambda_{k+1} = \min\left\{\frac{\int_\Omega \|\nabla u\|^2 dx}{\int_\Omega \|u\|^2 dx}; u \in H - \{0\},\right.$$

$$\left.\int_\Omega u\phi_i dx = 0 \text{ for } i = 1, \ldots, k\right\},$$

where H is the subspace of the Sobolev space (see [**1**]) of square integrable functions having first order partial derivatives in $L^2(\Omega)$ and that satisfy the corresponding boundary condition in (7). From (9) the following properties of eigenfunctions follow.

LEMMA 2.1. *If ϕ_1 is an eigenfunction corresponding to the eigenvalue λ_1 then ϕ_1 does not change sign in Ω. If ϕ_2 is an eigenfunction corresponding to the eigenvalue λ_2 then ϕ_2 changes sign exactly once, that is, $\{x \in \Omega; \phi_2(x) > 0\}$ is connected and $\{x \in \Omega; \phi_2(x) < 0\}$ is connected.*

The proof of Lemma 2.1 uses strongly that if $u \in H$ then $|u| \in H$ (see [**39**]).

When L is given by the wave operator $L(u) \equiv \Box(u) = u_{tt} - u_{xx}$ acting on functions defined on $[0, \pi] \times \mathbf{R}$ satisfying the Dirichlet-periodic boundary condition given by

$$(10) \quad u(0,t) = u(\pi, t) = 0, \quad u(x,t) = u(x, t+T), \quad \text{for all } x \in [0,\pi], \ t \in \mathbf{R},$$

the spectrum of L is the closure of

$$(11) \qquad \{k^2 - \frac{4\pi^2}{T^2}; k = 1, 2, \ldots, j = 0, 1, \ldots\}.$$

Unlike the elliptic case ($L = -\Delta$), now the spectrum of L accumulates both to $+\infty$ and $-\infty$. If T is a rational multiple of π or of the form $\sqrt{n}\pi$, with n a positive integer, then $\sigma(L)$ is a discrete subset of $\mathbf{R}$. It is easy to verify, for example, that when $T = 2\pi$ that all the eigenvalues are of finite multiplicity, except for 0 which has infinite multiplicity. In fact functions of the form $\sin(kx)\sin(kt)$, $k = 1, 2, \ldots$ are mutually orthogonal eigenfunctions corresponding to the eigenvalue 0. When $T = \sqrt{n}\pi$ all eigenvalues have infinite multiplicity (see [**10**]). This implies that if $c \notin \sigma(L)$ then $L - cI$ is invertible but $(L - cI)^{-1}$ is not compact. When T is not a rational multiple of 2π this set can be very complicated. Its study requires delicate tools from number theory (see [**46**]).

3. Critical point theory

Let $H = H_0^{1,2}(\Omega)$ be the Sobolev space of square integrable functions with first order partial derivatives in $L^2(\Omega)$ and vanishing on $\partial\Omega$ (see [**1**]). Finding solutions to (2) is equivalent to finding critical points of the functional of the functional $J : H \to \mathbf{R}$ defined by

$$J(u) = \int_\Omega \{\frac{1}{2}|\nabla u|^2 - F(x,u)\}\, dx, \tag{12}$$

where $F(x,u) = \int_0^u f(x,s)ds$. The gradient and Hessian of J are given by

$$\langle \nabla J(u), v\rangle = \int_\Omega \{\nabla u \cdot \nabla v - f(x,u)v\} dx \tag{13}$$

and

$$\langle D^2 J(u)v, w\rangle = \int_\Omega \{\nabla v \cdot \nabla w - \frac{\partial f}{\partial u}(x,u)vw\} dx. \tag{14}$$

If u is a critical point of J, we define its **the Morse index** to be the dimension of the largest subspace in which $D^2J(u)$ is negative definite. Also for future reference we note that if $\{\phi_1, \phi_2, \ldots,\}$ is a set of eigenfunctions corresponding to the eigenvalues given in (8), $X = \text{span}\ \{\phi_1, \ldots \phi_k\}$, and $Y = \text{span}\ \{\phi_{k+1}, \ldots\}$ then from (9)

$$\begin{aligned} \lambda_k \textstyle\int_\Omega u^2(x)dx &\geq \textstyle\int_\Omega \|\nabla u(x)\|^2 dx \text{ for } u \in X \\ \lambda_{k+1} \textstyle\int_\Omega u^2(x)dx &\leq \textstyle\int_\Omega \|\nabla u(x)\|^2 dx \text{ for } u \in Y, \end{aligned} \tag{15}$$

the latter are known as *Poincaré's inequalities*.

The solutions to the equation

$$\begin{cases} \Box(u) + f(x,t,u) = 0 \qquad (x,t) \in (0,\pi)\times\mathbf{R} \\ u(0,t) = u(\pi,t) = 0, \quad u(x,t) = u(x,t+T), \quad \text{for all } x \in [0,\pi],\ t \in \mathbf{R} \end{cases} \tag{16}$$

may also be studied as critical points of a functional. In this case H is given by the space of functions $u : [0,\pi]\times\mathbf{R} \to \mathbf{R}$ such that $u(0,t) = u(\pi,t) = 0$ for all $t \in \mathbf{R}$, $u(x,t) = u(x,t+T)$ for all $(x,t) \in (0,\pi)\times\mathbf{R}$, $u, u_x, u_t \in L^2((0,\pi)\times[0,T])$. The functional whose critical points are the solutions of (16) is given by

$$W(u) = \int_\Omega \{\frac{1}{2}(u_t^2 - u_x^2) - F(x,t,u)\}\, dx, \tag{17}$$

where $\Omega = (0,\pi)\times[0,T]$, and $F(x,t,u) = \int_0^u f(x,t,s)ds$. Results for J do not immediately transfer to W. A key difference between J and W is that while the integrand in J depending on the partial derivatives of u is positive definite, the one in W is positive definite in an infinite dimensional subspace, negative definite in an infinite dimensional subspace, and may be identically equal to zero in an infinite dimensional subspace. This creates many difficulties when applying arguments based on *compactness*.

Another class of semilinear equations that may be studied via critical point theory are *Hammerstein Integral Equations*. These are equations of the form

$$u = \psi(x) + \int_\Omega K(x,y) f(x, u(x))dx, \tag{18}$$

where $\Omega \subset \mathbf{R}^N$ is a bounded region, f is as above, and $K \in L^2(\Omega \times \Omega)$ with $K(x,y) = K(y,x)$ for all $x, y \in \Omega$. Since the operator $\mathcal{K}(u)(y) \equiv \int_\Omega K(x,y)u(x)dx$ defines a self-adjoint compact operator in $L^2(\Omega)$, there exist closed subspaces $X, Y \subset L^2(\Omega)$ such that $X \oplus Y = L^2(\Omega)$, $\mathcal{K}(X) \subset X$, $\mathcal{K}(Y) \subset Y$, $\langle \mathcal{K}(u), u\rangle \geq 0$ for $u \in X$, and $\langle \mathcal{K}(v), v\rangle \leq 0$ for $v \in Y$. Assuming further that $\mathcal{K}$ is positive definite in X and negative definite on Y one can find self-adjoint positive definite operators $K_1 : X \to X$, and $K_2 : Y \to Y$ such that $K_1(K_1(u)) = \mathcal{K}(u)$ for all $u \in X$, and $K_2(K_2(v)) = -\mathcal{K}(v)$ for all $v \in Y$. Let F be as in (12). Now the functional $I : X \oplus Y = L^2(\Omega) \to R$ defined by

$$I(u+v) = \int_\Omega \{\frac{1}{2}(u^2 + v^2) - F(x, K_1 u - K_2 v) - \psi(x)(u+v)\}\, dx, \tag{19}$$

is such that $u + v$ is a critical point of I if and only if $K_1 - K_2 v$ is a solution to (18) (see [**11**].)

Hamiltonian systems provide also an important class of semilinear problems with discrete spectrum whose solutions are critical points (see [**53**].)

4. Equations without resonance

Assuming that $\rho(L)$ contains an interval [a, b] with $a < b$, we can give a condition on N (from (1)) that guarantees the existence and uniqueness of solution to (1). In fact we have the following result.

THEOREM 4.1. *Let H be real Hilbert space, and $N : H \to H$. Suppose that $a < b$ are real numbers and that $[a, b] \subset \mathbf{R} - \sigma(L)$. For each $c \in (a, b)$, there exist $d \in (0, \infty)$ such that if $\|N(u) - cu - (N(v) - cv)\| \leq d\|u - v\|$ for all $u, v \in H$, then the equation (1) has a unique solution. Moreover if L is selfadjoint then d may be any number in $(0, \min\{c - a, b - c\})$.*

The proof of this theorem follows from the contraction mapping principle. This is because the equation (1) is equivalent to the fixed point equation

$$u = (L - cI)^{-1}(N(u) - cu). \tag{20}$$

When $(L - cI)^{-1}$ is compact, by appealing to the Schauder fixed point theorem (see [**26**]), we can further relax the hypotheses on N in the previous theorem. For example, the following holds.

THEOREM 4.2. *Let H be real Hilbert space. Suppose that $a < b$ are real numbers and that $[a, b] \subset \mathbf{R} - \sigma(L)$. If there exist $c \in (a, b)$ and $M \in \mathbf{R}$ such that $\|N(u) - cu\| \leq M$ for all $u \in H$, and $(L - cI)^{-1}$ is compact then (1) has a solution. This solution may not be unique (see Theorem 4.5 below).*

From the properties of the properties of the spectrum of $-\Delta$ or the wave operator $\Box$ stated in the previous section, it follows that Theorem 4.1 applies to (2) and to (16). In fact the reader is invited to verify the following.

THEOREM 4.3. *Let L be the linear operator defined in (2) or (16). If f is continuous function, and $(\partial f/\partial u)(y,u) \in (a,b)$ for all (y,u), and $[a,b] \cap \sigma(L) = \emptyset$, then the corresponding problem has a unique solution.*

While Theorem 4.2 applies to (2), it does not apply to (16). In fact, as noted in previous section the inverse of the linear operator defined by $-\Delta + cI$ subject to the boundary conditions in (2) is compact while that corresponding to $\Box - cI$, with the boundary condition in (16), is not so. Applying Theorem 4.2 to (2) the following may be established.

THEOREM 4.4. *Let $f(x,u) = cu + h(x,u)$ be continuous with $\lim_{|u|\to\infty} h(x,u) = 0$, uniformly for $x \in \Omega$. If $c \in (a,b)$, and $[a,b] \cap \sigma(L) = \emptyset$, then the problem (2) has a solution.*

In [**22**] and [**10**] sufficient conditions for the existence of solutions to (16) under assumptions related to those Theorem 4.4. See also [**58**].

Unlike Theorem 4.3, the solution provided in Theorem 4.4 need not be unique. In fact the issue of how many solutions (2) has under the hypotheses even stronger than those in Theorem 4.4 is a matter of many open questions. In fact, let us now assume that $f(x,u) = g(u)$ with

(21)
$$g(0) = 0, \ g'(0) \in (\lambda_k, \lambda_{k+1}), \ \text{and} \ \lim_{u\to+\infty} g'(u), \lim_{u\to-\infty} g'(u) \in (\lambda_j, \lambda_{j+1}),$$

where the $\lambda_i's$ are as in (8) and $\lambda_0 = -\infty$. When Ω is an interval in $\mathbf{R}$, it can be shown that (21) implies that (2) has at least $2|k-j|+1$ solutions. This, however, does not extend to higher dimensions. In [**31**], N. Dancer provided an example of where Ω is the unit ball in $\mathbf{R}^N$, $|k-j| = N-1$ and yet (2) has only three solutions. In [**44**] it is shown that if $k \neq j$ then (2) has at least two solutions. Also, building on the results of [**21**], [**16**], and [**13**] the following theorem was established in [**15**].

THEOREM 4.5. *Suppose (21) holds with $k = 1$ and $j \geq 2$, and that $rg'(r) > 0$ for $r \in \mathbf{R} - \{0\}$. If there exists $a < \lambda_{j+1}$ such that $g'(r) \leq a$ for all $r \in \mathbf{R}$, then (2) has at least five solutions. Two of these solutions change sign.*

As stated in section 2, the solutions to

$$\begin{cases} -\Delta u = g(u) & \text{en } \Omega \\ u = 0 & \text{en } \partial\Omega, \end{cases} \tag{22}$$

are given by the critical points of

$$J(u) = \int_\Omega \{\frac{1}{2}|\nabla u|^2 - G(u)\}\, dx, \tag{23}$$

where $G(t) = \int_0^t g(s)\,ds$. Since, trivially, $u \equiv 0$ is a solution to (22) we see that $u \equiv 0$ is a critical point of J. From (14) we see that

$$
\begin{aligned}
< D^2J(0)v, v > &= \int_\Omega [\|\nabla v(x)\|^2 - g'(0)v^2]dx \\
&\geq (1 - \frac{g'(0)}{\lambda_1}) \int_\Omega \|\nabla v(x)\|^2 dx,
\end{aligned}
\tag{24}
$$

where $\lambda_1 > 0$ is as in (8) for the boundary condition $u = 0$ on $\partial\Omega$. From (24) we conclude that $u \equiv 0$ is a local minimum. Moreover, its Morse index is 0, and its *local degree* is $+1$.

Let $g_+(x) = \max\{g(x), 0\}$. Let J_+ be the functional whose critical points are the solutions to (22) when g is replaced by g_+. Now the set of critical points of ∇J_+ is bounded and the Leray-Schauder degree of ∇J_+ on any open set containing all its critical points is equal to zero. Hence the Leray-Schauder degree of ∇J restricted to any open set containing all positive solutions is -1. Similar argument shows that the Leray-Schauder degree of ∇J restricted to any open set containing all negative solutions to (22) is -1.

Following the arguments of [**16**] one sees that (22) has a solution w that changes sign exactly once. Moreover such a solution is an isolated solution, then the local Leray-Schauder degree if ∇J is $+1$. From (21) one sees that the set of critical points of J is bounded and that the Leray-Schauder degree of ∇J is large balls centered at 0 is $(-1)^j$. Assuming that (22) has one only solution that changes sign we have then that

$$(-1)^j = 1 + (-1) + (-1) + 1, \tag{25}$$

where 1 accounts for the local Leray-Schauder of 0, (-1) for the Leary-Schauder degree of the set of positive solutions to (22), similarly the set of negative solutions to (22), and the last term in (25) accounts for the local Leray-Schauder degree of w. Since (25) is contradictory we see that (22) must have a second solution that changes sign. This is concludes a sketch of the proof of Theorem 4.5.

If $j = k + 1$ and $rg''(r) \geq 0$, then (22) has exactly three solutions (see [**21**]). Also if $j = 2$ then, under the assumptions of Theorem 4.5, (22) has exactly two solutions that change sign (see [**15**]).

Thus it remains open determining the number of solutions (2) must have in terms of $|k-j|$. It is tempting to conjecture that as $|k-j|$ tends to infinity, then number of solutions must tend to infinity. For a version Theorem 4.5 for the hyperbolic problem (16) the reader is referred to [**8**]. Even when T is a rational multiple of π additional hypotheses on the monotonicity of f are needed due to the infinite multiplicity of the eigenvalue 0.

5. Resonance and the Landesman-Lazer condition

In the hypotheses of the previous section the terms

$$\frac{\partial f}{\partial u}(x, -\infty) \equiv \lim_{u \to -\infty} \frac{\partial f}{\partial u}(x, u), \quad \text{and} \quad \frac{\partial f}{\partial u}(x, +\infty) \equiv \lim_{u \to +\infty} \frac{\partial f}{\partial u}(x, u) \tag{26}$$

stay in an interval contained in $\rho(L)$. Thus we can take advantage of the invertibility of $L - cI$. In contrast, in [**38**] the authors considered the case

$$f(x,u) = \lambda_k u + g(u) + p(x), \tag{27}$$

with g bounded and monotone. For the sake of simplicity let us assume that g is monotonically increasing. Let $g(-\infty) \equiv \lim_{u\to-\infty} g(u) < \lim_{u\to+\infty} g(u) \equiv g(+\infty)$. When (27) holds we say that the nonlinearity is **in resonance** with the eigenvalue λ_k. Surprisingly the following was established in [**38**].

THEOREM 5.1. *Suppose λ_k is a simple eigenvalue of $-\Delta$ subject to the Dirichlet boundary condition (see (7). Let ϕ_k be an eigenfunction associated with the eigenvalue λ_k. Let ϕ_+, ϕ_- be the positive and negative parts of ϕ_k, respectively. If*

$$\begin{aligned} g(-\infty)\textstyle\int_\Omega \phi_+(x)dx \;+\; g(+\infty)\int_\Omega \phi_-(x)dx \;&<\; \textstyle\int_\Omega p(x)\phi_k(x)dx \\ &<\; g(-\infty)\textstyle\int_\Omega \phi_-(x)dx + g(+\infty)\int_\Omega \phi_+(x)dx, \end{aligned} \tag{28}$$

then (2) has a solution. Moreover, if $g(-\infty) < g(u) < g(+\infty)$ for all $u \in \mathbf{R}$, then (28) is a necessary condition for the solvability of (2).

Theorem 5.1 not only pioneers the solvability of nonlinear problems with non invertible linear part ($L = -\Delta + \lambda_k I$), but also motivates important developments in variational methods. In fact, the **saddle point principle** proved by P. Rabinowitz in [**52**] owes its origin to a variational interpretation of the hypothesis (28), also known as Landesman-Lazer condition.

Let us state a simplified version of the saddle point principle. Let X, Y of H such that $H = X \oplus Y$, and $dim X < \infty$. Suppose that $\lim_{\|x\|\to\infty, x\in X} I(x) = -\infty$, $\lim_{\|y\|\to\infty, y\in Y} I(y) = \infty$, and I is bounded below on Y. Let $r > 0$ be so large that

$$\begin{aligned} \max\{I(x); x \in X, \|x\| = r\} \;&<\; \inf\{I(y); y \in Y\}, \text{ and} \\ \inf\{I(y); y \in Y, \|y\| = r\} \;&>\; \max\{I(x); \|x\| = r, x \in X\}. \end{aligned} \tag{29}$$

THEOREM 5.2. *Let H be a Hilbert space and $I : H \to \mathbf{R}$ a functional of class C^1 and r be as in (29). If*

(PS) *every sequence $\{u_j\}$ in H such that $\{I(u_j)\}$ is bounded and $\lim_{j\to\infty} \nabla I(u_j) = 0$ has a converging subsequence,*

then

$$c = \inf_{x\in\Sigma} \max_{\|x\|=r} I(s(x)), \tag{30}$$

is a critical value, where $\Sigma = \{s : \{x \in X; \|x\| \le r\} \to H; s$ is continuous, and $s(x) = x$ for $\|x\| = r\}$.

To derive Theorem 5.1 from the Saddle Point Principle we let X and Y be as in (15). From the first inequality in (15) it is easily seen that

$$J(u) \to -\infty \text{ as } \|u\| \to \infty, \tag{31}$$

for $u \in X_1 \equiv \text{span}\{\phi_1, \ldots, \phi_{k-1}\}$. Here we use strongly that $\lambda_{k-1} < \lambda_k$. Similarly, $J(u) \to +\infty$ as $\|u\| \to \infty$ for $u \in Y$, and J is bounded below on Y. Thus applying Theorem 5.2 only requires checking the hypothesis (PS), known as Palais-Smale condition, and that (31) is satisfied in all of X. The assumption (28) implies that $J(c\phi_k) \to -\infty$ as $|c| \to \infty$. This and (31) imply that $J(x) \to \infty$ as $\|x\| \to \infty$ for $x \in X$. Thus (29) is satisfied. The reader is referred to [**52**] or [**53**] for the detail of that part of the proof.

For extensions of the saddle point principle, and other critical point methods, and their applications to the solvability of boundary value problems the reader is referred to [**55**].

6. Jumping nonlinearities

The term *jumping nonlinearity* was introduced by S. Fucik in [**36**] to refer to cases where

$$\frac{\partial f}{\partial u}(x, -\infty) \neq \frac{\partial f}{\partial u}(x, +\infty).$$

In order to state the seminal result concerning problems with **jumping nonlinearities** let us assume that $f(x, u) = g(u) + p(x)$ with

$$-\infty < g'(-\infty) \equiv \lim_{u \to -\infty} g'(u) < \lambda_1 < \lim_{u \to \infty} g'(u) \equiv g'(+\infty) < \lambda_2. \tag{32}$$

We recall that the eigenvalue λ_1 is simple and that eigenfunctions corresponding to λ_1 do not change sign (see Lemma 2.1). Thus from now on we denote by ϕ_1 an eigenfunction corresponding to λ_1 and we assume that $\phi_1(x) > 0$ for all $x \in \Omega$, and that $\int_\Omega (\phi_1(x))^2 dx = 1$. In [**3**] A. Ambrosetti and G. Prodi proved the following fundamental result.

THEOREM 6.1. *Let g' be a nondecreasing function. If (32) holds, then for each $q \in L^2(\Omega)$ such that $\int_\Omega q(x)\phi_1(x)dx = 0$ there exists $C(q) \in \mathbf{R}$ such that*

$$\begin{cases} -\Delta u = g(u) + q(x) - c\phi_1(x) & \text{in } \Omega \\ \quad u = 0 \quad \text{on } \partial\Omega, \end{cases} \tag{33}$$

has no solution if $c < C(q)$, exactly one solution if $c = C(q)$, and exactly two solutions if $c > C(q)$.

This result motivated intense study of problems with jumping nonlinearities. Hypothesis (32) implies that $g(u) - \lambda_1 u$ is bounded below. Hence if we assume that $u \in H$ is a solution to $-\Delta u = g(u) + q + c\phi_1$, then multiplying by ϕ_1 and integrating on Ω one has

$$\begin{aligned} \lambda_1 \int_\Omega u\phi_1 dx &= \int_\Omega u(-\Delta\phi_1)dx = \int_\Omega(-\Delta u)\phi_1 dx \\ &= \int_\Omega (g(u) + q(x) + c\phi_1(x))\phi_1(x)dx \\ &\geq \int_\Omega (\lambda_1 u + M + q(x) - c\phi_1(x))\phi_1(x)dx, \end{aligned} \tag{34}$$

where M is a lower bound for $g(u) - \lambda_1 u$. Hence only if $c \geq M|\Omega|$ may (33) have a solution. This explains the nonexistence of solutions in Theorem 6.1. That (32) can have at most two solutions is strongly related to the nodal properties of eigenfunctions corresponding to the eigenvalue λ_2 (see Lemma 2.1).

If we replace in (32) the assumption $g'(+\infty) \in (\lambda_1, \lambda_2)$ by $g'(+\infty) \in (\lambda_k, \lambda_{k+1})$, one is tempted to conjecture that for large $c > 0$ the equation (33) has $2k$ solutions. This conjecture turns out to be correct when Ω is one dimensional, and even when Ω is a ball in $\mathbf{R}^N$ and one considers only radial solutions (see [**17**]) and [**28**]). E. N. Dancer in [**33**] provides counter examples to this general conjecture. To date, in general, all we can claim is that if $g'(-\infty) < \lambda_1 < \lambda_k < g'(+\infty) < \lambda_{k+1}$ with $k \geq 2$, then for $c > 0$ the problem (33) has at least four solutions. For early results on jumping nonlinearities the reader is referred to [**31**], [**40**], [**41**], [**42**], [**36**], among others. For a survey on problems with jumping nonlinearities and their applications to the design of suspension bridges see [**43**]. For recent results in problems with jumping nonlinearities see [**34**], [**50**], and [**56**].

In [**36**], S. Fucik noticed the relevance of understanding the set

$$\{(a, b);\ \text{for some } u \in H - \{0\},\ -\Delta u = au_+ bu_-\} \tag{35}$$

in the solvability of (33). This set is known as the Fucik spectrum of $-\Delta$ in Ω. For additional information on the Fucik spectrum see [**31**] and [**56**]. Full understanding of the Fucik spectrum remains open. For the one dimensional case, a full characterization of the Fucik spectrum may be found in [**36**] for the case of Dirichlet boundary condition or periodic solutions.

After thirty years of Theorem 6.1 there remains the conjecture that, for $c > 0$ large, the number of solutions to (33) is large when k is large and $\lambda_k < g'(+\infty) < \lambda_{k+1}$.

7. Superlinear problems

In this section we consider the *superlinear* case. That is when

$$\lim_{|u|\to\infty} \frac{\partial f}{\partial u}(x, u) = \infty. \tag{36}$$

This occurs, for example, if $f(x, u) \equiv g(u) = |u|^\alpha u$ and $\alpha > 0$. Because the Sobolev space H considered in Section 3 is not always contained in $L^\infty(\Omega)$, the functional J need not be well defined in H. However, since H is imbedded in $L^p(\Omega)$ for any $p \in (1, 2^*)$ with $2^* = (N+2)/(N-2)$ for $N > 2$, and $2^* = \infty$ for $N \leq 2$ (see [**1**]), for J to be well defined it is sufficient that

$$|f(x, u)| \leq A(|u|^p + 1), \tag{37}$$

for some $p \in (1, 2^* - 1)$. When (37) is satisfied we say that f has *subcritical growth*, or simply that f is *subcritical.* Similarly we will say that f is *supercritical* if there exist $p > 2^*$ and $A > 0$ such that

$$|f(x, u)| \geq A(|u|^p - 1) \quad \text{for all } (x, u) \in \Omega \times \mathbf{R}. \tag{38}$$

Finally we will say that f has *critical growth* if

$$\lim_{|u|\to\infty} \frac{|f(x,u)|}{|u|^{2^*}} \in (0,+\infty). \tag{39}$$

This guarantees that the functional J is of class C^1 (see [**53**]). When f is odd in the second variable, $f(x,u) = -f(x,-u)$, arguments that go back to the Liusternik-Schnirelmann theory show that (2) has infinitely many solutions if, in addition, there exist $\theta > 2$ and $\tau > 0$ such that

$$uf(x,u) \geq \theta F(x,u) \text{ for } |u| \geq \tau, x \in \Omega. \tag{40}$$

This condition plays a central role in proving that the functional J satisfies the Palais-Smale condition. For f not necessarily odd and $N = 1$, (2) also has infinitely many solutions. If Ω is a ball in $\mathbf{R}^N$, f is radial in the first variable ($f(x,u) = f(y,u)$ for $\|x\| = \|y\|$) but not necessarily odd in the variable u, under additional *structure conditions* on f it may also be shown that (2) has infinitely many solutions (see [**18**]). For the role of these structure conditions see Theorem 7.1 below. In fact, infinitely many solutions may be found even when f exceeds the growth condition (37). For general $f's$ proving the existence of infinitely many solutions for superlinear problems has remained elusive. The following theorem, due to A. Bahri and P. L. Lions (see [**5**]), obtains existence of solutions to (2) when f is a perturbation of $f_1(u) \equiv |u|^{p-1}u$ with $p \in (1, 2^*)$. More precisely, let $f(x,u) = |u|^{p-1}u - h(x,u)$ with

$$\begin{aligned} |h(x,t)| &\leq p(x) + C|t|^q \text{ for some } C \geq 0,\ p \in L^r_+(\Omega) \\ |H(x,t)| &\leq a(x) + b(x)|t|^\alpha \text{ for some } \alpha \in (0,2), \end{aligned} \tag{41}$$

where $H(x,t) = \int_0^t h(x,s)ds$, $a \in L^1_+(\Omega)$, $q = (N+2)/(N-2)$ ($q < \infty$ if $N \leq 2$), $r = 2N/(N+2)$ ($r > 1$ if $N \leq 2$), $b \in L^\beta_+(\Omega)$ with $\beta > 1$, $\beta' < 2N/(N-2)(1/\alpha)$ ($\beta > 1$ if $N \leq 2$).

THEOREM 7.1. *Let* $f(x,u) = |u|^{p-1}u + h(x,u)$. *If* $p \in (1, (N+2(1-\alpha))/(N-2)$, *then (2) has infinitely many solutions.*

One notes that, for example, if $f(x,u) = |u|^\alpha u + p(x)$ for $u \geq 0$, $f(u) = |u|^\beta u + p(x)$ for $u \leq 0$ and $\alpha \neq \beta$ Theorem 7.1 does not apply. Neither does it apply if $h \equiv 0$ and $p \in (N/(N-2), (N+2)/(N-2))$. These are just some of the cases that remain open in the case of superlinear problems, even when f is subcritical and (40) holds.

8. Critical exponents

In [**51**], S. Pohozaev noted that if u is a solution to

$$\begin{cases} -\Delta u = g(u) & x \in \Omega \\ u(x) = 0 & x \in \partial\Omega, \end{cases} \tag{42}$$

then

$$\frac{1}{2}\int_{\partial\Omega}\|\nabla u\|^2(x\cdot\nu)ds = N\int_{\Omega}G(u)dx - \frac{N-2}{2}\int_{\Omega}ug(u)dx. \tag{43}$$

In the latter equations ν denotes the outward unit normal, and ds the volume element on $\partial\Omega$. This identity, known as *Pohozaev's identity*, is particularly useful. For example, when Ω is a starlike region with respect to 0 the quantity $x\cdot\nu$ is positive, which forces $2NG(t)-(N-2)tg(t)$ to be nonnegative somewhere on the range of any solution u. Since, for example, $2NG(t)-(N-2)tg(t)<0$ for $t\neq 0$ when $g(t)=|t|^{p-1}t$ with $p>2^*$, we see that for such g's the equation (42) has no nontrivial solution. With the help of the maximum principle on can also see that (42) has no nontrivial solution on starlike regions when $p=2^*$. This is in contrast with the case $p\in(1,2^*)$ where the problem has infinitely many solutions.

Despite the negative result of [**51**], in [**9**] H. Brezis and L. Nirenberg discovered the existence of solutions to (42) when $g(t)=\lambda t+|u|^{2^*-1}u$, and $N\geq 4$. In fact they proved that if $\lambda\in(0,\lambda_1)$, the problem (42) has a positive solution. These results in turn led for further understanding of the solvability of semilinear equations with critical growth (see (39)). The reader is referred to [**7**], [**19**], [**20**], and [**24**] for further developments in this respect. It must be noted that most of these results pertain to radially symmetric solutions. For general bounded regions the solvability of (42) is yet to be adequately explored.

9. Qualitative properties of solutions

While the above developments only state the existence, or multiplicity of solutions to semilinear problems, the methods of proof carry information on the nature of the solutions. For example when f is nonnegative, owing to *sub-supersolution methods* one may determine that solutions to (2) are actually positive functions. The reader is referred to [**45**] for extensive information on this topic. Positive solutions may also exist when $f(x,0)<0$ and $f(x,u)$ is positive for $u>0$ large. Equations of this type are known as semipositone equations (see [**23**]).

The study of solutions that change sign is relatively new. Motivated by the results of [**57**] the following was established in [**14**].

THEOREM 9.1. *If $g(0)=0$, $g'(0)<\lambda_1$, $tg'(t)>g(t)$ for $t\neq 0$, g satisfies (40), and for some $p\in(1,(N+2)/(N-2)$, $A\in(0,\infty)$*

$$|g'(t)|\leq A(|t|^{p-1}+1) \text{ for all } t\in\mathbf{R}, \tag{44}$$

then (42) has a solutions that changes sign exactly once (that is, $\{x\in\Omega; u(x)>0\}$ is connected and $\{x\in\Omega; u(x)<0\}$ is connected.)

The existence of the solution in this theorem includes a variational characterization. Such a characterization has been skillfully used by J. M. Neuberger (see [**47**] and [**27**]) to approximate such solution. The existence of solution that changes sign has also been used to establish the existence of other sign changing solutions (see [**16**]) and exact number of solutions that change sign (see [**15**]). For additional results in this direction, the reader is referred to [**6**] and [**24**].

While the assumption that g grow subcritically is essential in the proof of Theorem [**14**], results for regions where g grows at the critical rate have been recently found for regions with some symmetry (see [**12**] and [**37**]).

In case g grows subcritically, say satisfying the assumption of Theorem 7.1, and Ω is a ball one may conclude the existence of infinitely many solutions either from the variational arguments of Theorem 7.1 or from one-dimensional arguments as in [**18**]. In [**2**] it is shown that infinitely many of the solutions provided by Theorem 7.1 are nonradial. The proof is based on the analysis of the growth rate of *energy levels* of radial solutions that mimics the asymptotic expansion of radial versus nonradial eigenvalues.

Motivated by equations coming from, among others, biological models, extensive work has been devoted to finding nonnegative solutions to the Neumann problem (see [**48**])

$$\begin{cases} -\epsilon^2 \Delta u = g(u) & x \in \Omega \\ \dfrac{\partial u}{\partial \eta}(x) = 0 & x \in \partial\Omega. \end{cases} \tag{45}$$

Here a key issue is the number and location of *spikes* of the solutions, and their behavior as ϵ tends to zero.

10. Extensions and Applications

The above results beg for extensions to cases where, for example, L is non-selfadjoint, L is quasilinear, or L is given by a system of linear operators. Little is known when the operator $-\Delta$ is replaced in the above by a second order non-selfadjoint operator. Since most of the results have been derived not from the potential J but rather from the vector field generated by the operator $u \to -\Delta u - f(x, u)$ on H, new ideas are needed to deal with the nonselfadjoint case. At this point one finds no reason to believe that, for example, if $-\Delta$ is perturbed by a first order operator then the above theorems extend to the new equation. That (5) has no nontrivial solution suggests, for example, that Theorem 7.1 does not extend to nonselfadjoint problems. For extensions to quasilinear problems the reader is referred to [**35**] and [**54**] and references therein.

The equation (1) models multiple phenomena. In fact the motivating examples may be derived from physical examples that go back to classical mathematical physics. For more recent applicability of semilinear models, the reader is referred to [**25**], [**43**], and [**49**].

References

[1] R. Adams, *Sobolev Spaces*, New York, Academic Press (1975).

[2] H. Aduen and A. Castro, *Infinitely many nonradial solutions to a superlinear Dirichlet Prob- lem,* Proc. Amer. Math. Soc. 131 (2003), pp. 835-843.
[3] A. Ambrosetti and G. Prodi, *On the inversion of some differentiable mappings with singularities between Banach spaces,* Ann. Mat. Pura Appl. 93(1972), 231-247.
[4] F. Atkinson, H. Brezis and L. Peletier, *Solutions d'équations elliptiques avec exposant de Sobolev critique que changent de signe*, C. R. Acad. Sci. Paris Serie I, (1988), vol.306, pp. 711-714.
[5] A. Bahri and P. L. Lions, *Morse index of some min-max critical points. I. Application to multiplicity results*, Comm. Pure Appl. Math. 41 (1988), no. 8, pp. 1027-1037.
[6] T. Bartsch, K.-C. Chang and Z.-Q. Wang *On the Morse indices of sign changing solutions of nonlinear elliptic problems*, Math. Z. 233 (2000), pp. 655-677.
[7] R. Benguria, J. Dolbeault, and M. Esteban, *Classification of the solutions of semilinear elliptic problems in a ball,* J. Differential Equations 167 (2000), no. 2, pp. 438-466.
[8] H. Brezis and M. Coron, *Periodic solutions of nonlinear wave equations and Hamiltonian systems,* Amer. J. Math. 103 (1981), no. 3, pp. 559-570.
[9] H. Brezis and L Nirenberg, *Positive solutions of nonlinear elliptic equations involving critical Sobolev exponents,* Comm. Pure Appl. Math. 36 (1983), no. 4, pp. 437-477.
[10] J. Caicedo and A. Castro, *A semilinear wave equation with derivative of nonlinearity containing multiple eigenvalues of infinite multiplicity,* . Contemp. Math., 208, pp. 111-132.
[11] A. Castro, *Hammerstein integral equations with indefinite kernel,* Internat. J. Math. Math. Sci. 1 (1978), no. 2, pp. 187-201.
[12] A. Castro and M. Clapp, *The effect of the domain topology on the number of minimal nodal solutions of an elliptic equation at critical growth in a symmetric domain*, to appear in Nonlinearity.
[13] A. Castro and J Cossio, *Multiple solutions for a nonlinear Dirichlet problem,* SIAM J. Math. Anal. 25 (1994), no. 6, pp. 1554-1561.
[14] A. Castro, J. Cossio, and J. M. Neuberger, *A sign-changing solution for a superlinear Dirichlet problem,* Rocky Mountain J. Math. 27 (1997), no. 4, pp. 1041-1053.
[15] A. Castro, J. Cossio, and J. M. Neuberger, *On multiple solutions of a nonlinear Dirichlet problem,* Nonlinear Anal. 30 (1997), no. 6, pp. 3657-3662.
[16] A. Castro, J. Cossio, and J. M. Neuberger, *A minmax principle, index of the critical point, and existence of sign-changing solutions to elliptic boundary value problems,* Electron. J. Differential Equations 1998, No. 02, 18 pp. (electronic).
[17] A. Castro and S. Gadam, *The Lazer Mckenna Conjecture for Radial Solutions in the* $\mathbf{R}^N$ *Ball*, Electron. J. Diff. Eqns. Vol. 1993(1993), No. 07, pp. 1-6.
[18] A. Castro and A. Kurepa, *Infinitely many solutions to a superlinear Dirichlet problem in a ball*, Proc. Amer. Math. Soc., (1987), vol. 101 No. 1, pp. 57-64.
[19] A. Castro and A. Kurepa, *Radially symmetric solutions to a superlinear Dirichlet problem with jumping nonlinearities,* Trans. Amer. Math. Soc., (1989), vol. 315, pp. 353-372
[20] A. Castro and A. Kurepa, *Radially symmetric solutions to a Dirichlet problem involving critical exponents,* Trans. Amer. Math. Soc., 348 (1996), no. 2, pp. 781-798.
[21] A. Castro and A. C. Lazer, *Critical point theory and the number of solutions of a nonlinear Dirichlet problem,* Ann. Mat. Pura Appl. (4) 120 (1979), pp. 113-137.
[22] A. Castro and S. Unsurangsie, *A semilinear wave equation with nonmonotone nonlinearity,* Pacific J. Math. 132 (1988), no. 2, pp. 215-225.
[23] A. Castro, C. Maya, and R. Shivaji, *Nonlinear eigenvalue problems with semipositone structure,* Electron. J. Differ. Equ., Conf. 5, (2000), pp. 33-49 (electronic).
[24] G. Cerami, S. Solimini, M. Struwe, *Some existence results for superlinear elliptic boundary value problems involving critical exponents,* J. Funct. Anal. **69** (1986), 289-306.

[25] G. Chen, Z. Ding, C. Hu, W. Ni, and J. Zhou *A note on the elliptic sine-Gordon equation,* (preprint).

[26] S. N. Chow and J. K. Hale, *Methods of bifurcation theory* Springer-Verlag, New York-Berlin, (1982).

[27] D. Costa, Z. Ding, and J. M. Neuberger, John M. *A numerical investigation of sign-changing solutions to superlinear elliptic equations on symmetric domains,* J. Comput. Appl. Math. 131 (2001), no. 1-2, pp. 299-319.

[28] D. G. Costa and D. G. de Figueiredo, *Radial solutions for a Dirichlet problem in a ball,* J. Diff. Equations 60 (1985), pp. 80-89.

[29] R. Courant and D. Hilbert, *Methods of Mathematical Physics, Volume I,* New York, John Wiley (1989).

[30] M. Crandall and P. Rabinowitz, *Bifurcation from simple eigenvalues,* J. Funct. Anal. 8(1971), 321-340.

[31] Dancer, E. N. *On the ranges of certain weakly nonlinear elliptic partial differential equations,* J. Math. Pures Appl. (9) 57 (1978), no. 4, 351–366. (Abstract version of amb-prodi.)

[32] E. N. Dancer, *A counter example to the Lazer-McKenna conjecture,* Nonlinear Anal., T.M.A., 13(1989), 19-22.

[33] E. N. Dancer, *Counterexamples to some conjectures on the number of solutions of nonlinear equations. Math. Ann.,* 272 (1985), no. 3, pp. 421-440.

[34] E. N. Dancer, *Some results for jumping nonlinearities,* Topological Methods Nonlinear Anal. 19 (2002), no. 2, 221–235. (35J60 (35J65 47Jxx 58E05).)

[35] P. Drabek, and S. Robinson, *On the generalization of the Courant nodal domain theorem,* J. Differential Equations 181 (2002), no. 1, pp. 58-71.

[36] S. Fucik, *Boundary value problems with jumping nonlinearities,* Casopis Pest. Mat. 101 (1976), no. 1, pp. 69-87.

[37] P. Girao, and M. Ramos, *Sign changing solutions for elliptic equations with critical growth in cylinder type domains,* ESAIM Control Optim. Calc. Var. 7 (2002), pp. 407-419.

[38] E. Landesman and A. C. Lazer, *Nonlinear perturbations of linear elliptic boundary value problems at resonance,* J. Math. Mech. 19 1969/1970, pp. 609-623.

[39] D. Kinderlehrer-G. Stampacchia, *Introduction to Variational Inequalities and Their Applications,* New York, Academic Press (1979).

[40] A. C. Lazer and P. J. McKenna, *On the number of solutions of a nonlinear Dirichlet problem,* J. Math. Anal. Appl. 84 (1981), no. 1, 282–294. (Existence of at least four solutions when crossing from below the first eigenvalue to one larger than the second.)

[41] A. C. Lazer and P. J. McKenna, Multiple of solutions of nonlinear boundary value problems with nonlinearities crossing several higher eigenvalues, J. Reine Angew. Math. 368 (1986), 184–200. (existence of our solutions.)

[42] A. C. Lazer and P. J. McKenna, *Critical point theory and boundary value problems with nonlinearities crossing multiple eigenvalues,* II. Comm. Partial Differential Equations 11 (1986), no. 15, 1653–1676. (Existence of j solutions (j in (1,3)) for c large and 4-j solutions for c large negative.)

[43] A. C. Lazer and P. J. McKenna, *Large-amplitude periodic oscillations in suspension bridges: some new connections with nonlinear analysis.* SIAM Rev. 32 (1990), no. 4, pp. 537-578.

[44] A. C. Lazer and S. Solimini, *Nontrivial solutions of operator equations and Morse indices of critical points of min-max type,* Nonlinear Anal. 12 (1988), no. 8, pp. 761-775.

[45] P. L. Lions, *On the existence of positive solutions of semilinear elliptic equations,* SIAM Rev. 24 (1982), no. 4, pp. 441-467.

[46] P. J. McKenna, *On solutions of a nonlinear wave equation when the ratio of the period to the length of the interval is irrational,* Proc. Amer. Math. Soc. 93 (1985), no. 1, pp. 59-64.

[47] J. M. Neuberger, *A numerical method for finding sign-changing solutions of superlinear Dirichlet problems,* Nonlinear World 4 (1997), no. 1, 73–83.

[48] W. Ni, I. Takagi, and J. Wei, *On the location and profile of spike-layer solutions to a singularly perturbed semilinear Dirichlet problem: intermediate solutions,* Duke Math. J. 94 (1998), no. 3, pp. 597-618.

[49] S. Oruganti, J. Shi, and R. Shivaji, *Diffusive logistic equation with constant yield harvesting. I. Steady states,* Trans. Amer. Math. Soc. 354 (2002), no. 9, pp. 3601-3619 (electronic.)

[50] K. Perera, *Resonance problems with respect to the Fučik spectrum of the p-Laplacian,* Electron. J. Differential Equations 2002, No. 36, 10 pp. (electronic.)

[51] S. I. Pohozaev, *On the eigenfunctions of the equation* $\Delta u + \lambda f(u) = 0$. Dokl. Akad. Nauk SSSR 165 (965), pp. 36-39.

[52] P. H. Rabinowitz, *Some minimax theorems and applications to nonlinear partial differential equations,* Nonlinear analysis (collection of papers in honor of Erich H. Rothe), Academic Press, New York, (1978), pp. 161-177.

[53] P. H. Rabinowitz, *Minimax methods in critical point theory with applications to differential equations,* CBMS Regional Conference Series in Mathematics, No. 65. Published for the Conference Board of the Mathematical Sciences, Washington, DC; and the American Mathematical Society, Providence, RI, (1986).

[54] V. Shapiro, *Singular quasilinearity and higher eigenvalues,* Mem. Amer. Math. Soc. 153 (2001), no. 726.

[55] M. Schechter, *Linking Methods in Critical Point Theory,* Springer Verlag, New York, (1999).

[56] H. Therani, *A multiplicity result for the jumping nonlinearity problem,* preprint.

[57] Z. Q. Wang, *On a Superlinear Elliptic Equation,* Ann. Inst. H. Poincare Analyse Non Lineaire 8 (1991), 43-57.

[58] M. Willem, *Variational methods and almost solvability of semilinear equations,* Lecture Notes in Math., 846, Springer Verlag, Berlin, (1981).

E-mail address: `Castro@math.hmc.edu`

Contemporary Mathematics
Volume **357**, 2004

Semilinear Elliptic Boundary Value Problems with Nonlinear Oblique Boundary Conditions, a Boundary Element Monotone Iteration Approach

Goong Chen, Yuanhua Deng, Wei-Ming Ni, and Jianxin Zhou

ABSTRACT. Boundary element computation of an elliptic boundary value problem is studied. The governing equation is semilinear. The boundary condition, called a regular oblique derivative nonlinear type, contains linear normal and tangential derivative terms plus semilinear ones as well. The main thrust of the paper is to analyze the interaction between the spatial nonlinearity from the governing equation with the boundary nonlinearity from the boundary condition. This analysis leads to the formulation of condition $[H]$ and gives a contraction rate for the monotone iteration scheme. Boundary elements are then used to discretize the problem, whose error estimates are determined. Concrete examples of equations with multiple solutions are computed on the unit disk to experimentally examine the error estimates and to visualize the solution profiles.

1. Introduction

In a series of papers [**CD, CNZ, CNPZ, DC**], we have successfully utilized the boundary element methods (BEM) to compute numerical solutions of semilinear elliptic partial differential equations. Nevertheless, the boundary conditions in those papers were all linear. Here in this paper, the *main thrust* is to study the numerical analysis and computation of the *interaction between the nonlinearity in the governing equation with that in the boundary condition* for elliptic boundary value problems (BVP) using a boundary element monotone iteration scheme.

Boundary element analysis and computation of BVP containing certain nonlinearities have been studied by several researchers. First, if the governing equation is linear such as the potential equation and if *nonlinearity appears only in the boundary condition*, then we may first cite the work of Banerjee [**B**] where he discussed the computation of $\Delta w = f$ with nonlinear boundary conditions. Error analysis for such problems was first done by Ruotsalainen and Wendland [**RW**] for a nonlinear

2000 *Mathematics Subject Classification.* Primary 31B20, 35J65, 65N30.

Key words and phrases. Semilinear elliptic PDE, regular oblique nonlinear boundary condition, boundary elements, error analysis.

G. Chen and J. Zhou were supported in part by NSF Grants DMS 9610076 and NATO Grant CRG 940369. W. M. Ni was supported in part by NSF Grant DMS 9988635.

boundary element Galerkin approximation on a smooth boundary where the nonlinearity is monotone dissipative. Atkinson and Chandler [**AC**] studied the same problem based on the Nyström method. The collocation method was considered by Ruotsalainen and Saranen [**RS2**], Ganesh [**G**], and by Doucette [**Dou**] where he treated polygonal boundaries. See also a more recent paper by Efendiev, Schmitz and Wendland [**ES**]. When the governing equation(s) (single or coupled) contain semilinear terms but the boundary condition(s) are linear, then we may refer to [**CD, DC**] for the boundary element analysis of such problems. Nevertheless, when *both the governing equation and the boundary condition contain nonlinearities*, no results on the *boundary element error analysis* are available, to the best of our knowledge. Thus, this is the focus of our work here. Even though, quite naturally, the problem studied by us here may also be computed by *finite elements* or *finite differences* instead of boundary elements, we have not yet found many references in the literature dealing with error analysis for those computational methods, either.

The nonlinearity in the governing equation, commonly called *spatial* or *distributed nonlinearity*, interacts with the *boundary nonlinearity* (due to the nonlinearity in the boundary condition). These two types of nonlinearities in general do not correspond to a nonlinear monotone dissipative operator and, therefore, one does not have the useful linear or nonlinear coerciveness to utilize at allfor the purpose of error estimation. However, in this study we have observed that if either one of these nonlinearities is "bad" (see Remark 3.6 in §3) in the sense that it causes the associated linearized elliptic operator to lose coerciveness, then the "goodness" (see (3.35) and (3.37)) of the other nonlinearity can compensate to a certain extent that "badness". This is perhaps the most central idea of this paper which is formalized by us in §3, as can be seen particularly in the formulation of the basic assumption [H] therein. Here, in our consideration of nonlinear boundary conditions, we consider a type somewhat more general than those treated in the previously cited work [**AC, B, Dou, G, RS2**] by including *linear tangential derivative terms* in the nonlinear boundary condition, leading to a so-called *regular oblique derivative nonlinear boundary condition.* (A colleague of ours has communicated that a regular oblique derivative nonlinear boundary condition may arise in fracture mechanics where the boundary tangential derivatives are related to shear forces in elasticity. Nevertheless, we could not provide a precise reference for this in the literature.) Such nonlinear boundary conditions somehow provide weaker regularity properties for the linearized problem than those treated by us earlier in [**CD, DC**]. Therefore, a considerable amount of additional efforts is required to establish a key property such as (5.31) to ensure that the iterative numerical solutions stay within a small error bound of the given super- and subsolutions. Incidentally, we may also note that the occurrence of the tangential boundary derivative prevents the usual Ritz minimization – variational formulation for a regular oblique derivative BVP. So more work needs to be done in the *finite element error analysis* for such BVP. See the contributions made by the Czech school of numerical analysts.

The main error estimates are given in Theorem 5.1. Concrete examples with two multiple solutions on the unit disk are computed and discussed in §6.

2. Monotone Iteration Scheme for Regular Oblique Derivative Nonlinear Boundary Conditions

Throughout, we consider the following model problem of a semilinear elliptic equation with a nonlinear boundary condition:

$$\begin{cases} \Delta w(x) = f(x, w(x)), & x \in \Omega, \\ \dfrac{\partial w(x)}{\partial n} + a(x) \cdot \overline{\nabla} w(x) + g(x, w(x)) = 0, & x \in \partial\Omega, \end{cases} \tag{2.1}$$

where Ω is bounded domain in $\mathbb{R}^N$, $N \geq 2$, with boundary $\partial\Omega$. The operator Δ is the Laplacian, and $\overline{\nabla}$ denotes the (hyper-) *surface gradient* defined on $\partial\Omega$:

$$\overline{\nabla} = \nabla - n\frac{\partial}{\partial n}, \qquad (n = \text{ the unit outward pointing normal on } \partial\Omega) \tag{2.2}$$

i.e., for any smooth function h defined in a neighborhood of $\partial\Omega$, $\overline{\nabla} h$ is the projection of ∇h in the tangent space $T(\partial\Omega)$. The vector-valued function $a\colon\ \partial\Omega \to \mathbb{R}^N$ is a tangent vector field on $\partial\Omega$ and, thus, $a(x) \cdot \overline{\nabla} w(x)$ denotes a tangential derivative of the function w on $\partial\Omega$. (Note that even though (2.2) seems to require that $\overline{\nabla} h$ be defined only for functions h that are defined in a *neighborhood of* $\partial\Omega$, through an extension and continuity argument we can easily make $\overline{\nabla} h$ well-defined for functions h *defined on* $\partial\Omega$ *only*.) The two nonlinear maps f and g in (2.1) satisfy

$$\begin{aligned} &f\colon\ \Omega \times I \to \mathbb{R}, \quad g\colon\ \partial\Omega \times I \to \mathbb{R}, \\ &I = [\alpha, \beta] \subset \mathbb{R} \quad \text{for certain given} \quad \alpha, \beta. \end{aligned}$$

All of the data in (2.1), including f, a, g and $\partial\Omega$, are assumed to be sufficiently smooth. For most of our purposes in this paper, $C^{m,\alpha}$-smoothness of these data for some $m \geq 2$, $0 < \alpha < 1$, suffices.

The boundary condition $(2.1)_2$ is a *regular oblique derivative nonlinear boundary condition.* Let $W^{r,p}(M)$ denote the usual Sobolev space of order $r \in \mathbb{R}$ of power p on a manifold M. For $p = 2$, we write $H^r(M)$ instead of $W^{r,2}(M)$. The $H^r(M)$-norm is denoted as $\|\ \|_{r,M}$. For a *linear* regular oblique derivative elliptic boundary value problem, we have the following regularity and continuity with respect to data.

PROPOSITION 2.1. *Let* $b_1(x) \geq 0$ *and* $b_2(x) > 0$ *on* $\partial\Omega$. *Assume that* $a(\cdot)$ *and* $b_i(\cdot)$, $i = 1, 2$, *and* $\partial\Omega$ *are sufficiently smooth. Consider the linear elliptic boundary value problem*

$$\begin{cases} \Delta w(x) - b_1(x)w(x) = h_1(x) & \text{on } \Omega, \\ \dfrac{\partial w(x)}{\partial n} + a(x) \cdot \overline{\nabla} w(x) + b_2(x)w(x) = h_2(x), & \text{on } \partial\Omega. \end{cases} \tag{2.3}$$

(i) *For* $h_1 \in C^{0,\alpha}(\overline{\Omega})$ *and* $h_2 \in C^{1,\alpha}(\partial\Omega)$, *where* $0 < \alpha \leq 1$, *(2.3) has a unique solution* $w \in C^{2,\alpha}(\overline{\Omega})$ *such that*

$$C_1(\|h_1\|_{C^{0,\alpha}(\overline{\Omega})} + \|h_2\|_{C^{1,\alpha}(\partial\Omega)}) \leq \|w\|_{C^{2,\alpha}(\overline{\Omega})} \leq C_2(\|h_1\|_{C^{0,\alpha}(\overline{\Omega})} + \|h_2\|_{C^{1,\alpha}(\partial\Omega)}), \tag{2.4}$$

where $C_1, C_2 > 0$ *are independent of* h_1 *and* h_2.

(ii) *For* $h_1 \in W^{r_1,p}(\Omega)$ *and* $h_2 \in W^{r_2,p}(\partial\Omega)$ *with* $p > 1$, $r_1 \geq -1$ *and* $r_2 \in \mathbb{R}$, *(2.3) has a unique solution* $w \in W^{s,p}(\Omega)$ *with* $s = \min\left(2 + r_1, \frac{3}{2} + r_2\right)$

such that

$$C_3(\|h_1\|_{W^{s-2,p}(\Omega)} + \|h_2\|_{W^{s-\frac{3}{2},p}(\partial\Omega)}) \leq \|w\|_{W^{s,p}(\Omega)}$$
$$\leq C_4(\|h_1\|_{W^{r_1,p}(\Omega)} + \|h_2\|_{W^{r_2,p}(\partial\Omega)}), \tag{2.5}$$

where $C_3, C_4 > 0$ are independent of $h_1(\cdot)$ and $h_2(\cdot)$.

PROOF. The Schauder estimates (2.4) are given in [**GT**, Sect. 6.7] (in particular, see Theorems 6.30 and 6.31, and equation (6.80), p. 128, therein) through an argument making the boundary condition $(2.3)_2$ homotopic to the usual Neumann boundary condition $\partial w/\partial n + w = h_2$ on $\partial\Omega$.

For the $W^{s,p}$ estimates, there do not seem to be many references readily available. However, we may quote the (now standard) pseudodifferential operator approach for elliptic systems (see [**T**, Chap. V] and [**K**, Chap. 6]) to check that the Lopatinski-Shapiro condition [**K**, p. 195] is satisfied and therefore the overall system (2.3) is *elliptic*, which then yields the usual shifts of regularity (under the assumption of smooth data). But the ellipticity is already evident in the Schauder estimates (2.4). A much more detailed discussion involving the evaluation of the Lopatinski determinant [**K**, (1.15), p. 195] would constitute a digression from the primary numerical interests of this paper, without gaining significant new knowledge. We therefore will not pursue this issue any further. □

We now proceed to discuss the monotone iteration scheme for (2.1). A pair of $C^2(\overline{\Omega})$- functions $\underline{u}(x)$ and $\bar{u}(x)$ are said to be, respectively, a *subsolution* and a *supersolution* for (2.1) if they satisfy

$$\begin{cases} \Delta\underline{u}(x) - f(x, \underline{u}(x)) \geq 0, & \text{on } \Omega, \\ \dfrac{\partial \underline{u}(x)}{\partial n} + a(x) \cdot \overline{\nabla}\underline{u}(x) + g(x, \underline{u}(x)) \leq 0, & \text{on } \partial\Omega, \end{cases}$$

$$\begin{cases} \Delta\bar{u}(x) - f(x, \bar{u}(x)) \leq 0, & \text{on } \Omega, \\ \dfrac{\partial \bar{u}(x)}{\partial n} + a(x) \cdot \overline{\nabla}\bar{u}(x) + g(x, \bar{u}(x)) \geq 0, & \text{on } \partial\Omega, \end{cases}$$

and

$$\underline{u}(x) \leq \bar{u}(x), \qquad x \in \Omega.$$

We define the *sector*

$$\Theta = \{u \in C(\overline{\Omega}) \mid \underline{u}(x) \leq u(x) \leq \bar{u}(x)\}, \tag{2.6}$$

and let its trace set be

$$\partial\Theta = \{u|_{\partial\Omega} \mid u \in \Theta\}. \tag{2.7}$$

Now choose $\lambda > 0$, $\mu > 0$. Write

$$f_\lambda(x, w) \equiv \lambda w - f(x, w), \quad g_\mu(x, w) \equiv \mu w - g(x, w).$$

We assume that λ and μ are positive and sufficiently large such that

$$\lambda - \frac{\partial f(x, u)}{\partial u} > 0, \quad \forall x \in \overline{\Omega}, u \in \Theta; \qquad \mu - \frac{\partial g(x, u)}{\partial u} > 0, \quad \forall x \in \partial\Omega, u \in \partial\Theta. \tag{2.8}$$

THEOREM 2.2 (The Monotone Iteration Scheme). *Assume that a pair of subsolution and supersolution $(\underline{u}, \bar{u})$ exist for (2.1). Then the mapping*

$$\left.\begin{array}{l} T\colon \ \phi \mapsto w, \quad w = T\phi, \\ \phi \in C^2(\overline{\Omega}) \cap \Theta, \end{array}\right\} \tag{2.9}$$

where $w(x)$ is the unique solution to the linear BVP

$$(2.10)\qquad \begin{cases} \Delta w(x) - \lambda w(x) = -f_\lambda(x, \phi(x)) & \text{on } \Omega, \\ \dfrac{\partial w(x)}{\partial n} + a(x) \cdot \overline{\nabla} w(x) + \mu w(x) = g_\mu(x, \phi(x)) & \text{on } \partial\Omega, \end{cases}$$

is monotone, i.e., for any ϕ_1, ϕ_2 satisfying (2.9) and $\phi_1 \leq \phi_2$, we have

$$(2.11)\qquad T\phi_1, T\phi_2 \text{ satisfy (2.10), and } T\phi_1 \leq T\phi_2 \text{ on } \Omega.$$

Consequently, the iterations

$$(2.12)\qquad \begin{cases} \bar{u}_0(x) = \bar{u}(x), \\ (\Delta - \lambda)\bar{u}_{m+1}(x) = -f_\lambda(x, \bar{u}_m(x)) & \text{on } \Omega, m = 0, 1, 2, \ldots, \\ \dfrac{\partial}{\partial n}\bar{u}_{m+1}(x) + a(x) \cdot \overline{\nabla}\bar{u}_{m+1}(x) + \mu\bar{u}_{m+1}(x) \\ = g_\mu(x, \bar{u}_m(x)) & \text{on } \partial\Omega, \end{cases}$$

and

$$(2.13)\qquad \begin{cases} \underline{u}_0(x) = \underline{u}(x), \\ (\Delta - \lambda)\underline{u}_{m+1}(x) = -f_\lambda(x, \underline{u}_m(x)) & \text{on } \Omega, m = 0, 1, 2, \ldots, \\ \dfrac{\partial}{\partial n}\underline{u}_{m+1}(x) + a(x) \cdot \overline{\nabla}\underline{u}_{m+1}(x) + \mu\underline{u}_{m+1}(x) \\ = g_\mu(x, \underline{u}_m(x)) & \text{on } \partial\Omega, \end{cases}$$

yield iterates $\underline{u}_m$ and $\bar{u}_m$ satisfying

$$(2.14)\qquad \underline{u} = \underline{u}_0 \leq \underline{u}_1 \leq \ldots \leq \underline{u}_m \leq \ldots \leq \bar{u}_m \leq \ldots \leq \bar{u}_1 \leq \bar{u}_0 = \bar{u},$$

so that the limits

$$(2.15)\qquad \underline{u}_\infty(x) \equiv \lim_{m \to \infty} \underline{u}_m(x), \quad \bar{u}_\infty(x) \equiv \lim_{m \to \infty} \bar{u}_m(x)$$

exist in $C^2(\overline{\Omega})$; they both are solutions to (2.1). We have

(i) *$\underline{u}_\infty \leq \bar{u}_\infty$ on $\overline{\Omega}$;*

(ii) *$\underline{u}_\infty$ and $\bar{u}_\infty$ are, respectively, stable from below and above;*

(iii) *if $\underline{u}_\infty \not\equiv \bar{u}_\infty$, and both $\underline{u}_\infty$ and $\bar{u}_\infty$ are asymptotically stable, then there exists an unstable solution $\phi \in C^2(\overline{\Omega})$ such that $\underline{u}_\infty \leq \phi \leq \bar{u}_\infty$.*

PROOF. Let us adapt the proofs in [**N, P**] for (2.11). First, note that (2.10) is uniquely solvable because of Proposition 2.1. Let $w_i = T\phi_i$ for $i = 1, 2$. Since w_i and ϕ_i satisfy (2.10) with $w = w_i$ $\phi = \phi_i$ therein for $i = 1, 2$, letting $v(x) \equiv w_2(x) - w_1(x)$ and subtracting, we obtain

$$\Delta v(x) - \lambda v(x) = -\{\lambda[\phi_2(x) - \phi_1(x)] - [f(x, \phi_2(x)) - f(x, \phi_1(x))]\} \leq 0 \text{ on } \Omega,$$

$$\begin{aligned} \frac{\partial v(x)}{\partial v} &+ a(x) \cdot \overline{\nabla} v(x) + \mu v(x) = \mu[\phi_2(x) - \phi_1(x)] \\ &- [g(x, \phi_2(x)) - g(x, \phi_1(x))] \geq 0 \text{ on } \partial\Omega, \end{aligned}$$

both due to (2.8). Now assume that $\min_{\overline{\Omega}} v < 0$. Let $v(P) = \min_{\overline{\Omega}} v$, where $P \in \overline{\Omega}$. There are two cases:

(a) $P \in \Omega$. Then

$$0 \geq (\Delta - \lambda)v(x)|_{x=P} \geq 0 - \lambda v(P) > 0,$$

a contradiction.

(b) $P \in \partial\Omega$. Then since $a(x) \cdot \overline{\nabla} v(x)$ is a tangential differentiation along $\partial\Omega$, we have $a(x) \cdot \overline{\nabla} v(x)|_{x \in P} = 0$. Therefore

$$\left[\frac{\partial v(x)}{\partial n} + \mu v(x)\right]\Big|_{x=P} \geq 0,$$

and we conclude $\frac{\partial v(P)}{\partial n} > 0$, again a contradiction because v is decreasing from Ω toward $P \in \partial\Omega$ along the direction n at P.

Therefore $v(x) \geq 0$ on $\overline{\Omega}$ and (2.11) is verified. (Actually, using Hopf's Boundary Point Lemma, one can show that either $v \equiv 0$ or $v > 0$ on Ω.) The results (i) and (ii) follow. Part (iii) will not be needed subsequently and its proof is omitted. □

Let F be a tangential vector field defined on $\partial\Omega$. We define the *surface divergence* of $F, \overline{\nabla} \cdot F$, as in [**M**]; [**CK**, p. 60]. Then it is known that the Gauss Divergence Theorem can be generalized to $\partial\Omega$ in the following form:

$$\int_{\partial\Omega} F \cdot \overline{\nabla}\phi \, d\sigma = -\int_{\partial\Omega} (\overline{\nabla} \cdot F)\phi \, d\sigma \tag{2.16}$$

for all sufficiently smooth function ϕ defined on $\partial\Omega$.

Proposition 2.3. *Let $\alpha, f_1, f_2, : \partial\Omega \to \mathbb{R}$ and let F be a tangential vector field defined on $\partial\Omega$. Then we have*

(2.17)
(i)
$$\overline{\nabla}(f_1(x)f_2(x)) = f_1(x)\overline{\nabla} f_2(x) + f_2(x)\overline{\nabla} f_1(x),$$

(2.18)
(ii)
$$\overline{\nabla} \cdot (\alpha(x)F(x)) = \overline{\nabla}\alpha(x) \cdot F(x) + \alpha(x)\overline{\nabla} \cdot F(x).$$

Proof. Property (i) easily follows from (2.2) by extending the functions f_1 and f_2 smoothly to a neighborhood of $\partial\Omega$. Property (ii) can be easily obtained from (i) and (2.16) using a denseness argument. □

Corollary 2.4. *Let F be a $C^1(\partial\Omega)$-smooth tangential vector field defined on $\partial\Omega$. Then*

$$\int_{\partial\Omega} \overline{\nabla} \cdot F \, d\sigma = 0. \tag{2.19}$$

Proof. Use $\phi = 1$ in (2.16). □

Corollary 2.4 indicates that $\overline{\nabla} \cdot F$ *will never be of one sign* on $\partial\Omega$.

As a final comment for this sectin, we note that the presence of the tangential derivative terms in the nonlinear boundary condition $(2.1)_2$ prevents the existence of a variational functional (i.e., a first integral) for which w in (2.1) is a solution of the variational problem whether in the form of max, min or minimax, locally or globally. The same comment also applies to the linear problems (2.3), (2.12) and (2.13).

3. A Contraction Convergence Rate for the Monotone Iteration Scheme

In general, the rate of monotone convergence (2.14) cannot be determined. Obviously, additional assumptions are required in order to achieve this. In this section, we will explore the interaction between the distributed nonlinearity f in $(2.1)_1$ with the boundary nonlinearity g in $(2.1)_2$ to obtain a contraction rate for convergence.

Denote $\lambda_D^* > 0$ the smallest eigenvalue of $-\Delta$ on Ω corresponding to the zero Dirichlet boundary condition. We first present two lemmas.

LEMMA 3.1. *For any $\nu_1 \in \mathbb{R}$, we have*

$$-\infty < \lambda_1^*(\nu_1) \equiv \inf_{\substack{\phi \in H^1(\Omega) \\ \phi \not\equiv 0}} \frac{\int_\Omega |\nabla\phi|^2 dx + \nu_1 \int_{\partial\Omega} \phi^2 \, d\sigma}{\int_\Omega \phi^2 \, dx} < \infty \tag{3.1}$$

for some $\lambda_1^*(\nu_1) \in \mathbb{R}$; $\lambda_1^*(\cdot)$ satisfies

$$\lambda_1^*(0) = 0, \quad \lambda_1^*\colon (-\infty,\infty) \to (-\infty, \lambda_D^*) \text{ strictly monotonically increasing and onto.} \tag{3.2}$$

The infimum in (3.1) is attained by a nontrivial function ϕ^* satisfying

$$\begin{cases} \Delta\phi^* + \lambda_1^*(\nu_1)\phi^* = 0 & \text{on } \Omega, \\ \dfrac{\partial\phi^*}{\partial n} + \nu_1\phi^* = 0 & \text{on } \partial\Omega. \end{cases} \tag{3.3}$$

PROOF. If $\nu_1 > 0$, then $\int_\Omega |\nabla\phi|^2 \, dx + \nu_1 \int_{\partial\Omega} \phi^2 \, d\sigma$ defines an equivalent norm in $H^1(\Omega)$ [**DL**]. It is routine to show that $\lambda_1^*(\nu_1)$ exists and is positive such that (3.3) is satisfied by some $\phi^* \not\equiv 0$. If $\nu_1 = 0$, then we can also show that $\lambda_1^*(0) = 0$ and $\phi^* \equiv 1$ satisfies (3.3). So let us consider the case $\nu_1 < 0$. Thus $-\nu_1 > 0$. By the proof of the Trace Theorem (see [**E**, pp. 258–259], e.g.), we infer that for any $(-\nu_1) > 0$ there exists some $K > 0$ (depending only on $-\nu_1$) sufficiently large such that

$$(-\nu_1) \int_{\partial\Omega} \phi^2 \, d\sigma \le \int_\Omega |\nabla\phi|^2 \, dx + K \int_\Omega \phi^2 \, dx, \qquad \forall \phi \in H^1(\Omega).$$

Therefore

$$\frac{\int_\Omega |\nabla\phi|^2 dx + \nu_1 \int_{\partial\Omega} \phi^2 \, dx}{\int_\Omega \phi^2 \, dx} \ge \frac{(-K)\int_\Omega \phi^2 \, dx}{\int_\Omega \phi^2 \, dx} = -K > -\infty,$$

and the infimum in (3.1) is bounded from below by $-K$. It is trivial to see that the infimum in (3.1) is bounded from above by 0 if we try $\phi = 1$. Therefore $\lambda_1^*(\nu_1)$ is finite and negative when $\nu_1 < 0$. The existence of a nontrivial eigenfunction ϕ^* satisfying (3.3) follows from the standard calculus of variations arguments. The monotonicity property (3.2) can also be verified without too much difficulty; see [**Ch**]; [**CZ2**, §3.1], e.g. □

LEMMA 3.2. *For any $\nu_2 \in \mathbb{R}$, consider the minimization problem*

$$\lambda_2^*(\nu_2) \equiv \inf_{\substack{\phi \in H^1(\Omega) \\ \phi \not\equiv 0 \text{ on } \partial\Omega}} \frac{\int_\Omega [|\nabla\phi|^2 + \nu_2\phi^2]dx}{\int_{\partial\Omega} \phi^2 \, d\sigma}. \tag{3.4}$$

Then we have

$$\text{(i)} \quad 0 = \lambda_2^*(0) < \lambda_2^*(\alpha_1) < \lambda_2^*(\alpha_2) < \infty \text{ if } 0 < \alpha_1 < \alpha_2; \tag{3.5}$$

$$\lim \lambda_2^*(\nu_2) \to \infty \text{ as } \nu_2 \to \infty;$$

$$\text{(ii)} \quad -\infty < \lambda_2^*(\alpha_1) < \lambda_2^*(\alpha_2) < 0 \quad \text{if} \quad -\lambda_D^* < \alpha_1 < \alpha_2 < 0; \tag{3.6}$$

$$\text{(iii)} \quad \lambda_2^*(\nu_2) = -\infty \quad \text{if} \quad \nu_2 < -\lambda_D^*. \tag{3.7}$$

In cases (i) and (ii) above, the infimum in (3.4) is attained by a nontrivial function ψ^* satisfying

$$\begin{cases} \Delta\psi^* - \nu_2\psi^* = 0 & \text{on } \Omega, \\ \dfrac{\partial\psi^*}{\partial n} - \lambda_2^*(\nu_2)\psi^* = 0 & \text{on } \partial\Omega. \end{cases} \tag{3.8}$$

PROOF. If $\nu_2 \geq 0$, then the arguments for (3.5) and (3.8) go essentially in the same way as in the proof of Lemma 3.1. So let us consider only the case $\nu_2 < 0$. Thus $-\nu_2 > 0$. First, let $0 < -\nu_2 < \lambda_D^*$. By (3.2), the function $\lambda_1^*(\cdot)\colon (0,\infty) \to (0,\lambda_D^*)$ is strictly monotonic, therefore for each $-\nu_2 \in (0,\lambda_D^*)$ there exists a unique $\nu_1 > 0$ such that $\lambda_1^*(\nu_1) = -\nu_2$. By this and (3.1), we have

$$(-\nu_2)\int_\Omega \phi^2 \, dx = \lambda_1^*(\nu_1)\int_\Omega \phi^2 \, dx \leq \int_\Omega |\nabla\phi|^2 \, dx + \nu_1 \int_{\partial\Omega} \phi^2 \, dx, \qquad \forall \phi \in H^1(\Omega).$$

Therefore

$$\frac{\int_\Omega |\nabla\phi|^2 \, dx + \nu_2 \int_\Omega \phi^2 \, dx}{\int_{\partial\Omega} \phi^2 \, d\sigma} \geq \frac{-\nu_1 \int_{\partial\Omega} \phi^2 \, d\sigma}{\int_{\partial\Omega} \phi^2 \, d\sigma} = -\nu_1 > -\infty, \qquad \forall \phi \in H^1(\Omega),$$

and we have established (3.6). Now consider the case $\nu_2 < -\lambda_D^*$. Let $\hat\phi \in H^1(\Omega)$ satisfy

$$0 < \int_\Omega |\nabla\hat\phi|^2 dx < (-\nu_2)\int_\Omega \hat\phi^2 \, dx, \quad \hat\phi|_{\partial\Omega} \equiv 0,$$

which is possible because $-\nu_2 > \lambda_D^*$ and thus we can find some $\hat\phi$ violating the Poincaré inequality. Choose a sequence $\hat\phi_m \in H^1(\Omega)$ such that

$$\hat\phi_m \to \hat\phi \quad \text{in} \quad H^1(\Omega), \quad \hat\phi_m|_{\partial\Omega} \not\equiv 0.$$

Then obviously

$$\lim_{m\to\infty} \frac{\int_\Omega |\nabla\hat\phi_m|^2 dx + \nu_2 \int_\Omega \hat\phi_m^2 \, dx}{\int_{\partial\Omega} \hat\phi_m^2 d\sigma} = -\infty$$

because the denominator is vanishing and the numerator is becoming negative. Therefore (3.7) has been confirmed. □

REMARK 3.3. Comparing (3.3) with (3.8), we note the correspondence

$$\left.\begin{aligned} \lambda_1^{*^{-1}}(-\nu_2) &= -\lambda_2^*(\nu_2), \quad \forall \nu_2 \in (-\lambda_D^*, \infty), \\ \lambda_2^{*^{-1}}(-\nu_1) &= -\lambda_1^*(\nu_1), \quad \forall \nu_1 \in (-\infty, \lambda_D^*), \end{aligned}\right\} \tag{3.9}$$

where $\lambda_1^{*^{-1}}(\cdot)$ and $\lambda_2^{*^{-1}}(\cdot)$ are, respectively, the inverse functions of $\lambda_1^*(\cdot)$ and $\lambda_2^*(\cdot)$.□

THEOREM 3.4 (A Contraction Rate for the Monotone Iteration Scheme). *Assume that (2.1) possesses a pair of subsolution and supersolution $(\underline{u}, \bar{u})$. Assume further that there exists a $\nu_1 \in \mathbb{R}$ such that*

$$[H] \quad \begin{cases} \nu_1 + \dfrac{1}{2}(\overline{\nabla} \cdot a) < \dfrac{\partial g(x, u(x))}{\partial u}, & \forall x \in \partial\Omega, u \in \partial\Theta; \\ \lambda_1^*(\nu_1) > -\dfrac{\partial f(x, u(x))}{\partial u}, & \forall x \in \Omega, u \in \Theta. \end{cases}$$

Consider the monotone iteration (2.12) where $\lambda > 0, \mu > 0$ are chosen so large that

$$\lambda + \lambda_1^*(\nu_1) > 0, \quad \mu - \nu_1 - \frac{1}{2}(\overline{\nabla} \cdot a) > 0 \quad \text{on} \quad \partial\Omega \tag{3.10}$$

and that (2.8) is satisfied. Then there exists $C > 0$ and $0 < \theta < 1$ independent of m such that

$$|||\bar{e}_m||| \le C\theta^m |||\bar{e}_0|||, \qquad m = 1, 2, 3, \ldots, \tag{3.11}$$

where $\bar{e}_m = \bar{u}_\infty - \bar{u}_m$ is the error, and

$$|||\bar{e}_m|||^2 \equiv \int_\Omega \bar{e}_m^2 \, dx + \int_{\partial\Omega} \bar{e}_m^2 \, d\sigma.$$

The same also holds for the monotone iteration (2.13):

$$|||\underline{e}_m||| \le C\theta^m |||\underline{e}_0|||; \quad \underline{e}_m = \underline{u}_\infty - \underline{u}_m, \qquad m = 1, 2, 3, \ldots .$$

PROOF. We write (2.12) or (2.13) as

$$\begin{cases} -\Delta u_{m+1}(x) + \lambda u_{m+1}(x) = f_\lambda(x, u_m(x)) & \text{on } \Omega, \\ \dfrac{\partial u_{m+1}(x)}{\partial n} + a(x) \cdot \overline{\nabla} u_{m+1}(x) + \mu u_{m+1}(x) = g_\mu(x, u_m(x)) & \text{on } \partial\Omega, \end{cases} \tag{3.12}$$

where u_{m+1} is $\bar{u}_{m+1}$ (resp., $\underline{u}_{m+1}$). Since $\bar{u}_\infty$ (resp., $\underline{u}_\infty$) is a solution to (2.1), we have

$$\begin{cases} -\Delta u_\infty(x) + \lambda u_\infty(x) = f_\lambda(x, u_\infty(x)) & \text{on } \Omega, \\ \dfrac{\partial u_\infty(x)}{\partial n} + a(x) \cdot \overline{\nabla} u_\infty(x) + \mu u_\infty(x) = g_\mu(x, u_\infty(x)) & \text{on } \partial\Omega, \end{cases} \tag{3.13}$$

where $u_\infty = \bar{u}_\infty$ (resp., $\underline{u}_\infty$). Subtracting (3.12) from (3.13) and applying the Mean Value Theorem, we obtain

$$\begin{cases} -\Delta e_{m+1}(x) + \lambda e_{m+1}(x) = \left[\lambda - \dfrac{\partial f(x, \tilde{u}_m(x))}{\partial u}\right] e_m(x) & \text{on } \Omega, \\ \dfrac{\partial e_{m+1}(x)}{\partial n} + a(x) \cdot \overline{\nabla} e_{m+1}(x) + \mu e_{m+1}(x) = \left[\mu - \dfrac{\partial g(x, \tilde{\tilde{u}}_m(x))}{\partial u}\right] e_m(x) & \text{on } \partial\Omega, \end{cases} \tag{3.14}$$

for some $C^2(\overline{\Omega})$-functions $\tilde{u}_m, \tilde{\tilde{u}}_m$ such that $u_\infty \le \tilde{u}_m, \tilde{\tilde{u}}_m \le u_m$ (resp., $u_m \le \tilde{u}_m, \tilde{\tilde{u}}_m \le u_\infty$). Multiply e_{m+1} to both sides of $(3.14)_1$, integrate by parts, and substitute $(3.14)_2$ in for $\partial e_{m+1}/\partial n$:

$$-\int_{\partial\Omega} \frac{\partial e_{m+1}}{\partial n} e_{m+1} d\sigma + \int_{\Omega} [|\nabla e_{m+1}|^2 + \lambda e_{m+1}^2] dx = \int_{\Omega} \left[\lambda - \frac{\partial f(\cdot, \tilde{u}_m)}{\partial u}\right] e_m e_{m+1}\, dx,$$

$$\int_{\partial\Omega} [(a \cdot \overline{\nabla} e_{m+1}) e_{m+1} + \mu e_{m+1}^2] d\sigma + \int_{\Omega} [|\nabla e_{m+1}|^2 + \lambda e_{m+1}^2] dx$$

(3.15)

$$= \int_{\partial\Omega} \left[\mu - \frac{\partial g(\cdot, \tilde{\tilde{u}}_m)}{\partial u}\right] e_{m+1} e_m \, d\sigma + \int_{\Omega} \left[\lambda - \frac{\partial f(\cdot, \tilde{u}_m)}{\partial u}\right] e_{m+1} e_m \, dx.$$

But by (2.17) $(a \cdot \overline{\nabla} e_{m+1}) e_{m+1} = a \cdot \overline{\nabla}(e_{m+1}^2)/2$ and by (2.16), we have

$$\int_{\partial\Omega} (a \cdot \overline{\nabla} e_{m+1}) e_{m+1} \, d\sigma = -\frac{1}{2} \int_{\partial\Omega} (\overline{\nabla} \cdot a) e_{m+1}^2 \, d\sigma. \tag{3.16}$$

Substituting (3.16) into (3.15) and introducing $\nu_1 \in \mathbb{R}$ in (3.15) as follows:

$$\int_{\partial\Omega} \left[\mu - \frac{1}{2}(\overline{\nabla} \cdot a) - \nu_1\right] e_{m+1}^2 \, d\sigma + \left[\int_{\Omega} |\nabla e_{m+1}|^2 \, dx + \nu_1 \int_{\partial\Omega} e_{m+1}^2 \, d\sigma\right]$$

$$+ \lambda \int_{\Omega} e_{m+1}^2 \, dx$$

$$= \int_{\partial\Omega} \left[\mu - \frac{\partial g(\cdot, \tilde{\tilde{u}}_m)}{\partial u}\right] e_{m+1} e_m \, d\sigma + \int_{\Omega} \left[\lambda - \frac{\partial f(\cdot, \tilde{u}_m)}{\partial u}\right] e_{m+1} e_m \, dx, \tag{3.17}$$

from (3.1) we now obtain

$$\int_{\partial\Omega} \left[\mu - \nu_1 - \frac{1}{2}(\overline{\nabla} \cdot a)\right] e_{m+1}^2 \, d\sigma + [\lambda + \lambda_1^*(\nu_1)] \int_{\Omega} e_{m+1}^2 \, dx$$

$$\leq \int_{\partial\Omega} \left[\mu - \frac{\partial g(\cdot, \tilde{\tilde{u}}_m)}{\partial u}\right] |e_{m+1}|\, |e_m| \, d\sigma + \int_{\Omega} \left[\lambda - \frac{\partial f(\cdot, \tilde{u}_m)}{\partial u}\right] |e_{m+1}|\, |e_m| \, dx. \tag{3.18}$$

We choose $\lambda > 0$, $\mu > 0$ sufficiently large such that (3.10) holds and now invoke assumption $[H]$ and (2.8) to obtain

$$\left.\begin{array}{ll} \mu - \nu_1 - \frac{1}{2}(\overline{\nabla} \cdot a(x)) > \mu - \dfrac{\partial g(x, u(x))}{\partial u} > 0, & \forall x \in \partial\Omega, u \in \partial\Theta, \\ \lambda + \lambda_1^*(\nu_1) > \lambda - \frac{\partial f(x,u(x))}{\partial u} > 0, & \forall x \in \overline{\Omega}, u \in \Theta, \end{array}\right\} \tag{3.19}$$

and, as a consequence, there exists a $\tilde{\theta}$: $0 < \tilde{\theta} < 1$, such that

$$\left.\begin{array}{ll} 0 < \dfrac{\mu - \frac{\partial g(x,u(x))}{\partial u}}{\mu - \nu_1 - \frac{1}{2}\overline{\nabla} \cdot a(x)} \leq \tilde{\theta}, & \forall x \in \partial\Omega, u \in \partial\Theta, \\ 0 < \dfrac{\lambda - \frac{\partial f(x,u(x))}{\partial u}}{\lambda + \lambda_1^*(\nu_1)} \leq \tilde{\theta}, & \forall x \in \overline{\Omega}, u \in \Theta. \end{array}\right\} \tag{3.20}$$

The inequalities in (3.19) suggest the following norm: for u such that $|||u||| < \infty$, define

$$|||u|||^{\wedge} = \left\{ \int\limits_{\partial\Omega} \left[\mu - \nu_1 - \frac{1}{2}(\overline{\nabla} \cdot a)\right] u^2 \, d\sigma + [\lambda + \lambda_1^*(\nu_1)] \int\limits_{\Omega} u^2 \, dx \right\}^{1/2}. \tag{3.21}$$

This norm depends on λ, μ, ν_1 and a. By (3.10) there exist $\widehat{K}_1$ and $\widehat{K}_2 > 0$ such that

$$\widehat{K}_1|||u||| \le |||u|||^{\wedge} \le \widehat{K}_2|||u|||. \tag{3.22}$$

Continuing from (3.18) and using (3.20), we obtain

$$\begin{aligned}
|||e_{m+1}|||^{\wedge^2} \le & \int\limits_{\partial\Omega} \left[\frac{\mu - \frac{\partial g}{\partial u}(\cdot, \tilde{\tilde{u}}_m)}{\mu - \nu_1 - \frac{1}{2}(\overline{\nabla} \cdot a)}\right] \left[\mu - \nu_1 - \frac{1}{2}(\overline{\nabla} \cdot a)\right] |e_{m+1}| \, |e_m| \, d\sigma \\
& + \int\limits_{\Omega} \left[\frac{\lambda - \frac{\partial f}{\partial u}(\cdot, \tilde{u}_m)}{\lambda + \lambda_1^*(\nu_1)}\right] [\lambda + \lambda_1^*(\nu_1)] |e_{m+1}| \, |e_m| \, dx \\
& \le \tilde{\theta} \left\{ \int\limits_{\partial\Omega} \left[\mu - \nu_1 - \frac{1}{2}(\overline{\nabla} \cdot a)\right] |e_{m+1}| \, |e_m| \, d\sigma \right. \\
& \left. + [\lambda + \lambda_1^*(\nu_1)] \int\limits_{\Omega} |e_{m+1}| \, |e_m| \, dx \right\} \\
& \le \frac{\tilde{\theta}}{2} \left\{ \int\limits_{\partial\Omega} \left[\mu - \nu_1 - \frac{1}{2}(\overline{\nabla} \cdot a)\right] e_{m+1}^2 \, d\sigma + \int\limits_{\partial\Omega} \left[\mu - \nu_1 - \frac{1}{2}(\overline{\nabla} \cdot a)\right] e_m^2 \, d\sigma \right. \\
& \left. + [\lambda + \lambda_1^*(\nu_1)] \int\limits_{\Omega} e_{m+1}^2 \, dx + [\lambda + \lambda_1^*(\nu_1)] \int\limits_{\Omega} e_m^2 \, dx \right\}.
\end{aligned}$$

Hence

$$|||e_{m+1}|||^{\wedge^2} \le \theta^2 |||e_m|||^{\wedge^2} \tag{3.23}$$

where $0 < \theta^2 \equiv \tilde{\theta}/(2 - \tilde{\theta}) < 1$. Therefore

$$|||e_m|||^{\wedge} \le \theta^m |||e_0|||^{\wedge}. \tag{3.24}$$

Using (3.22) in (3.24), we have completed the proof of (3.11). □

COROLLARY 3.5. *Let $\Omega \subset \mathbb{R}^N$ for $N = 2, 3$ or 4. Under the same assumptions as in Theorem 3.4, we have*

$$\|e_m\|^2_{s_1,\Omega} \le C_1\theta^{2m}|||e_0|||^2 \le C_2\theta^{2m}\|e_0\|^2_{\frac{3}{2},\Omega}, \quad \text{for} \quad s_1 \le 3/2, \tag{3.25}$$

$$\|e_m\|^2_{s_2,\partial\Omega} \le C_3\theta^{2m}|||e_0|||^2, \quad \text{for} \quad s_2 \le 1, \tag{3.26}$$

$$\|e_m\|^2_{s_1,\Omega} + \|e_m\|^2_{s_2,\partial\Omega} \le C_4\theta^{2m}|||e_0|||^2, \qquad 0 \le s_1 \le 3/2, \quad 0 \le s_2 \le 1, \tag{3.27}$$

$$\begin{aligned} &\|e_m\|_{W^{r,p}(\Omega)} \le C_5\theta^m|||e_0|||, \quad r = 3/2, p \in \left[1, \frac{2N-2}{N-2}\right] \\ &\qquad \text{if } N = 3, 4; p \in [1, \infty) \text{ if } N = 2; \end{aligned} \tag{3.28}$$

$$\begin{aligned} &\|e_m\|_{C^{0,\alpha}(\overline{\Omega})} \le C_6\theta^m|||e_0|||; \quad \alpha \in \left(0, \frac{-N^2+5N-3}{2(N-1)}\right] \\ &\qquad \text{if } N = 3, 4; \alpha \in (0, 1) \text{ if } N = 2. \end{aligned} \tag{3.29}$$

for some $C_i > 0$, $i = 1, 2, \ldots, 6$, independent of m, but may depend on s_1, s_2, r, p and α.

PROOF. Apply the right inequality in (2.5) to (3.14) with $r_1 = r_2 = 0$ and $p = 2$ to obtain $e_{m+1} \in H^{\frac{3}{2}}(\Omega)$ and

$$\|e_{m+1}\|^2_{\frac{3}{2},\Omega} \le \widetilde{C}_1|||e_m|||^2,$$

where $\widetilde{C}_1 > 0$ is independent of m. Then apply (3.11) to obtain the left inequality in (3.25). The right inequality in (3.25) follows from the Trace Theorem.

Inequality (3.26) follows from the left inequality in (3.25) by using $s_1 = 3/2$ therein and the Trace Theorem.

Inequality (3.27) is a consequence of the left inequality in (3.25) and inequality (3.26).

Let Ω have dimensionality N with N as large as possible. We now consider (3.28). First, note from (3.27) that $e_m \in H^{3/2}(\Omega)$ and $e_m|_{\partial\Omega} \in H^1(\partial\Omega)$, by the Sobolev Imbedding Theorem ([**Ad**]) we have

(3.30)

(i) for $N > 3$: $e_m \in L^{p_1}(\Omega), e_m|_{\partial\Omega} \in L^{p_2}(\partial\Omega)$; $\quad p_1 \in \left[1, \frac{2N}{N-3}\right], p_2 \in \left[1, \frac{2N-2}{N-2}\right]$;

(ii) for $N = 3$: $e_m \in L^{p_1}(\Omega), e_m|_{\partial\Omega} \in L^{p_2}(\partial\Omega)$; $\quad p_1 \in [1, \infty), p_2 \in [1, 4]$;

(iii) for $N = 2$: $e_m \in C^{0,\alpha}(\overline{\Omega}), e_m|_{\partial\Omega} \in L^{p_2}(\partial\Omega)$; $\quad \alpha \in (0, 1/2], p_2 \in [1, \infty)$.

Thus, let us only consider the case $N > 3$ because the other cases are easier and they follow from the same argument. From $(3.30)_1$, choose $\hat{p}_1 = (2N)/(N-3)$, $\hat{p}_2 = (2N-2)/(N-2)$. Then $\hat{p}_2 < \hat{p}_1$ for all $N \ge 2$. The Sobolev Imbedding Theorem [**Ad**] and (3.27) gives

$$\|e_m\|_{L^{\hat{p}_2}(\Omega)} + \|e_m\|_{L^{\hat{p}_2}(\partial\Omega)} \le \widetilde{C}_2(\|e_m\|_{\frac{3}{2},\Omega} + \|e_m\|_{1,\partial\Omega}) \le \widetilde{C}_3\theta^m|||e_0|||. \tag{3.31}$$

Applying (3.31) and the right inequality in (2.5) with $r_1 = r_2 = 0$ and $p = \hat{p}_2$ to (3.14), we obtain

$$\begin{aligned} \|w\|_{W^{\frac{3}{2},\hat{p}_2}(\Omega)} &\le \widetilde{C}_4(\|e_m\|_{W^{0,\hat{p}_2}(\Omega)} + \|e_m\|_{W^{0,\hat{p}_2}(\partial\Omega)}) \\ &\le \widetilde{C}_5\theta^m|||e_0|||. \end{aligned} \tag{3.32}$$

Therefore (3.28) has been proved.

Finally, we consider (3.29). Let $N = 3$ or 4. Choose $r = 3/2$ and $\hat{p} = \frac{2N-2}{N-2}$ in (3.28) to get $e_m \in W_{\frac{3}{2},\hat{p}}(\Omega)$, and apply the Sobolev Imbedding Theorem [**Ad**] to obtain

$$\|e_m\|_{C^{0,\alpha}(\overline{\Omega})} \le \widetilde{C}_6 \|e_m\|_{W^{\frac{3}{2},\hat{p}}(\Omega)} \le \widetilde{C}_7 \theta^m |||e_0|||; \quad \alpha \in \left(0, \frac{-N^2+5N-3}{2(N-1)}\right]. \tag{3.33}$$

(Note that $-N^2 + 5N - 3 < 0$ if $N > 4$ and, therefore, N cannot go above 4.) If $N = 2$, then the same argument yields

$$\|e_m\|_{C^{0,\alpha}(\overline{\Omega})} \le \widetilde{C}_8 \theta^m |||e_0||| \quad \text{for any} \quad \alpha \in (0,1), \tag{3.34}$$

where $\widetilde{C}_8$ depends only on α. Therefore (3.29) is established. □

Inequality (3.29) is important in establishing a *crucial property (5.31)* below. Even though N in (3.29) can go as high as 4, due to other restrictions, see (5.29) and (5.30), so far we are only able to establish (5.31) for $N = 2$ and 3. As a consequence, the study in §5 covers only $N = 2$ and 3 but does not encompass $N = 4$.

REMARK 3.6. Let us examine the key assumption $[H]$ here. The nonlinearity $g(x, w(x))$ in the boundary condition $(2.1)_2$ is regarded as "*good*" if

$$\frac{\partial g(x,u)}{\partial u} \ge 0, \qquad x \in \partial\Omega, \tag{3.35}$$

because the above makes the boundary condition behave rather like

$$\frac{\partial u(x)}{\partial n} + \alpha(x)u(x) = h(x) \tag{3.36}$$

for some nonnegative function $\alpha(x)$. Ill-posedness of the boundary value problem is less likely to occur for (3.36) because it *reinforces the coerciveness* of the elliptic operator. All the nonlinearities g in [**RW**] necessarily satisfy (3.35) because of the *monotone dissipative* nature and assumption made therein. Conversely, if (3.35) is violated, then $\alpha(x)$ in (3.36) looks like a negative function which is known to impair the coerciveness of an elliptic operator; the more negative $\alpha(\cdot)$ in (3.36), the worse the loss of coerciveness. Thus it is "*bad*".

On the other hand, the nonlinearity of $f(x, w(x))$ in the governing equation $(2.1)_1$ is also regarded as "*good*" if

$$\frac{\partial f(x,u)}{\partial u} \ge 0. \tag{3.37}$$

For example, $f(x, w) = w^3$ is "good" because the equation $\Delta w = w^3$ (with the homogeneous Dirichlet boundary condition, e.g.) would correspond to a monotone dissipative operator and thus reinforces the coerciveness, while $f(x, w) = -w^3$, which is the opposite of (3.37), is considered as "*bad*" because it usually damages the coerciveness of the elliptic operator. Multiplicity (i.e., nonuniqueness) of solutions is well understood for the well-known *Lane-Emden equation* $\Delta w + w^3 = 0$ (see [**N**], e.g.).

Our condition $[H]$ allows either f or g to be "*bad*". For example, if (3.35) is violated, we are allowed to choose a negative ν_1 to make $[H]_1$ satisfied. (Note that one cannot enlist the help of the term $\frac{1}{2}(\overline{\nabla} \cdot a)$ because *it is never of one sign.* See the comment after Corollary 2.4.) In that case by (3.2) we have $\lambda_1^*(\nu_1) < 0$. To make $[H]_2$ satisfied, we must have

$$\frac{\partial f(x,u)}{\partial u} \geq -\lambda_1^*(\nu_1) > 0,$$

i.e., the nonlinearity f must be "rather good" in order to compensate the badness from the nonlinearity g.

One can also observe that if f is "bad", then g must be "good" in order for $[H]$ to be satisfied. Our gist here is that *the spatial nonlinearity f and the boundary nonlinearity g can mutually help each other.* □

REMARK 3.7. Since a major emphasis of this paper is to study the interaction between the spatial nonlinearity f with the boundary nonlinearity g, let us review the proof of Theorem 3.4, where the *key step* leading to the said interaction occurs in (3.17) through the introduction of ν_1 therein and an application of Lemma 3.1 afterwards. An alternative approach goes as follows: instead of introducing ν_1 one can introduce $\nu_2 \in (-\lambda_D^*, \infty)$ below to replace (3.17):

$$\int_{\partial\Omega} \left[\mu - \frac{1}{2}(\overline{\nabla} \cdot a)\right] e_{m+1}^2 \, d\sigma + \left[\int_{\Omega} |\nabla e_{m+1}|^2 \, dx + \nu_2 \int_{\Omega} e_{m+1}^2 \, dx\right]$$
$$+ (\lambda - \nu_2) \int_{\Omega} e_{m+1}^2 \, dx = \text{RHS of (3.17)}.$$

Then we apply Lemma 3.2 to obtain

$$\int_{\partial\Omega} \left[\mu - \frac{1}{2}(\overline{\nabla} \cdot a) + \lambda_2^*(\nu_2)\right] e_{m+1}^2 \, d\sigma + (\lambda - \nu_2) \int_{\Omega} e_{m+1}^2 \, dx \leq \text{RHS of (3.18)}.$$

The rest of the arguments in the proof of Theorem 3.4 also follow through if we replace $[H]$ by

$$[H]' \quad \begin{cases} \text{There exists a } \nu_2 \in (-\lambda_D^*, \infty) \text{ such that} & \\ \qquad \nu_2 < \dfrac{\partial f(x,u)}{\partial u}, & \forall x \in \Omega, \quad u \in \Theta; \\ \dfrac{1}{2}(\overline{\nabla} \cdot a) - \lambda_2^*(\nu_2) < \dfrac{\partial g(x,u)}{\partial u}, & \forall x \in \partial\Omega, \quad u \in \partial\Theta. \end{cases}$$

Therefore (3.11) is again obtained.

Note that $[H]'$ and $[H]$ are equivalent via (3.9). □

4. Properties of Boundary Integral Operators and Equations

Let $E(x, \lambda)$ be the fundamental solution of the Helmholtz operator $-\Delta + \lambda$:

$$-\Delta E(x,\lambda) + \lambda E(x,\lambda) = \delta(x), \qquad \lambda > 0. \tag{4.1}$$

We inherit the operators as well as the ensuing properties from [**CD, DC**]:

(i) $\widetilde{V}_\lambda$ denotes the volume (Newtonian) potential operator: for $\phi\colon\ \Omega \to \mathbb{R}$,

(4.2)

$$(\widetilde{V}_\lambda \phi)(x) = \int_{\Omega} E(x-y;\lambda)\phi(y)dy, \qquad x \in \Omega; \widetilde{V}_\lambda\colon\ H^s(\Omega) \to H^{s+2}(\Omega), \quad \forall s \geq -1;$$

(ii) $\widetilde{S}_\lambda$ denotes the simple-layer potential operator: for $\eta\colon\ \partial\Omega \to \mathbb{R}$,

(4.3)
$$(\widetilde{S}_\lambda \eta)(x) = \int_{\partial\Omega} E(x-y;\lambda)\eta(y)d\sigma_y, \qquad x \in \overline{\Omega};\, \widetilde{S}_\lambda\colon\ H^s(\partial\Omega) \to H^{s+\frac{3}{2}}(\Omega), \forall s \in \mathbb{R};$$

(iii) S_λ and V_λ are the trace of $\widetilde{S}_\lambda$ and $\widetilde{V}_\lambda$:

(4.4) $$S_\lambda \eta = (\widetilde{S}_\lambda \eta)|_{\partial\Omega}; \quad S_\lambda\colon\ H^s(\partial\Omega) \to H^{s+1}(\partial\Omega) \text{ isomorphically}, \forall s \in \mathbb{R};$$

(4.5) $$V_\lambda \phi = (\widetilde{V}_\lambda \phi)|_{\partial\Omega}; \quad V_\lambda\colon\ H^s(\Omega) \to H^{s+3/2}(\partial\Omega), \qquad \forall s \geq -1;$$

and V'_λ is the normal derivative of V_λ:

(4.6) $$V'_\lambda \phi = \frac{\partial}{\partial n}\widetilde{V}_\lambda \phi|_{\partial\Omega}; \quad V'_\lambda\colon\ H^s(\Omega) \to H^{s+1/2}(\partial\Omega), \quad \forall s \geq -1;$$

(iv) K_λ is the boundary integral operator defined by

$$(K_\lambda \eta)(x) = \int_{\partial\Omega} \frac{\partial E(x-y;\lambda)}{\partial n_x}\eta(y)d\sigma_y, \qquad x \in \partial\Omega;$$

(4.7) $$K_\lambda\colon\ H^s(\partial\Omega) \to H^{s+1}(\partial\Omega), \quad \forall s \in \mathbb{R};$$

see [**CD**, Theorem 3.1].

An important new feature in this paper involves the tangential derivative $a\cdot\overline{\nabla}w$ in the boundary condition $(2.1)_2$, for which we need to introduce some additional integral operators.

LEMMA 4.1. *Assume that $\partial\Omega$ and $a(\cdot)$ are C^∞-smooth. Let $\phi\colon\ \Omega \to \mathbb{R}$ and $\eta\colon\ \partial\Omega \to \mathbb{R}$. Define*

(4.8) $$\left.\begin{array}{l} (V'_{\lambda,a}\phi)(x) = a(x)\cdot\overline{\nabla}\widetilde{V}_\lambda\phi(x)|_{\partial\Omega}, \\ (K_{\lambda,a}\eta)(x) = a(x)\cdot\overline{\nabla}S_\lambda(x)|_{\partial\Omega}, \end{array}\right\} \quad x \in \partial\Omega.$$

Then

(4.9)

(i) $$\left.\begin{array}{ll} (V'_{\lambda,a}\phi)(x) = \int_\Omega a(x)\cdot\overline{\nabla}_x E(x-y;\lambda)\phi(y)dy, & x \in \partial\Omega, \\ V'_{\lambda,a}\colon\ H^s(\Omega) \to H^{s+1/2}(\partial\Omega) \textit{ is continuous}, & \forall s \geq -1; \end{array}\right\}$$

(4.10)

(ii) $$\left.\begin{array}{ll} (K_{\lambda,a}\eta)(x) = \int_{\partial\Omega} a(x)\cdot\overline{\nabla}_x E(x-y;\lambda)\eta(y)d\sigma_y, & x \in \partial\Omega, \\ K_{\lambda,a}\colon\ H^s(\partial\Omega) \to H^s(\partial\Omega) \textit{ is continuous}, & \forall s \in \mathbb{R}. \end{array}\right\}$$

(iii) *The boundary integral operator $A_{\lambda,\mu} \equiv \frac{1}{2}I + K'_\lambda + K_{\lambda,a} + \mu S_\lambda\colon\ H^s(\partial\Omega) \to H^s(\partial\Omega)$ is a strongly-elliptic pseudodifferential operator with principal part P of order 0 satisfying Gårding's inequality: for any $t \in \mathbb{R}$,*

(4.11) $$Re\langle A_{\lambda,\mu}\psi,\psi\rangle_{H^s(\partial\Omega)} \geq C_1\|\psi\|^2_{H^s(\partial\Omega)} - C_2\|\psi\|^2_{H^t(\partial\Omega)}, \qquad \forall \psi \in C^\infty(\partial\Omega),$$

for some $C_1, C_2 > 0$ independent of ψ. Consequently, $A_{\lambda,\mu}$ is Fredholm on $H^s(\partial\Omega)$. Indeed, $A_{\lambda,\mu}$ is an isomorphism on $H^s(\partial\Omega)$ if $\lambda,\mu > 0$.

PROOF. Properties in (4.9) follow from (4.2) and the Trace Theorem, while property (4.10) follows from [**CZ1**, (6.61) and (6.62)], [**CK**].

It is known from [**HW**]; [**CZ1**, §4.4] the principal symbol of $\frac{1}{2}I + K'_\lambda$ is $1/2$ while that of μS_λ is $\mu/(2|\xi|)$, where $\xi \in \mathbb{R}^{N-1}$ is the Fourier transform variable on $\partial\Omega$. Therefore we need only determine the principal symbol of $K_{\lambda,a}$. Following [**CZ1**, §4.4] by flatting out $\partial\Omega$ locally into $\mathbb{R}^{N-1}$, noting that $K_{\lambda,a}\eta = a \cdot \overline{\nabla}(S_\lambda\eta)$ and the principal symbol of $S_\lambda\eta$ is $1/(2|\xi|)$, the Fourier transform immediately gives us the symbol of $K_{\lambda,a}$ as $ia(x) \cdot \xi/(2|\xi|)$ modulo a symbol $\mathcal{O}(|\xi|^{-1})$ due to the geometrical factors contributed by locally flatting out of the boundary $\partial\Omega$ into $\mathbb{R}^{N-1}$. Therefore the local principal symbol of $A_{\lambda,\mu}$ is $[1+ia(x)\cdot\xi/|\xi|]/2$, and $A_{\lambda,\mu}$ has order zero. Because the part $ia(x)\cdot\xi/(2|\xi|)$ is *purely imaginary*, for $\xi \in \mathbb{R}^{N-1}$, we have

$$\operatorname{Re}\left(\frac{1}{2} + i\frac{1}{2}a(x)\cdot\xi/|\xi|\right) = \frac{1}{2} = \frac{1}{2}|\xi|^0 \geq \frac{1}{2}|\xi|^0, \qquad \forall\xi \in \mathbb{R}^{N-1}.$$

Therefore, by [**T**, Theorem 8.1, p. 55], we conclude (4.11). As a consequence of the Gårding inequality (4.11), we obtain that $A_{\lambda,\mu}$ is the sum of an invertible operator plus a compact operator. Therefore $A_{\lambda,\mu}$ is Fredholm on $H^s(\partial\Omega)$.

To conclude that $A_{\lambda,\mu}$ is an isomorphism, we now need only show that $A_{\lambda,\mu}$ is 1-1. Assume that the null space $\mathcal{N}(A_{\lambda,\mu})$ is nontrivial, say $\eta_0 \in \mathcal{N}(A_{\lambda,\mu})$ and $\eta_0 \not\equiv 0$. Consider the simple-layer potential

$$u(x) = \int_{\partial\Omega} E(x-y;\lambda)\eta_0(y)d\sigma_y = (\widetilde{S}_\lambda\eta_0)(x) \qquad x \in \overline{\Omega}. \tag{4.12}$$

Then u is a solution of the elliptic BVP

$$\begin{cases} (\Delta-\lambda)u = 0 & \text{on } \Omega, \\ \dfrac{\partial}{\partial n}u + a\cdot\overline{\nabla}u + \mu u = A_{\lambda,\mu}\eta_0 = 0 & \text{on } \partial\Omega. \end{cases} \tag{4.13}$$

By Proposition 2.1, the unique solution u to (4.13) is $u \equiv 0$. Therefore $S_\lambda\eta_0 = \widetilde{S}_\lambda\eta_0|_{\partial\Omega} = 0$. Since S_λ is an isomorphism by (4.4), we conclude that $\eta_0 = 0$, a contradiction. □

We are now in a position to formulate the boundary integral equations (BIE) for (2.1) through (2.12) or (2.13). Write the ansatz

$$\begin{aligned} u_{m+1}(x) &= \int_{\partial\Omega} E(x-y;\lambda)\eta_{m+1}(y)d\sigma_y - \int_{\Omega} E(x-y;\lambda)f_\lambda(y,u_m(y))dy, \\ &= (\widetilde{S}_\lambda\eta_{m+1})(x) - (\widetilde{V}_\lambda f_\lambda(\cdot,u_m))(x), \qquad x\in\Omega, \end{aligned} \tag{4.14}$$

where u_{m+1} represents either $\bar{u}_{m+1}$ or $\underline{u}_{m+1}$, depending on whether $u_0 = \bar{u}$ or $\underline{u}$, respectively. Then obviously u_{m+1} satisfies $(2.12)_2$ or $(2.13)_2$. Assume that u_m in (4.14) is known. Then the only unknown is the simple-layer density η_{m+1} in (4.14), which must be solved by substituting (4.14) into $(2.12)_3$ or $(2.13)_3$, yielding

$$(A_{\lambda,\mu}\eta_{m+1})(x) = G_{\lambda,\mu}(x,u_m(x)), \quad x\in\partial\Omega, \quad m=0,1,2,\ldots, \tag{4.15}$$

where $G_{\lambda,\mu}(\cdot,u_m)$ is defined by

$$\begin{aligned} G_{\lambda,\mu}(x,v(x)) &\equiv g_\mu(x,v(x)) + (V'_\lambda f_\lambda(\cdot,v))(x) + (V'_{\lambda,a}f_\lambda(\cdot,v))(x) + \mu(V_\lambda f_\lambda(\cdot,v))(x) \\ &\equiv g_\mu(x,v(x)) + \widetilde{V}_{\lambda,\mu}f_\lambda(\cdot,v), \end{aligned} \tag{4.16}$$

with $v = u_m$.

According to Lemma 4.1, (4.15) has a unique solution η_{m+1}. The boundary layer density η_{m+1} preserves the (anti)-monotonicity property of the monotone iteration scheme as follows.

THEOREM 4.2. *Let $\underline{\phi}, \bar{\phi} \in \Theta$ such that $\underline{\phi} \leq \bar{\phi}$, and let $\underline{\eta}$ and $\bar{\eta}$ be, respectively, the unique solutions of the BIE*

$$(A_{\lambda,\mu}\eta)(x) = G_{\lambda,\mu}(x, \phi(x))$$

for $(\eta, \phi) = (\underline{\eta}, \underline{\phi})$ and $(\bar{\eta}, \bar{\phi})$. Then we have $\bar{\eta} \leq \underline{\eta}$.

PROOF. The extra new ingredients here involve several tangential derivative terms on $\partial\Omega$. However, one can easily adapt the proofs in [**DC**, Theorem 3.1] and [**CD**, Theorem 3.3] to make them work and then obtain $\bar{\eta} \leq \underline{\eta}$. Details are omitted. □

5. Boundary Element Error Estimates

Throughout this section, we let $N = 2$ or 3. From the sector Θ in (2.6), let us define a slightly enlarged sector

$$\Theta^* = \langle \underline{u} - \delta, \bar{u} + \delta \rangle, \tag{5.1}$$

where δ is a sufficiently small positive number; $\partial\Theta^*$ is defined in the same way as (2.7). Since f and g are at least $C^{1,\alpha}$-smooth, it is easily seen from $[H_1]$ that the following $[H]^*$ is satisfied:

$$[H]^* \quad \begin{cases} \text{There exists a } \nu_1 \in \mathbb{R} \text{ such that} & \\ \nu_1 + \dfrac{1}{2}(\overline{\nabla} \cdot a) < \dfrac{\partial g(x, u(x))}{\partial u}, & \forall x \in \partial\Omega, u \in \partial\Theta^*; \\ \lambda_1^*(\nu_1) > -\dfrac{\partial f(x, u(x))}{\partial u}, & \forall x \in \overline{\Omega}, u \in \Theta^*. \end{cases}$$

We now choose $\lambda, \mu > 0$ sufficiently large so that

$$\left.\begin{aligned} \mu - \nu_1 - \frac{1}{2}(\overline{\nabla} \cdot a(x)) > \mu - \frac{\partial g(x, u(x))}{\partial u} > 0, \quad & \forall x \in \partial\Omega, u \in \partial\Theta^*, \\ \lambda + \lambda_1^*(\nu_1) > \lambda - \frac{\partial f(x, u(x))}{\partial u} > 0, \quad & \forall x \in \overline{\Omega}, u \in \Theta^*. \end{aligned}\right\} \tag{5.2}$$

From (5.2), we also obtain the following Lipschitz properties: for $u, v \in \Theta^*$,

$$\left.\begin{aligned} |g_\mu(x, u(x)) - g_\mu(x, v(x))| \leq K_\mu^* |u(x) - v(x)|, \quad & x \in \partial\Omega, \\ |f_\lambda(x, u(x)) - f_\lambda(x, v(x))| \leq M_\lambda^* |u(x) - v(x)|, \quad & x \in \Omega, \end{aligned}\right\} \tag{5.3}$$

$$\left.\begin{aligned} |g(x, u(x)) - g(x, v(x))| \leq K^* |u(x) - v(x)|, \quad & x \in \partial\Omega, \\ |f(x, u(x)) - f(x, v(x))| \leq M^* |u(x) - v(x)|, \quad & x \in \Omega, \end{aligned}\right\} \tag{5.4}$$

for some $K^*, K_\mu^*, M^*, M_\lambda^* > 0$ independent of u and v. Note that (5.3) and (5.4) are *local Lipschitz* properties.

As in [**CD, DC**], we let $\mathcal{S}_h$: $0 < h \leq h_0$ be a 1-parameter family of finite-dimensional boundary element approximation spaces that form an $(\bar{\ell}, \overline{m})$-system $\mathcal{S}_h^{\bar{\ell},\overline{m}}$, with $\bar{\ell}, \overline{m} \in \mathbb{Z}^+ \equiv \{0, 1, 2, \ldots\}$, and $\bar{\ell} \geq \overline{m} + 1$, on $\partial\Omega$, in the sense of Babuška and Aziz [**BA**]. The Galerkin boundary element scheme to solve (4.15) is as follows:

(5.5)

(i) Initialization: Set $\qquad u_h^{(0)}(x) = u_0(x) = \underline{u}(x)$ or $\bar{u}(x)$.

Note that $u_h^{(0)}(x)$ also can be chosen to be an arbitrary sufficiently smooth function in Θ^* other than those in (5.5).

(ii) Choose $\lambda, \mu > 0$ satisfying (5.2).

(iii) For each $m = 0, 1, 2, \ldots$, and h: $0 < h \le h_0$, find $\eta_h^{(m+1)} \in \mathcal{S}_h$ such that

$$\langle A_{\lambda,\mu}\eta_h^{(m+1)}, \phi_h\rangle = \langle G_{\lambda,\mu}(\cdot, u_h^{(m)}), \phi_h\rangle, \qquad \forall \phi_h \in \mathcal{S}_h, \tag{5.6}$$

where $\langle\ ,\ \rangle$ signifies the $L^2(\partial\Omega)$ inner product, with

$$u_h^{(m+1)}(x) \equiv (\widetilde{S}_\lambda \eta_h^{(m+1)})(x) - (\widetilde{V}_\lambda f_\lambda(\cdot, u_h^{(m)}))(x), \qquad x \in \Omega, m = 0, 1, 2, \ldots, \tag{5.7}$$

$G_{\lambda,\mu}(x, u_h^{(m)})$ is defined by (4.16) with $v = u_h^{(m)}$ therein.

As in [**CD, DC**], we assume that the quadrature for the volume potential $\widetilde{V}_\lambda f_\lambda(\cdot, u_h^{(m)}))$ is *exact*.

Now, for the purpose of error estimation, let us also define an intermediate simple-layer density $\zeta_h^{(m+1)}$, $m = 0, 1, 2, \ldots$, to be the unique solution of the BIE

$$A_{\lambda,\mu}\zeta_h^{(m+1)} = G_{\lambda,\mu}(\cdot, u_h^{(m)}), \qquad \text{on } \partial\Omega, \tag{5.8}$$

with the ensuing

$$w_h^{(m+1)}(x) \equiv (\widetilde{S}_\lambda \zeta_h^{(m+1)})(x) - (V_\lambda f_\lambda(\cdot, u_h^{(m)}))(x), \qquad x \in \Omega, m = 0, 1, 2, \ldots. \tag{5.9}$$

This $w_h^{(m+1)}$ is the unique solution to the following linear elliptic BVP

$$\left.\begin{aligned} &-\Delta w_h^{(m+1)}(x) + \lambda w_h^{(m+1)}(x) = f_\lambda(x, u_h^{(m)}(x)), && x \in \Omega,\\ &\frac{\partial}{\partial n} w_h^{(m+1)}(x) + a(x)\cdot \overline{\nabla} w_h^{(m+1)}(x) + \mu w_h^{(m+1)}(x) = g_\mu(x, u_h^{(m)}(x)), && x \in \partial\Omega. \end{aligned}\right\} \tag{5.10}$$

Let u_∞ be either the solution $\underline{u}_\infty$ or $\bar{u}_\infty$ as guaranteed by (2.15) (according to whether we choose $u_h^{(0)} = \underline{u}$ or $\bar{u}$ in (5.5), respectively), and let η_∞ be the unique solution of the BIE

$$A_{\lambda,\eta}\eta_\infty = G_{\lambda,\mu}(\cdot, u_\infty).$$

We can now state the main theorem on error estimates.

THEOREM 5.1. **(Error estimates for the boundary element monotone iteration scheme of the nonlinear regular oblique derivative semilinear BVP)**

(1) *Let $\{\mathcal{S}_h \mid 0 < h \le h_0\}$ be a family of $(\bar{\ell}, \overline{m})$-systems on $\partial\Omega$ with $\bar{\ell} \ge 1$, $\overline{m} \ge 0$, and $\bar{\ell} \ge \overline{m}+1$, and let $N = 2$ or 3. Assume $[H]$ so that there exists a small $\delta > 0$ for which $[H]^*$ holds. Then there exists an $\bar{h}_0$: $0 < \bar{h}_0 \le h_0$ such that for any r: $0 \le r \le 3/2$, we have $u_h^{(m)} \in \Theta^*$ and there are positive constants K_1, K_2, L^*, and θ: $0 < \theta < 1$ such that*

$$\|u_h^{(m)} - u_\infty\|_{r,\Omega} \le K_1\theta^m|||e_0||| + K_2h^{2-r}[L^* + |||e_0||| + \|u\|_{2,\Omega}]. \tag{5.11}$$

Therefore $u_h^{(m)}$ converges to u_∞ as $m \to \infty$ and $h \downarrow 0$. Also, for s: $-\frac{3}{2} \le s \le \frac{1}{2}$, there are constants $K_{3,s} > 0$ and $K_{4,5} > 0$ such that

$$\|\eta_h^{(m)} - \eta_\infty\|_{s,\partial\Omega} \le K_{3,s}\theta^m|||e_0||| + K_{4,s}h^{\frac{1}{2}-s}[L^* + |||e_0||| + \|u\|_{2,\infty}]. \tag{5.12}$$

(2) *If $N = 2$ and if $\{S_h^d \mid 0 < h \le h_0\}$ is a family of S_h^d spaces ([**RS2**]) of smoothest splines of degree $d \ge 0$ with respect to a quasiuniform mesh on $\partial\Omega$, then (5.11) holds for $0 \le r \le 3/2$, and (5.12) holds for s: $-\frac{3}{2} \le s \le 1/2$.* □

We now proceed to prove Theorem 5.1. In comparison with the proofs given by us earlier in [**CD, DC**], we note that the new ingredients here are:

(i) the strongly elliptic pseudodifferential operator $A_{\lambda,\mu}$ of order 0 on $\partial\Omega$;
(ii) the operator $K_{\lambda,a}$ (see (4.10)) appearing in $A_{\lambda,\mu}$;
(iii) the nonlinearity g included in $G_{\lambda,\mu}$ on the RHS of (4.15);
(iv) the term $V'_{\lambda,a}f(\cdot,v)$ included in $G_{\lambda,\mu}$ on the RHS of (4.15);
(v) the contraction rate as given in the forms of (3.11) and (3.25).

Among these, (iii) and (v) are the most significant. We will present arguments as concisely as possible leading to the proof of Theorem 5.1.

LEMMA 5.2. *Assume that $u_h^{(m)} \in \Theta^*$, and let $\{\mathcal{S}_h \mid 0 < h \le h_0\}$ be an $(\bar{\ell}, \overline{m})$-system as in Theorem 5.1 for $N = 2$ or 3. Then for each r given in the indicated range below, there exist, respectively, constants C_i, $i = 1, 2, 3$ such that*

$$r\colon\ 0 \le r \le \min\left\{2, m + \frac{3}{2}\right\};\ \|u_h^{(m+1)} - w_h^{(m+1)}\|_{r,\Omega} \le C_1 h^{2-r}[\|g_\mu(\cdot, u_h^{(m)})\|_{1/2,\partial\Omega} + \|f_\lambda(\cdot, u_h^{(m)})\|_{0,\Omega}], \tag{5.13}$$

$$r\colon\ 0 \le r \le \min\left\{\frac{3}{2}, m + 1\right\};\ \|u_h^{(m+1)} - w_h^{(m+1)}\|_{r,\partial\Omega} \le C_2 h^{\frac{3}{2}-r}[\|g_\mu(\cdot, u_h^{(m)})\|_{1/2,\partial\Omega} + \|f_\lambda(\cdot, u_h^{(m)})\|_{0,\Omega}], \tag{5.14}$$

$$r\colon\ 0 \le r \le 1;\ \|u_h^{(m+1)} - w_h^{(m+1)}\|_{r,\partial\Omega} \le C_3 h^{1-r}[|||u_h^{(m)} - u_m||| + \theta^m |||e_0||| + \|u_\infty\|_{2,\Omega}]. \tag{5.15}$$

PROOF. Subtracting (5.9) from (5.7) and using (4.2) and Lemma 4.1, we obtain

$$\begin{aligned}
&\|u_h^{(m+1)} - w_h^{(m+1)}\|_{r,\Omega} = \|\widetilde{S}_\lambda \eta_h^{(m+1)} - \widetilde{S}_\lambda \zeta_h^{(m+1)}\|_{r,\Omega} \\
&\le \|\widetilde{S}_\lambda\|\ \|\eta_h^{(m+1)} - \zeta_h^{(m+1)}\|_{r-\frac{3}{2},\partial\Omega} \\
&\le C_1\left(\frac{1}{2}, r - \frac{3}{2}\right) \|\widetilde{S}_\lambda\| h^{2-r} \|\zeta_h^{(m+1)}\|_{\frac{1}{2},\partial\Omega} \\
&\quad \left(\text{for } r \in \left[-\ell + \frac{3}{2}, m + \frac{3}{2}\right], \text{ by [\textbf{HW}, Cor. 2.1]}\right) \\
&\le C_1\left(\frac{1}{2}, r - \frac{3}{2}\right) \|\widetilde{S}_\lambda\| h^{2-r} \|A_{\lambda,\mu}^{-1}\| [\|g_\mu(\cdot, u_h^{(m)})\|_{\frac{1}{2},\partial\Omega} + \|\widetilde{\mathcal{V}}_{\lambda,\mu}\|\ \|f_\lambda(\cdot, u_h^{(m)})\|_{0,\Omega}],
\end{aligned}$$

which is (5.13); (5.14) can be proved in a nearly identical way using S_λ instead of $\widetilde{S}_\lambda$ in the above.

We now show (5.15) below:

$$\|u_h^{(m+1)} - w_h^{(m+1)}\|_{r,\partial\Omega} = \|S_\lambda \eta_h^{(m+1)} - S_\lambda \zeta_h^{(m+1)}\|_{r,\partial\Omega}$$
$$\le \|S_\lambda\| \cdot \|\eta_h^{(m+1)} - \zeta_h^{(m+1)}\|_{r-1,\partial\Omega}$$
$$\le \|S_\lambda\| C_2(0, r-1) h^{1-r} \|\zeta_h^{(m+1)}\|_{0,\partial\Omega} \qquad \text{(by [**HW**, Cor. 2.1])}$$
$$\le h^{1-r} \cdot C_2(0, r-1) \|S_\lambda\| \; \|A_{\lambda,\mu}^{-1}\| [\|g_\mu(\cdot, u_h^{(m)})\|_{0,\partial\Omega} + \|\widetilde{\mathcal{V}}_{\lambda,\mu}\| \; \|f_\lambda(\cdot, u_h^{(m)})\|_{-\frac{1}{2},\Omega}]$$
$$\le h^{1-r} \cdot C_2(0, r-1) \|S_\lambda\| \; \|A_{\lambda,\mu}^{-1}\| \{ [\|g_\mu(\cdot, u_h^{(m)}) - g_\mu(\cdot, u_m)\|_{0,\partial\Omega}$$
$$+ \|g_\mu(\cdot, u_m) - g_\mu(\cdot, u_\infty)\|_{0,\partial\Omega}$$
$$+ \|g_\mu(\cdot, u_\infty)\|_{0,\partial\Omega}] + C_3 [\|f_\lambda(\cdot, u_h^{(m)}) - f_\lambda(\cdot, u_m)\|_{0,\Omega} + \|f_\lambda(\cdot, u_m) - f_\lambda(\cdot, u_\infty)\|_{0,\Omega}$$
$$+ \|f_\lambda(\cdot, u_\infty)\|_{0,\Omega}) \}$$
$$\le h^{1-r} \cdot C[|||u_h^{(m)} - u_m||| + \theta^m |||e_0||| + \|u_\infty\|_{2,\Omega}],$$

where in the last inequality, we have used the Lipschitz continuity of f_λ and g_μ, (3.11), and that u_∞ satisfies (2.10) with $w = u_\infty$ and $\phi = u_\infty$ therein. □

COROLLARY 5.3. *Assume that* $u_h^{(m)} \in \Theta^*$, *and let* $\{S_h^d \mid 0 < h \le h_0\}$ *be a family of* S_h^d *spaces of smoothest splines of degree* $d \ge 0$ *with respect to a quasiuniform mesh on* $\partial\Omega$ *with* $N = 2$. *Then*

$$\textit{(5.13) holds for } r\colon\ 0 \le r \le \min\{d + 2 - \varepsilon, 2\};$$
$$\textit{(5.14) holds for } r\colon\ 0 \le r \le \min\left\{\frac{3}{2} + d - \varepsilon, \frac{3}{2}\right\};$$
$$\textit{(5.15) holds for } r\colon\ 0 \le r \le 1,$$

where ε *is an arbitrarily small positive number.*

PROOF. Use [**RS1**, Cor. 4]. □

LEMMA 5.4. *Assume* [H] *and* $u_h^{(m)} \in \Theta^*$. *Let* λ, μ *and* ν_1 *be chosen in the same way as in the proof of Theorem 3.4. Then there exists a* $\bar{\theta}_{\lambda,\mu}\colon\ 0 < \bar{\theta}_{\lambda,\mu} < 1$ *such that*

$$|||w_h^{(m+1)} - u_{m+1}|||^\wedge \le \bar{\theta}_{\lambda,\mu} |||u_h^{(m)} - u_m|||^\wedge; \tag{5.16}$$

cf. (3.21) for $||| \cdot |||^\wedge$.

PROOF. From (3.12) and (5.10), by subtraction we obtain that $v = w_h^{(m+1)} - u_{m+1}$ satisfies

$$\begin{cases} -\Delta v(x) + \lambda v(x) = f_\lambda(x, u_h^{(m)}(x)) - f_\lambda(x, u_m(x)), & \text{on } \Omega, \\ \dfrac{\partial}{\partial n} v(x) + \mu v(x) = g_\mu(x, u_h^{(m)}(x)) - g_\mu(x, u_m(x)) & \text{on } \partial\Omega. \end{cases} \tag{5.17}$$

The rest of the proof is done in the same way as that of Theorem 3.4. □

LEMMA 5.5. *Assume that* $u_h^{(m)} \in \Theta^*$. *For each* $r\colon\ 0 \le r \le 3/2$, *there exists a constant* $C > 0$ *independent of* m *and* h *such that*

$$\|w_h^{(m+1)} - u_{m+1}\|_{r,\Omega} \le C |||u_h^{(m)} - u_m|||. \tag{5.18}$$

PROOF. Same as that of Cor. 3.5. □

By the Lipschitz continuity of $g(\cdot,\cdot)$, we have a constant L^* such that

$$|\nabla_x g(x,u)| \le L^* \qquad \forall (x,u) \in \partial\Omega \times \partial\Theta^*,$$

where ∇_x denotes the gradient with respect to x. Let $\alpha \in (0,1)$. From the properties of the Sloboditskii norm, for any u we have

$$\|u\|^2_{\alpha,\partial\Omega} = \|u\|^2_{0,\partial\Omega} + |u|^2_{\alpha,\partial\Omega}; \quad |u|^2_{\alpha,\partial\Omega} \equiv \int\limits_{\partial\Omega}\int\limits_{\partial\Omega} \frac{|u(x)-u(y)|^2}{|x-y|^{N-1+2\alpha}}\, d\sigma_x d\sigma_y,$$

and, thus,

$$\|g(\cdot,u)\|^2_{\frac12,\partial\Omega} = \|g(\cdot,u)\|^2_{0,\partial\Omega} + \int\limits_{\partial\Omega}\int\limits_{\partial\Omega} \frac{|g(x,u(x))-g(y,u(y))|^2}{|x-y|^{(N-1)+2\cdot(1/2)}} d\sigma_x d\sigma_y$$

$$\le \|g(\cdot,u)\|^2_{0,\partial\Omega} + \int\limits_{\partial\Omega}\int\limits_{\partial\Omega} \frac{|g(y,u(x))-g(y,u(y))|^2}{|x-y|^N} d\sigma_x d\sigma_y$$

$$+ \int\limits_{\partial\Omega}\int\limits_{\partial\Omega} \frac{|g(x,u(x))-g(y,u(x))|^2}{|x-y|^N} d\sigma_x d\sigma_y$$

(5.19)

$$\le \|g(\cdot,u)\|^2_{0,\partial\Omega} + K^{*^2} \int\limits_{\partial\Omega}\int\limits_{\partial\Omega} \frac{|u(x)-u(y)|^2}{|x-y|^N} d\sigma_x d\sigma_y + (L^*)^2 \int\limits_{\partial\Omega}\int\limits_{\partial\Omega} \frac{d\sigma_x d\sigma_y}{|x-y|^{N-2}}$$

(5.20)

$$\equiv \|g(\cdot,u)\|^2_{0,\partial\Omega} + K^{*^2}|u|^2_{1/2,\partial\Omega} + C_1(L^*)^2$$

where $C_1 > 0$ is independent of u. Note that the last integral in (5.19) is finite because the kernel is only *weakly singular* on $\partial\Omega \times \partial\Omega$.

LEMMA 5.6. *Assume that $u_h^{(m)} \in \Theta^*$. Let $\{\mathcal{S}_h \mid 0 < h \le h_0\}$ be a family of $(\bar\ell, \overline{m})$-systems on $\partial\Omega$ with $\bar\ell \ge 1$, $\overline{m} \ge 0$ and $\bar\ell \ge \overline{m}+1$, and let $N = 2$ or 3. Then*
(1) for r: $0 \le r \le \min\left(2, m+\frac32\right)$
(5.21)

$$\|u_h^{(m+1)} - w_h^{(m+1)}\|_{r,\Omega} \le C_1 h^{2-r}(L^* + |||u_h^{(m-1)} - u_{m-1}||| + \theta^{m-1}|||e_0||| + \|u_\infty\|_{2,\Omega});$$

(2) for r: $0 \le r \le \min\left(\frac32, m+1\right)$,
(5.22)

$$\|u_h^{(m+1)} - w_h^{(m+1)}\|_{r,\partial\Omega} \le C_2 h^{\frac32 - r}(L^* + |||u_h^{(m-1)} - u_{m-1}||| + \theta^{m-1}|||e_0||| + \|u_\infty\|_{2,\Omega}),$$

(3)

(5.23)
$$|||u_h^{(m+1)} - w_h^{(m+1)}||| \le C_3 h^{3/2}(L^* + |||u_h^{(m-1)} - u_{m-1}||| + \theta^{m-1}|||e_0||| + \|u_\infty\|_{2,\Omega}),$$

where $C_i > 0$, $i = 1,2,3$, are independent of h and m (and C_1 and C_2 are dependent only on r).

PROOF. Inequalities (5.21) and (5.22) will be obtainable from (5.13) and (5.14), respectively, if we can estimate $\|f_\lambda(\cdot, u_h^{(m)})\|_{0,\Omega}$ and $\|g_\mu(\cdot, u_h^{(m)})\|_{\frac12,\partial\Omega}$. Using (3.11) and (5.3), we have

$$\begin{aligned}\|f_\lambda(\cdot, u_h^{(m)})\|_{0,\Omega} &\le \|f_\lambda(\cdot, u_h^{(m)}) - f_\lambda(\cdot, u_m)\|_{0,\Omega} + \|f_\lambda(\cdot, u_m) - f_\lambda(\cdot, u_\infty)\|_{0,\Omega} \\ &\quad + \|f_\lambda(\cdot, u_\infty)\|_{0,\Omega} \\ &\le M_\lambda^* \|u_h^{(m)} - u_m\|_{0,\Omega} + C[M_\lambda^* \theta^m |||e_0||| + \|u_\infty\|_{2,\Omega}] \\ &\le M_\lambda^* \|u_h^{(m)} - u_m\|_{1/2,\Omega} + C[M_\lambda^* \theta^m |||e_0||| + \|u_\infty\|_{2,\Omega}]. \end{aligned} \tag{5.24}$$

Using (5.20), we obtain

$$\begin{aligned}\|g_\mu(\cdot, u_h^{(m)})\|_{\frac{1}{2},\partial\Omega} &\le C[\mu \|u_h^{(m)}\|_{\frac{1}{2},\partial\Omega} + \|g(\cdot, u_h^{(m)})\|_{0,\partial\Omega} + C_1(L^*) + K^* |u_h^{(m)}|_{1/2,\partial\Omega}] \\ &\le C_2[C_1 L^* + (\mu + K^*)\|u_h^{(m)} - u_m\|_{1/2,\partial\Omega} + (\mu + K^*)\|u_m - u_\infty\|_{\frac{1}{2},\partial\Omega} \\ &\quad + \|g(\cdot, u_\infty)\|_{0,\partial\Omega} + (\mu + K^*)\|u_\infty\|_{\frac{1}{2},\partial\Omega}].\end{aligned}$$

By (3.25), (5.15) and (5.16), we know

$$\begin{aligned}&\|u_m - u_\infty\|_{\frac{1}{2},\partial\Omega} \le C_3 \theta^m |||e_0|||, \\ &\|u_h^{(m)} - w_h^{(m)}\|_{\frac{1}{2},\partial\Omega} \le C_4 h^{1/2}[|||u_h^{(m-1)} - u_{m-1}||| + \theta^{m-1}|||e_0||| + \|u_\infty\|_{2,\Omega}], \\ &\|w_h^{(m)} - u_m\|_{\frac{1}{2},\partial\Omega} \le C_5 |||u_h^{(m-1)} - u_{m-1}|||.\end{aligned}$$

Through triangle inequalities, we therefore obtain a $C > 0$ such that

$$\|g_\mu(\cdot, u_h^{(m)})\| \le C(L^* + |||u_h^{(m-1)} - u_{m-1}||| + \theta^{m-1}|||e_0||| + \|u_\infty\|_{2,\Omega}). \tag{5.25}$$

The proofs of (5.21) and (5.22) are then complete by substituting (5.24) and (5.25) into (5.13) and (5.14).

The proof of (5.23) follows from adding (5.21) and (5.22) by using $r = 0$ therein. □

Note that (5.23) is of higher order in h than (5.15).

COROLLARY 5.7. *Assume that $u_h^{(m)} \in \Theta^*$. Let $\{S_h^d \mid 0 < h \le h_0\}$ be a family of smoothest splines of degree d on a quasiuniform mesh on $\partial\Omega$. Then*
(1) for $0 \le r \le \min\{d + 2 - \varepsilon, 2\}$,

$$\|u_h^{(m+1)} - w_h^{(m+1)}\|_{r,\Omega} \le C_1 h^{2-r}(L^* + |||u_h^{(m-1)} - u_{m-1}||| + \theta^{m-1}|||e_0||| + \|u_\infty\|_{2,\Omega}); \tag{5.26}$$

(2) for $0 \le r \le \min\left\{\frac{3}{2} + d - \varepsilon, \frac{3}{2}\right\}$,

$$\|u_h^{(m+1)} - w_h^{(m+1)}\|_{r,\partial\Omega} \le C_2 h^{\frac{3}{2}-r}(L^* + |||u_h^{(m-1)} - u_{m-1}||| + \theta^{m-1}|||e_0||| + \|u_\infty\|_{2,\Omega}), \tag{5.27}$$

hold for any small $\varepsilon > 0$, where C_1 and $C_2 > 0$ are independent of m and h. □

COROLLARY 5.8. *Let $u_h^{(m)} \in \Theta^*$, and let $\{S_h \mid 0 < h \le h_0\}$ and $\{S_h^d \mid 0 < h \le h_0\}$ be as in Lemma 5.6 and Corollary 5.7. Then*

$$\begin{aligned}&\|w_h^{(m+1)} - u_h^{(m+1)}\|_{1/2,\Omega} + \|w_h^{(m+1)} - u_h^{(m+1)}\|_{0,\partial\Omega} \\ &\quad \le C_1 h^{3/2}(L^* + |||u_h^{(m-1)} - u_{m-1}||| + \theta^{m-1}|||e_0||| + \|u_\infty\|_{2,\Omega}),\end{aligned} \tag{5.28}$$

and

(i) *for $N = 3, m \ge 1, \beta$: $0 < \beta \le \frac{1}{2} - \varepsilon_0$, where $\varepsilon_0 > 0$ is small,*

$$\|w_h^{(m+1)} - u_h^{(m+1)}\|_{C^{0,\beta}(\overline{\Omega})} \le C_2 h^{\varepsilon_0}(L^* + |||u_h^{(m-1)} - u_{m-1}||| + \theta^{m-1}|||e_0||| + \|u_\infty\|_{2,\Omega}), \tag{5.29}$$

(ii) *for* $N = 2, \beta$: $0 < \beta \le 1/2$,

(5.30)

$$\|w_h^{(m+1)} - u_h^{(m+1)}\|_{C^{0,\beta}(\overline{\Omega})} \le C_3 h^{1/2}(L^* + |||u_h^{(m-1)} - u_{m-1}||| + \theta^{m-1}|||e_0||| + \|u_\infty\|_{2,\Omega}),$$

for some $C_1, C_2, C_3 > 0$ *independent of* m *and* h. □

PROOF. We use (5.21) (resp. (5.26)) with $r = 1/2$ and add it with (5.22) (resp. (5.27)) with $r = 0$ to obtain (5.28).

Estimate (5.29) is derived from (5.21) by using $r = 2 - \varepsilon_0$ therein and the Sobolev Imbedding Theorem [**Ad**]; [**CZ1**, p. 30]. Estimate (5.30) follows likewise by using $r = \frac{3}{2}$ in (5.21). □

THEOREM 5.9. *Let* $\{\mathcal{S}_h \mid 0 < h \le h_0\}$ *or* $\{S_h^d \mid 0 < h \le h_0\}$ *be given as before such that* $\eta_h^{(m)} \in \mathcal{S}_h$ *or* S_h^d *(cf. (5.6)),* $m = 1, 2, \ldots,$ *and* $u_h^{(m)}$ *is defined by (5.7). Then there exists an* $\bar{h}_0$ *such that for* h: $0 < h \le \bar{h}_0$ *and* $m = 1, 2, \ldots,$

$$u_h^{(m)} \in \Theta^*. \tag{5.31}$$

PROOF. The situation is more complicated than that in [**DC**], so let us give some details here. From (5.17), we have

$$\|w_h^{(m+1)} - u_{m+1}\|_{3/2,\Omega} = \|v\|_{3/2,\Omega} \le \widetilde{C}_1 |||u_h^{(m)} - u_m||| \tag{5.32}$$

for some $\widetilde{C}_1 > 0$ independent of m and h. By the same kind of arguments as presented in (3.30) $\sim$ (3.34), with $N = 3$ we have, for $\beta = \frac{-N^2+5N-3}{2(N-1)} = \frac{3}{4}$,

$$\begin{aligned} \|w_h^{(m+1)} - u_{m+1}\|_{C^{0,\beta}(\overline{\Omega})} &\le \widetilde{C}_2 \|w_h^{(m+1)} - u_{m+1}\|_{W^{\frac{3}{2},\frac{2N-2}{N-2}}(\Omega)} \\ &\le \widetilde{C}_3 |||u_h^{(m)} - u_m|||, \qquad m = 0, 1, 2, \ldots, 0 < h \le \bar{h}_0, \end{aligned} \tag{5.33}$$

for some $\bar{h}_0$ sufficiently small. (For $N = 2$, we may just take $\beta = 1/2$ as in [**CD, DC**], e.g.)

We now estimate the RHS of (5.33). By (5.16), (5.23) and (3.22), we have

$$\begin{aligned} &|||u_h^{(m+1)} - u_{m+1}|||^\wedge \le |||u_{m+1} - w_h^{(m+1)}|||^\wedge + |||w_h^{(m+1)} - u_h^{(m+1)}|||^\wedge \\ &\le \hat{\theta}_{\lambda,\mu} |||u_h^{(m)} - u_m|||^\wedge + \widehat{K}_2 |||w_h^{(m+1)} - u_h^{(m+1)}||| \\ &\le \hat{\theta}_{\lambda,\mu} |||u_h^{(m)} - u_m|||^\wedge + C_3 \widehat{K}_2 h^{3/2}(L^* + |||u_h^{(m-1)} - u_{m-1}||| \\ &\quad + \theta^{m-1}|||e_0||| + \|u_\infty\|_{2,\infty}) \\ &= \hat{\theta}_{\lambda,\mu} |||u_h^{(m)} - u_m|||^\wedge + C_3 \widehat{K}_2 \widehat{K}_1^{-1} h^{3/2} |||u_h^{(m-1)} - u_{m-1}|||^\wedge \\ &\quad + C_3 \widehat{K}_2 h^{3/2}(L^* + \theta^{m-1}|||e_0||| + \|u_\infty\|_{2,\Omega}) \end{aligned}$$

The above is a second order recursive inequality, which can be recast into a 2×2 first order recursive inequality:

$$E_{m+1} \le \Lambda E_m + \theta^m A + B, \qquad m = 0, 1, 2, \ldots, \tag{5.34}$$

where in the above componentwise inequality is meant, with

(5.35)

$$\left.\begin{aligned} &E_m \equiv \begin{bmatrix} |||u_h^{(m)} - u_m|||^\wedge \\ |||u_h^{(m+1)} - u_{m-1}|||^\wedge \end{bmatrix}, \ \Lambda \equiv \begin{bmatrix} \hat{\theta}_{\lambda,\mu} & C_3 \widehat{K}_2 \widehat{K}_1^{-1} h^{3/2} \\ 1 & 0 \end{bmatrix}, \ E_0 = \begin{bmatrix} 0 \\ 0 \end{bmatrix}, \\ &A = \begin{bmatrix} C_3 \widehat{K}_2 h^{3/2} \theta^{-1} |||e_0||| \\ 0 \end{bmatrix}, \ B = \begin{bmatrix} C_3 \widehat{K}_2 h^{3/2}(L^* + \|u_\infty\|_{2,\Omega}) \\ 0 \end{bmatrix}. \end{aligned}\right\}$$

Iterating (5.34), we obtain

$$\begin{aligned} E_m &\le \Lambda^m E_0 + \sum_{j=0}^{m-1} (\theta^{m-1-j}\Lambda^j A + \Lambda^j B) \\ &\le \left(\sum_{j=0}^{\infty} \Lambda^j\right)(A+B). \end{aligned} \tag{5.36}$$

The eigenvalue of Λ are

$$\rho_{1,2} = \frac{1}{2}\left[\hat{\theta}_{\lambda,\mu} \pm \sqrt{\hat{\theta}^2_{\lambda,\mu} + 4C_3\widehat{K}_2\widehat{K}_1^{-1}h^{3/2}}\right] \in \mathbb{R};\ \text{“}+\text{” for } \rho_1,\ \text{“}-\text{” for } \rho_2. \tag{5.37}$$

Since $0 < \hat{\theta}_{\lambda,\mu} < 1$, we can choose h_0 sufficiently small such that

$$|\rho_i| < 1 \quad \text{for} \quad i = 1,2. \tag{5.38}$$

Therefore $\sum\limits_{j=0}^{\infty} \Lambda^j$ converges and

$$\sum_{j=0}^{\infty} \Lambda^j = (I-\Lambda)^{-1} = \begin{bmatrix} c_{11} & c_{12} \\ c_{21} & c_{22} \end{bmatrix}, \tag{5.39}$$

where each entry c_{ij} is positive.

Substituting (5.35) and (5.39) into (5.36) and taking the first component, we obtain

$$|||u_h^{(m)} - u_m|||^\wedge \le c_{11}\cdot C_3\widehat{K}_2 h^{3/2}[\theta^{-1}|||e_0||| + L^* + \|u_\infty\|_{2,\Omega}]. \tag{5.40}$$

Using (3.21), we therefore have

$$|||u_h^{(m)} - u_m||| \le c_{11}C_3\widehat{K}_2^2 h^{3/2}[L^* + \theta^{-1}|||e_0||| + \|u_\infty\|_{2,\Omega}], \qquad m = 1,2,3,\dots . \tag{5.41}$$

From (5.29), (5.30) and (5.33), using (5.41), for some β: $0<\beta<1/2$ and $\varepsilon > 0$, we have a $C > 0$ such that

$$\begin{aligned} \|u_h^{(m)} - u_m\|_{C^{0,\beta}(\overline{\Omega})} &\le \|u_h^{(m)} - w_h^{(m)}\|_{C^{0,\beta}(\overline{\Omega})} + \|w_h^{(m)} - u_h\|_{C^{0,\beta}(\overline{\Omega})} \\ &\le Ch^\varepsilon(L^* + |||e_0||| + \|u_\infty\|_{2,\Omega}), \quad \text{for all} \quad m = 1,2,\dots, h>0. \end{aligned} \tag{5.42}$$

By taking some $h_0 > 0$ sufficiently small, we therefore have

$$\|u_h^{(m)} - u_m\|_{C^{0,\beta}(\overline{\Omega})} < \delta/2, \quad \text{for all} \quad m = 1,2,\dots, 0 < h < h_0.$$

Therefore (5.31) has been established. □

PROOF OF THEOREM 5.1. Let $\{\mathcal{S}_h \mid 0 < h \le h_0\}$ be an $(\bar{\ell}, \overline{m})$-system. To establish (5.11), we note that for r: $-\bar{\ell} + \frac{3}{2} \le r \le \frac{3}{2}$,

$$\begin{aligned} \|u_h^{(m)} - u_\infty\|_{r,\Omega} &\le \|u_h^{(m)} - w_h^{(m)}\|_{r,\Omega} + \|w_h^{(m)} - u_m\|_{r,\Omega} + \|u_m - u_\infty\|_{r,\Omega} \\ &\le C_1' h^{2-r}(L^* + |||e_0||| + \|u_\infty\|_{2,\Omega}) \\ &\quad + C_2' h^{3/2}(L^* + |||e_0||| + \|u_\infty\|_{2,\Omega}) \\ &\quad + C_3'\theta^m|||e_0|||, \quad \text{by (5.21), (5.18), (5.41) and (3.25).} \end{aligned}$$

Therefore (5.11) has been proved. For an $\{\mathcal{S}_h^d \mid 0 < h \le h_0\}$ system when $N = 2$, the proof is the same.

The proof of (5.12) is partly evident in the arguments used in the proof of Lemma 5.2. We omit the details. □

Theorem 5.1 apparently gives the sharpest possible error estimates.

6. Numerical Examples

The numerical procedures on the computer are similar to the flow-chart in [**DC**, p. 971]. We wish to numerically confirm some of the estimates in Theorem 5.1. We choose the domain Ω as the unit open disk with the objective in mind that we want to visualize the solution profiles and also perhaps compare with some of those computed in [**DC**, §7].

EXAMPLE 6.1. *(A semilinear elliptic equation subject to a regular oblique derivative nonlinear boundary condition with at least two solutions)*
Let $\Omega = \{x = (x_1, x_2) \mid |x| < 1\}$ be the unit open disk in $\mathbb{R}^2$. Consider

$$\text{(6.1)} \qquad \begin{cases} \Delta w(x) = \dfrac{1}{2}\left(x_1 - \dfrac{1}{2}\right)^2 (w(x) - 2)\left(w(x) + \dfrac{3}{2}\right), & \text{on } \Omega, \\ \dfrac{\partial w(x)}{\partial r} + \dfrac{1}{2} x_1 x_2 \dfrac{\partial w(x)}{\partial \theta} + \dfrac{1}{2}(w(x) - 1)(w(x) + 1) = 0, & \text{on } \partial\Omega. \end{cases}$$

Note that on $\partial\Omega$ here we have $\frac{\partial}{\partial r} = \frac{\partial}{\partial n}$ and $\frac{\partial}{\partial \theta} = \overline{\nabla}$, where r and θ are, respectively, the radial and angular variables in polar coordinates. The surface divergence operator is simply $\overline{\nabla}\cdot = \frac{\partial}{\partial \theta}$ also. It is easy to check that (6.1) is *not monotone dissipative.* Now it is straightforward to verify that we have the following two pairs of constant super- and subsolutions:

$$\text{(6.2)} \qquad (\bar{u}, \underline{u}) = (2, 1); \quad (\bar{v}, \underline{v}) = (-1, -1.5).$$

Therefore we have the existence of at least two solutions obtainable by the monotone iteration scheme. There are likely more solutions other than these two solutions, which are of the *unstable* type and thus are unobtainable by the monotone iteration scheme. However, as we noted at the end of §2 that problem (6.1) does not admit a variational functional and therefore any such unstable solutions, if they indeed exist, cannot be readily determined by the standard form of the Mountain Pass Lemma of Ambrosetti and Rabinowitz [**AR**]; compare [**DC**, Example 7.3]. Further investigation is required in this regard.
Now define
(6.3)

$$\Theta_1^* = \{u \in C^2(\overline{\Omega}) \mid 1 \le u(x) \le 2\}, \quad \Theta_2^* = \{u \in C^2(\overline{\Omega}) \mid -1.5 \le u(x) \le -1\}.$$

Let us check condition $[H]^*$. We want to find a $\nu_1 \in \mathbb{R}$ such that

$$\begin{aligned} \nu_1 + \frac{1}{2}(\overline{\nabla} \cdot a) = \nu_1 + \frac{1}{2}\frac{\partial}{\partial \theta}\left[\frac{1}{2}\cos\theta \sin\theta\right] &< \frac{\partial g}{\partial u} \\ = \frac{\partial}{\partial u}\left[\frac{1}{2}(u-1)(u+1)\right] &= u \quad \text{on} \quad \partial\Theta_1^*, \end{aligned}$$

i.e.,

$$\text{(6.4)} \qquad \nu_1 + \frac{1}{4}\cos 2\theta < u \quad \text{on} \quad \partial\Theta_1^*,$$

and

$$\lambda_1^*(\nu_1) > -\frac{\partial f}{\partial u} = -\frac{\partial}{\partial u}\left[\frac{1}{2}\left(x_1 - \frac{1}{2}\right)^2 (u-2)\left(u+\frac{3}{2}\right)\right]$$

$$(6.5) \qquad = -\left(x_1 - \frac{1}{2}\right)^2\left(u - \frac{1}{4}\right) \quad \text{on} \quad \Theta_1^*.$$

Choose $\nu_1 = \varepsilon$ for a small $\varepsilon > 0$. Then $\lambda_1^*(\nu_1) > 0$, and we easily see that (6.4) and (6.5) are satisfied because

$$\min_{u \in \Theta_1^*} u = \min_{u \in \partial\Theta_1^*} u = 1.$$

In our numerical computation, we choose

$$(6.6) \qquad \lambda = 16, \quad \mu = 0.04, \quad u_h^{(0)}(x) = 1.5, \qquad x \in \Omega.$$

We then check again that (5.2) is also satisfied by this choice of λ and μ. The boundary $\partial\Omega$, which is the unit circle, is divided into piecewise constant boundary elements (i.e., a (1,0)-system $S_h^{(1,0)}$ as well as a S_h^0 system) using

$$(6.7) \qquad h = \frac{2\pi}{64}, \frac{2\pi}{128}, \frac{2\pi}{256}, \frac{2\pi}{512},$$

respectively.

First, fix $m = 3$ and let only h vary as in (6.7), we plot the relative logarithmic error for the simple layer density $\eta_h^{(m)}$:

$$(6.8) \qquad \ln \|\eta_h^{(m)} - \eta_{h/2}^{(m)}\|_{0,\partial\Omega} \quad \text{versus } \ln h, \text{ with } h \text{ given in (6.7).}$$

This is shown in Fig. 1. The slope is measured to be 0.4356, which is a little off the predicted rate $\frac{1}{2} - s = \frac{1}{2} - 0 = 0.5$ as given in (5.12). We attribute this small deviation to the quadrature errors in the volume integral $\widetilde{V}_\lambda f_\lambda$ in (5.7) (which was assumed to be exact).

Next, fix $h = 2\pi/256$ and let m vary, we plot the relative logarithmic error for the simple layer density $\eta_h^{(m)}$ again:

$$(6.9) \qquad \ln \|\eta_h^{(m)} - \eta_h^{(m+1)}\|_{0,\partial\Omega} \quad \text{versus} \quad m = 1, 2, \ldots, 7.$$

This is shown in Fig. 2. The dots are slightly scattered from a straight line, a fact we again attribute to the quadrature error in the evaluation of $\widetilde{V}_\lambda f_\lambda$ in (5.7). The slope of the "line" in Fig. 2 is averaged to be -1.53.

Using $h = 2\pi/512$ and iterating $m = 20$ times, we have obtained a positive convergent solution $w(x)$ as shown in Fig. 3.

The second negative solution (within Θ_2^* in (6.3)) is also computed in a similar manner. Its profile is plotted in Fig. 4. □

EXAMPLE 6.2 (The effect of the tangential derivative). Consider a problem nearly identical to (6.1), except that in the boundary condition $(6.1)_2$ the tangential term $\frac{1}{2}x_1x_2\frac{\partial w}{\partial\theta}$ is changed to $\frac{3}{2}x_1x_2\frac{\partial w}{\partial\theta}$, while all the rest remains the same. It is easy to check that (6.2) remains as two pairs of super- and subsolutions.

We compute the positive solution (within Θ_1^*) in a similar way as in Example 6.1, obtaining a convergent solution profile given in Fig. 5. By visually comparing the profiles of w in Figs. 4 and 5, we see that the difference is only about 5% relative magnitude. For a collection of other problems, we have also noticed

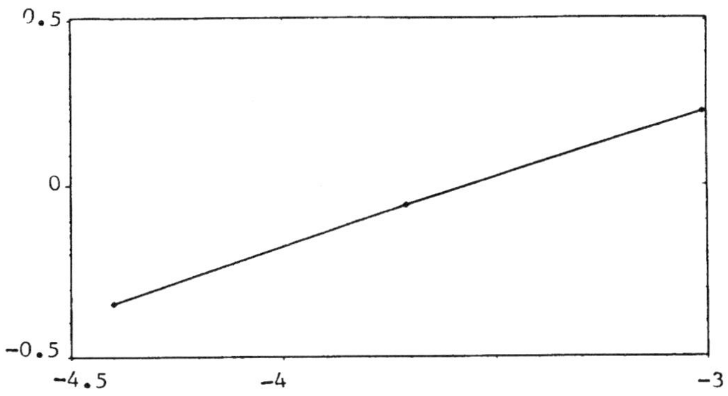

FIGURE 1. The plot for (6.8) in Example 6.1.

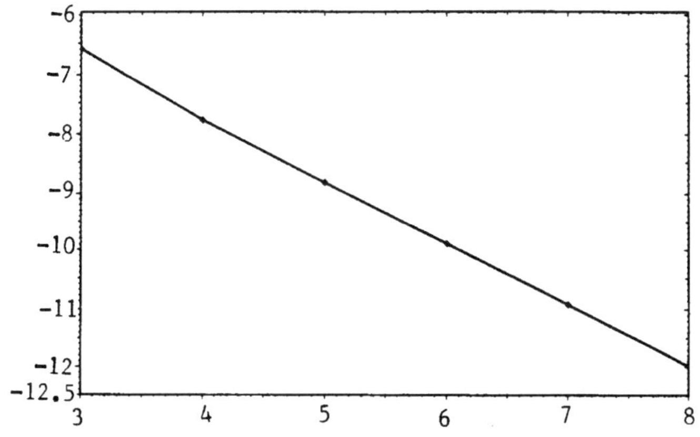

FIGURE 2. The plot for (6.9) in Example 6.1.

the phenomenon that *the effect of the change of tangential derivative terms in a regular oblique derivative boundary condition is rather minor as far as the solution profiles are concerned.* □

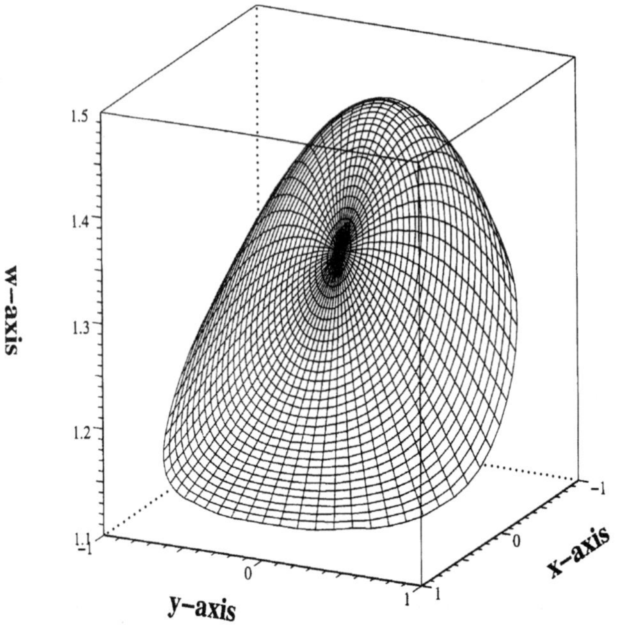

(a)

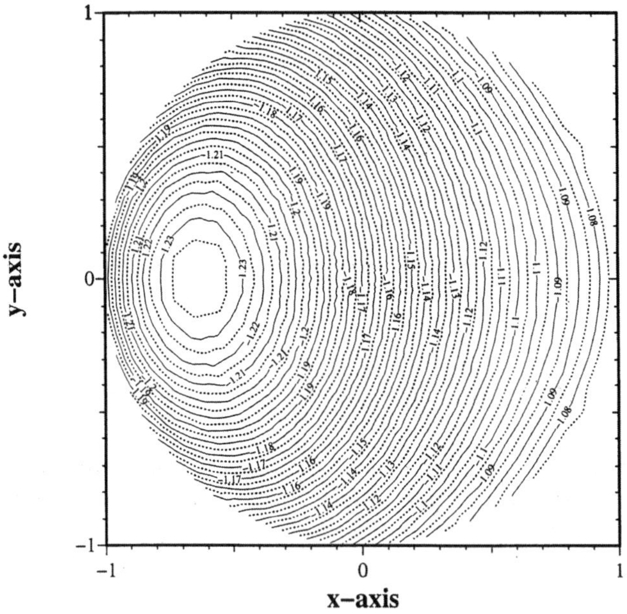

(b)

FIGURE 3. (a) The positive solution for Example 6.1; (b) its contour curves.

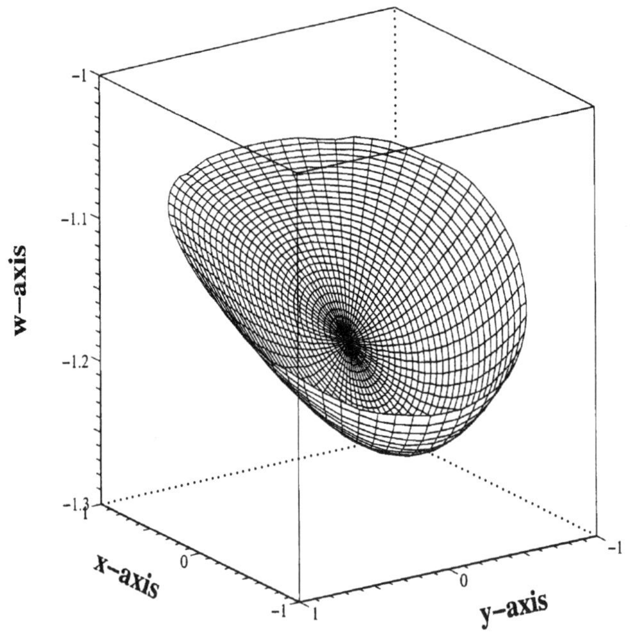

(a)

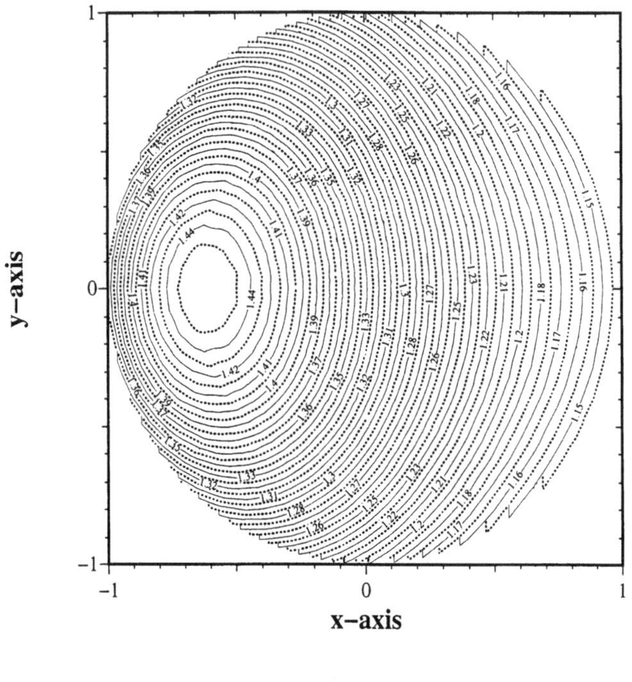

(b)

FIGURE 4. (a) The negative solution for Example 6.1; (b) its contour curves.

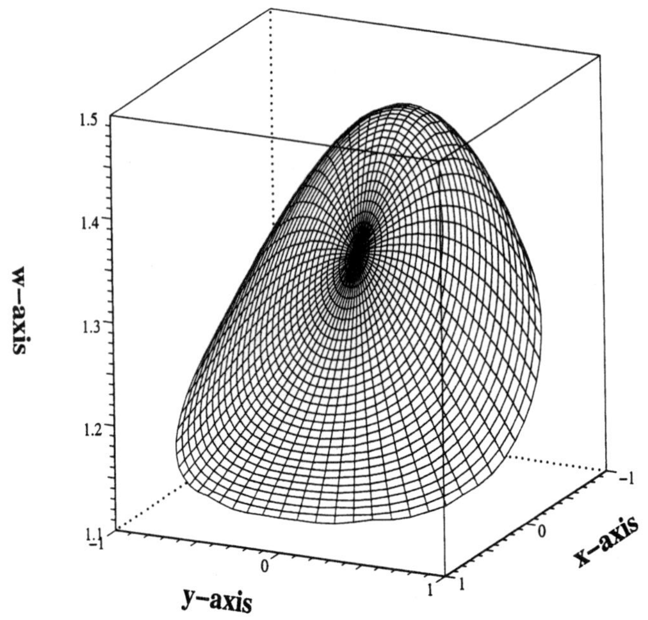

FIGURE 5. The positive solution for Example 6.2. Note that the difference between this and the profile in Fig. 4(a) is small.

References

[Ad] R. Adams, *Sobolev Spaces*, Academic Press, New York, 1975.

[AR] A. Ambrosetti and P. H. Rabinowitz, Dual variational methods in critical point theory and applications, J. Funct. Anal. **14** (1973), 349–381.

[AC] K. E. Atkinson and G. Chandler, Boundary integral equation methods for solving Laplace equation with nonlinear boundary conditions: the smooth boundary case, Math. Comp. **55** (1990), 451–472.

[B] P. K. Banerjee, Nonlinear problems of potential theory, in Developments in Boundary Element Methods Vol. 1, P. K. Banerjee and R. Butterfield (eds.), Appl. Sci. Publ., London, 1979, pp. 21–30.

[BA] I. Babuska and A. K. Aziz, *The Mathematical Foundations of the Finite Element Method with Applications to Partial Differential Equations*, Academic Press, New York, 1972.

[Ch] I. Chavel, *Eigenvalues in Riemannian Geometry*, Academic Press, New York, 1984.

[CD] G. Chen, Y,. Deng, W. M. Ni and J. Zhou, Boundary element monotone iteration scheme for semilinear elliptic partial differential equations, Part II: quasimonotone iteration for coupled 2×2 systems, Math. Comp. **69** (1999), 629–652.

[CZ1] G. Chen and J. Zhou, *Boundary Element Methods*, Academic Press, London-San Diego, 1992.

[CZ2] G. Chen and J. Zhou, *Vibration and Damping in Distributed Systems*, Vol. II: WKB and Wave Methods, Visualization and Experimentation, CRC Press, Boca Raton, Florida, 1993.

[CNZ] G. Chen, W. M. Ni and J. Zhou, Algorithms and visualization for solutions of nonlinear elliptic equations, Int. J. Bifurcation & Chaos **7** (2000), 1565–1612.

[CNPZ] G. Chen, W. M. Ni, A. Perronnet and J. Zhou, Algorithms and visualization for solutions of nonlinear elliptic equations, Part II: Dirichlet, Neumann and Robin boundary conditions and problems in 3D, Int. J. Bifurcation & Chaos **11** (2001), 1781–1799.

[CK] D. Colton and R. Kress, *Integral Equation Methods in Scattering Theory*, Wiley-Interscience, New York, 1983.

[DC] Y. Deng, G. Chen, W. M. Ni and J. Zhou, Boundary element monotone iteration scheme for semilinear elliptic partial differential equations, Math. Comp. **65** (1996), 943–982.

[DL] J. Deny and J. L. Lions, Les espaces du type de Beppo Levi, Ann. Institut Fourier (Grenoble), V (1953–54), 305–370.

[Dou] R. L. Doucette, A collocation method for the numerical solution of Laplace's equation with nonlinear boundary conditions on a polygon, SIAM J. Numer. Anal. **30** (1993), 717–732.

[ES] M. A. Efendiev, H. Schmitz and W. L. Wendland, On some nonlinear potential problems, Electronic J. Diff. Eq., **18** (1999), 1–17.

[E] L. C. Evans, *Partial Differential Equations*, Amer. Math. Soc., Providence, R.I., 1998.

[G] M. Ganesh, A BEI method for nonlinear BVP, J. Comp. Applied Math. **45** (1993), 299–308.

[GT] D. Gilbarg and N. S. Trudinger, *Elliptic Partial Differential Equations of the Second Order*, 2nd ed., Springer Verlag, New York, 1983.

[HW] G. C. Hsiao and W. L. Wendland, The Aubin-Nitsche lemma for integral equations, J. Integral Eq. **3** (1981), 299–315.

[K] H. Kumano-go, *Pseudodifferential Operators*, The MIT Press, Cambridge, MA, 1974.

[M] C. Müller, *Grundeprobleme der mathematischen Theorie electromagnetische Schwingungen*, Springer Verlag, Berlin, 1957.

[N] W. M. Ni, Some aspects of semilinear elliptic equations, Lectures Notes published by Institute of Mathematics, National Tsing Hua Univ., Hsinchu, Taiwan, Rep. of China, May, 1987.

[P] C. V. Pao, *Nonlinear Parabolic and Elliptic Equations*, Plenum Press, New York, 1992.

[RS1] K. Ruotsalainen and J. Saranen, On the convergence of Galerkin method for nonsmooth solutions of integral equations, Numer. Math. **54** (1988), 295–302.

[RS2] K. Ruotsalainen and J. Saranen, On the collocation method for a nonlinear boundary integral equation, J. Comp. Applied Math. **28** (1989), 339–348.

[RW] K. Ruotsalainen and W. L. Wendland, On the boundary element method for some nonlinear boundary value problems, Numer. Math. **53** (1988), 299–314.

[T] M. E. Taylor, *Pseudodifferential Operators*, Princeton Univ. Press, Princeton, 1981.

Department of Mathematics, Texas A&M University, College Station, TX 77843
E-mail address: gchen@math.tamu.edu

Northern Telecom, 2201 Lakeside Blvd., Richardson, TX 75082
E-mail address: ydeng@nortel.ca

School of Mathematics, University of Minnesota, Minneapolis MN 55455
E-mail address: ni@math.umn.edu

Department of Mathematics, Texas A&M University, College Station, TX 77843
E-mail address: jzhou@math.tamu.edu

Contemporary Mathematics
Volume **357**, 2004

A Note on the Elliptic Sine-Gordon Equation

Goong Chen, Zhonghai Ding, Chia-Ren Hu, Wei-Ming Ni, and Jianxin Zhou

ABSTRACT. The elliptic sine-Gordon equation originates from the static case of the hyperbolic sine-Gordon equation modelling the Josephson junction in superconductivity. However, the elliptic sine-Gordon boundary value problem as studied in the mathematical literature actually has an opposite sign in front of the sine nonlinearity; it models not the "usual" Josephson junction but rather the Josephson π-junction, which is of contemporary interest to physicists. We first furnish this physical backdrop that has motivated our study here. Then we aim to establish the existence of nonconstant solutions of the semilinear elliptic sine-Gordon equation subject to homogeneous Neumann and Dirichlet boundary conditions by using critical point theory. Positive numerical solutions of the Dirichlet case, which are global minima of the variational problem, are computed on a dumbbell-shaped 2D domain for visualization.

1. Origin of the model

The hyperbolic sine-Gordon equation

$$\phi_{xx} - \phi_{tt} = \sin\phi \tag{1.1}$$

describes the dynamics of many condensed matter systems. Examples include: a 1-D ferromagnet with planar anisotropy in the presence of a magnetic field perpendicular to the chain direction [**Mi**], spin dynamics of the A-phase of superfluid ^{3}He [**MK**], and a Josephson transmission line [**Sc**], etc., besides such classical examples as: a chain of coupled pendulums [**Dr**], and a classical model on 1-D dislocation [**La**].

The sine-Gordon equation may be derived from the Lagrangian:

$$\mathcal{L} = \int \left(\frac{1}{2}\phi_t^2 - \frac{1}{2}\phi_x^2 + \cos\phi - 1\right) dx, \tag{1.2}$$

2000 *Mathematics Subject Classification.* Primary: 35J20, 58E05; Secondary: 35Q20, 65N30.

Key words and phrases. Elliptic sine-Gordon equation, variational methods, numerical computation.

G. Chen was supported in part by a TITF initiative from Texas A&M University, C. R. Hu was supported in part by a grant from Texas Center for Superconductivity and Advanced Materials at the University of Houston, and W. M. Ni was supported in part by NSF Grant DMS 9988635.

or from the Hamiltonian:

$$\mathcal{H} = \int \left(\frac{1}{2}\phi_t^2 + \frac{1}{2}\phi_x^2 + 1 - \cos\phi \right) dx. \tag{1.3}$$

In the case of the Josephson transmission line, the variable ϕ above describes the *relative phase*, $\phi = \phi_I - \phi_{II}$, between the superconducting metals I and II, which causes the Josephson tunnelling current to flow across a very thin insulating barrier. If the Josephson transmission line is wide as well as long, the hyperbolic sine-Gordon equation must be generalized to two space dimensions and one in time:

$$\phi_{xx} + \phi_{yy} - \phi_{tt} = \sin\phi\,. \tag{1.4}$$

A somewhat detailed derivation of (1.4) may be found in [**RS**](p. 30); see (1.6)–(1.10) below. The static version of this equation reads

$$\phi_{xx} + \phi_{yy} = \sin\phi; \tag{1.5}$$

cf.[**RS**] (p. 31, (1.88)).

Now, consider however, a Josephson π-junction, which is formed by either inserting a ferromagnetic layer of a sufficient thickness inside the tunnelling barrier perpendicular to the current direction [**Ku**], or using d-wave high-T_c superconductors to form the two electrodes of the junction in an appropriate arrangement [**SR**]. Thus, the basic equations modelling the Josephson π-junction (based on the very recent papers [**Ku**] by T. Kontos *et al.* and Y. Blum *et al.*) now become

$$\frac{\partial\phi}{\partial t} = \frac{2eV}{\hbar}, \tag{1.6}$$

$$\frac{\partial\phi}{\partial x} = \left(\frac{2ed}{\hbar c}\right) H_2, \tag{1.7}$$

$$\frac{\partial\phi}{\partial y} = \left(-\frac{2ed}{\hbar c}\right) H_1, \tag{1.8}$$

$$j_3 = -j_0(x,y)\sin\phi, \tag{1.9}$$

where

$e =$ electron charge; $c =$ the speed of light in vacuo;

$d = \lambda_1 + \lambda_2 + b$, with

λ_1, λ_2 equal to, respectively, the London penetration depths for metals I and II;

$j_3(x,y) =$ the Josephson current per unit area across the barrier, with $j_0(x,y)$ depending on the properties of the barrier;

H_1, H_2 are, respectively, the x and y components of the magnetic field;

$\hbar =$ Planck's constant; and

$V(x,y,t) =$ the potential difference across the barrier.

See Fig. 1.

Substituting (1.6)–(1.9) into the Maxwell equation [**RS**] (p. 30, (1.84))

$$\frac{\partial H_2}{\partial x} - \frac{\partial H_1}{\partial y} - \frac{4\pi}{4}\cdot C\cdot\frac{\partial V}{\partial t} = \frac{4\pi}{c}j_3 \tag{1.10}$$

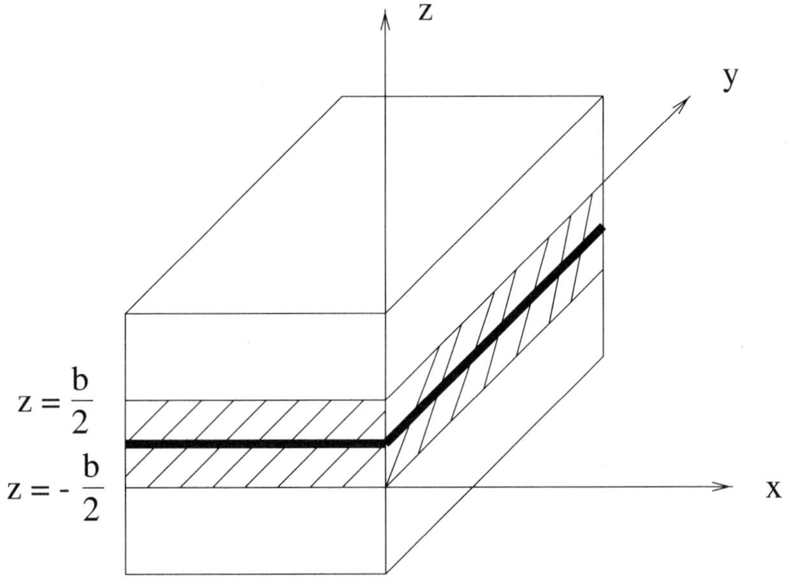

FIGURE 1. A Josephson π-junction. For $z > b/2$, there is superconducting metal I and for $z < -b/2$, superconducting metal II. The insulating barrier lying between $-b/2 < z < b/2$ (the light shaded area) is inserted with a thin layer (the dark area) of ferromagnetic material, causing j_3 in (1.10) to have a negative sign. (This is the namesake for the π-junction as it means the change of phase by π, yielding a negative sign for j_3.)

yields the barrier equation

$$\Delta\phi - \frac{1}{c_0^2}\frac{\partial^2\phi}{\partial t^2} = -\lambda_J^{-2}\sin\phi, \tag{1.11}$$

where $c_0 = c(4\pi CD)^{-1/2}$, $\lambda_J = c[\hbar/(8\pi ed j_0)]^{1/2}$, $\Delta = \partial^2/\partial x^2 + \partial^2/\partial y^2$, which, after normalization, gives

$$\phi_{xx} + \phi_{yy} - \phi_{tt} = -\sin\phi,$$

with static version

$$\phi_{xx} + \phi_{yy} = -\sin\phi. \tag{1.12}$$

This is the elliptic sine-Gordon equation we intend to investigate (rather than (1.5)) in this paper.

The boundary condition is the *natural* boundary condition

$$\frac{\partial\phi}{\partial n} = 0,$$

where n is the unit outward normal vector on the boundary. In order to secure the Dirichlet boundary condition $\phi = 0$ or the Robin condition $\phi + \alpha\partial\phi/\partial n = 0$, it is apparent that some devices or controls must be imposed thereupon. We have not yet found much discussion about how to carry out the design of such devices or controls in the literature, however.

In this paper, we present some theoretical and numerical results for the elliptic sine-Gordon boundary value problem. We will discuss both the homogeneous Dirichlet and Neumann boundary conditions, with the understanding that the discussion so far for the homogeneous Dirichlet condition is primarily of academic interest. Even though the existence of multiple solutions of a general class of nonlinear eigenvalue problems has already been discussed in [**Ra**] (in particular, Theorem 9.6), however, we must profess that for the elliptic sine-Gordon equation, the understanding of qualitative properties of the nontrivial solutions rather than the existence of multiple nontrivial solutions seems to be limited, at least as far as the authors are concerned. Furthermore, this paper should be viewed more as a research in progress rather than a major milestone on this subject.

2. The elliptic sine-Gordon equation (*I*): Neumann boundary condition

Let Ω be a bounded open domain in $\mathbb{R}^N$ with smooth boundary $\partial\Omega$. Let $H^s(\Omega)$ denote the usual Sobolev space of order $s \in \mathbb{R}$. Consider the boundary value problem (BVP)

$$(2.1)\qquad \begin{cases} -\Delta w(x) = \lambda \sin w(x), & x = (x_1, \ldots, x_N) \in \Omega, \quad \lambda > 0, \\ \dfrac{\partial w}{\partial n} = 0 & \text{on } \partial\Omega. \end{cases}$$

By making the change of variable

$$(2.2)\qquad w(x) = v(x) - \pi, \qquad x \in \Omega,$$

in (2.1), we obtain

$$(2.3)\qquad \begin{cases} -\Delta v = -\lambda \sin v, & \text{in } \Omega, \\ \dfrac{\partial v}{\partial n} = 0 & \text{on } \partial\Omega. \end{cases}$$

Thus, we see that as far as the Neumann BVP is concerned, mathematically speaking, the Josephson π-function and the usual Josephson junction are *equivalent.* However, for the homogeneous Dirichlet BVP, the change of variable (2.2) is inadmissible, and thus the equations for the Josephson π-junction and the usual Josephson junction are *not equivalent.*

We will study (2.3) instead of (2.1). Define $J_\lambda : H^1(\Omega) \to \mathbb{R}$ by

$$(2.4)\qquad J_\lambda(v) = \int_\Omega \left[\frac{1}{2}|\nabla v|^2 + \lambda(1 - \cos v)\right] dx, \qquad v \in H^1(\Omega).$$

One can verify easily that $J_\lambda \in C^1(H^1(\Omega), \mathbb{R})$; and the critical points of $J_\lambda(v)$ in $H^1(\Omega)$ are the weak solutions of (2.3) in $H^1(\Omega)$. For $v_0 \in H^1(\Omega)$ and $\varepsilon > 0$, define

$$B(v_0, \varepsilon) = \{v \in H^1(\Omega) \mid \|v - v_0\|_{H^1(\Omega)} < \varepsilon\}.$$

LEMMA 2.1. *Let $\lambda > 0$. Then $J_\lambda(v)$ has a local minimum at $v = 0$ and, consequently, at every $2n\pi$ for $n = \pm 1, \pm 2, \ldots$. Furthermore, for any $0 < \varepsilon < \varepsilon_0$ where $\varepsilon_0 = \pi\sqrt{|\Omega|}$, there exists a $\delta_0 > 0$ depending on ε such that*

$$J_\lambda(v) \geq \delta_0, \quad \forall v \in \partial B(0, \varepsilon).$$

PROOF. Let $\varepsilon_0 = \pi\sqrt{|\Omega|}$ and $0 < \varepsilon < \varepsilon_0$. For any $v \in B(0, \varepsilon)$, we have

$$J_\lambda(v) = \frac{1}{2}\int_\Omega |\nabla v|^2 dx + \lambda \int_\Omega (1 - \cos v) dx \geq 0 = J_\lambda(0).$$

Then 0 is a local minimum of $J_\lambda(v)$ in $B(0,\varepsilon)$. If $v \in H^1(\Omega)$ satisfies

$$\frac{1}{2}\int_\Omega |\nabla v|^2 dx + \lambda \int_\Omega (1-\cos v)dx = 0, \tag{2.5}$$

then

$$|\nabla v| = 0, \quad 1-\cos v = 0, \quad \text{in } \Omega.$$

By Corollary 3.27 of [**Ad**], we then have

$$v = 0,\ \pm 2\pi,\ \pm 4\pi, \cdots, \quad \text{in } \Omega.$$

For any integer $n \neq 0$, $2n\pi \notin \overline{B(0,\varepsilon)}$. Thus if $v \in \overline{B(0,\varepsilon)}$ satisfies (2.5), then $v = 0$. Therefore if $v \in \partial B(0,\varepsilon)$, then

$$J_\lambda(v) = \frac{1}{2}\int_\Omega |\nabla v|^2 dx + \lambda \int_\Omega (1-\cos v)dx > 0.$$

We claim that

$$\inf_{u\in\partial B(0,\varepsilon)} J_\lambda(u) = \delta_0 > 0.$$

Assume the contrary. Then there exists a sequence $\{v_n\} \subset \partial B(0,\varepsilon)$ such that

$$\lim_{n\to\infty} J_\lambda(v_n) = \lim_{n\to\infty}\left[\frac{1}{2}\int_\Omega |\nabla v_n|^2 dx + \lambda \int_\Omega (1-\cos v_n)dx\right] = 0. \tag{2.6}$$

By the weak compactness of $\overline{B(0,\varepsilon)}$ in $H^1(\Omega)$, there is a subsequence of $\{v_n\}$, still denoted by $\{v_n\}$, such that

$$\lim_{n\to\infty} v_n = v^* \in \overline{B(0,\varepsilon)} \quad \text{weakly in } H^1(\Omega).$$

By applying Theorem 1.6 in [**St**], $J_\lambda(v)$ is weakly lower semi-continuous. Then we have

$$0 = \lim_{n\to\infty}\left(\frac{1}{2}\int_\Omega |\nabla v_n|^2 dx + \lambda \int_\Omega (1-\cos v_n)dx\right) \geq \frac{1}{2}\int_\Omega |\nabla v^*|^2 dx + \lambda \int_\Omega (1-\cos v^*)dx.$$

Thus $v^* = 0$. Hence

$$\lim_{n\to\infty} v_n = 0 \quad \text{weakly in } H^1(\Omega).$$

By Sobolev imbedding [**Ad**], we have

$$\lim_{n\to\infty} v_n = 0 \quad \text{in } L^2(\Omega).$$

From (2.6), we have

$$\lim_{n\to\infty}\int_\Omega |\nabla v_n|^2 dx = 0.$$

Hence

$$\lim_{n\to\infty} \|v_n\|_{H^1(\Omega)} = 0,$$

which contradicts the assumption that $v_n \in \partial B(0,\varepsilon)$ for any n. Therefore

$$\inf_{u\in\partial B(0,\varepsilon)} J_\lambda(u) = \delta_0 > 0,$$

i.e.,

$$J_\lambda(v) \geq \delta_0,$$

for all $v \in \partial B(\pi, \varepsilon)$. □

Let α be a given positive number such that $\alpha \geq \lambda > 0$. Let us redefine the norm $\|\cdot\|$ of $H^1(\Omega)$ to be

$$\|v\|_{H^1(\Omega)} = \left[\int_\Omega |\nabla v|^2 dx + \alpha \int_\Omega v^2 dx\right]^{\frac{1}{2}}, \quad v \in H^1(\Omega);$$

$\|\cdot\|$ is equivalent to the usual norm of $H^1(\Omega)$. Rewrite (2.3) as

$$\text{(2.7)} \qquad \begin{cases} -\Delta v + \alpha v = \alpha v - \lambda \sin v, & \text{in } \Omega, \\ \dfrac{\partial v}{\partial n} = 0 & \text{on } \partial\Omega. \end{cases}$$

Define $f : \mathbb{R} \to \mathbb{R}$ by

$$f(t) = \begin{cases} 0, & t < 0, \\ \alpha t - \lambda \sin t, & 0 \leq t \leq 2\pi, \\ 2\pi\alpha, & t > 2\pi. \end{cases}$$

Note that $f \in C(\mathbb{R})$. Consider the following Neumann boundary value problem

$$\text{(2.8)} \qquad \begin{cases} -\Delta v + \alpha v = f(v), & \text{in } \Omega, \\ \dfrac{\partial v}{\partial n} = 0 & \text{on } \partial\Omega. \end{cases}$$

Note that $v = 0$, π and 2π are the only three constant solutions of (2.8).

LEMMA 2.2. *If v is a solution of (2.8), then $v \in H^2(\Omega) \bigcap C^{1,\nu}(\overline{\Omega})$ for some $\nu > 0$ and $0 \leq v(x) \leq 2\pi$ on Ω. Consequently, $v + 2n\pi$ are solutions of (2.3) for $n = 0, \pm 1, \pm 2, \pm 3, \cdots$.*

PROOF. If v is a solution of (2.8), then by elliptic regularity [**GT**], $v \in W^{2,p}(\Omega)$ for any $1 \leq p < \infty$. The Sobolev imbedding theorem [**Ad**] gives $v \in C^{1,\nu}(\overline{\Omega})$ for any $0 < \nu \leq 1$. Assume that

$$\Omega_{2\pi} = \{x \in \Omega \mid v(x) > 2\pi\},$$

and consider

$$\begin{cases} -\Delta v = f(v) - \alpha v & \text{in } \Omega_{2\pi}, \\ v = 2\pi & \text{on } \partial\Omega_{2\pi}. \end{cases}$$

Since $f(v) - \alpha v < 0$ on $\Omega_{2\pi}$, by the maximum principle [**GT**], we have $v < 2\pi$ on $\Omega_{2\pi}$, which is a contradiction. Thus $v(x)$ does not cross 2π. By a similar argument, $v(x)$ does not cross below 0. Thus $0 \leq v(x) \leq 2\pi$ on Ω. □

Let

$$F(t) = \int_0^t f(s) ds = \begin{cases} 0, & t < 0, \\ \dfrac{1}{2}\alpha t^2 - \lambda(1 - \cos t), & 0 \leq t \leq 2\pi, \\ 2\pi\alpha(t - \pi), & t > 2\pi. \end{cases}$$

Since $\alpha \geq \lambda$, we have

$$\text{(2.9)} \qquad F(t) \leq \frac{1}{2}\alpha t^2 - \lambda(1 - \cos t), \quad t \in \mathbb{R}.$$

Define $I_\lambda : H^1(\Omega) \to \mathbb{R}$ by

$$I_\lambda(v) = \frac{1}{2}\int_\Omega |\nabla v|^2 dx + \frac{\alpha}{2}\int_\Omega v^2 dx - \int_\Omega F(v)dx.$$

Since $f \in C(\mathbb{R})$, we have $I_\lambda \in C^1(H^1(\Omega), \mathbb{R})$ and the critical points of $I_\lambda(v)$ in $H^1(\Omega)$ correspond to the weak solutions of (2.8). By using (2.9), we have

$$I_\lambda(v) \geq J_\lambda(v), \quad \forall v \in H^1(\Omega).$$

Thus, by Lemma 2.1 we have the following.

LEMMA 2.3. *$I_\lambda(v)$ has a local minimum at $v = 0$.* □

The Palais–Smale (PS) condition is a key technical condition prevalent in critical point theory. We say that $I_\lambda(v)$ satisfies the (PS) condition if a sequence $\{v_n\} \subset H^1(\Omega)$ satisfies $|I_\lambda(v_n)| \leq M$ for some $M > 0$ and $I'_\lambda(v_n) \to 0$ as $n \to \infty$, then $\{v_n\}$ possesses a convergent subsequence in $H^1(\Omega)$. To show that $I_\lambda(v)$ satisfies the the (PS) condition, we need the following.

LEMMA 2.4. *$I_\lambda(v)$ is coercive in $H^1(\Omega)$.*

PROOF. To verify that $I_\lambda(v)$ is coercive in $H^1(\Omega)$, we need to prove $I_\lambda(v) \to \infty$ as $\|v\|_{H^1(\Omega)} \to \infty$. For any $v \in H^1(\Omega)$, define

$$\begin{aligned}
\Omega_1 &= \{x \in \Omega \mid v(x) < 0\}, \\
\Omega_2 &= \{x \in \Omega \mid 0 \leq v(x) \leq 2\pi\}, \\
\Omega_3 &= \{x \in \Omega \mid v(x) > 2\pi\}.
\end{aligned}$$

Then

$$\begin{aligned}
\int_\Omega F(v)dx &= \int_{\Omega_1} F(v)dx + \int_{\Omega_2} F(v)dx + \int_{\Omega_3} F(v)dx \\
&= \int_{\Omega_2} \left[\frac{\alpha}{2}v^2 - \lambda(1-\cos v)\right] dx + 2\pi\alpha\int_{\Omega_3}(v-\pi)dx \\
&\leq 2\pi\alpha\int_\Omega |v|dx + 4\pi^2\alpha|\Omega| + 2\lambda|\Omega| \\
&\leq 2\pi\alpha\left(\int_\Omega v^2 dx\right)^{\frac{1}{2}} |\Omega|^{\frac{1}{2}} + 4\pi^2\alpha|\Omega| + 2\lambda|\Omega|.
\end{aligned}$$

Thus

$$\begin{aligned}
I_\lambda(v) =& \frac{1}{2}\int_\Omega |\nabla v|^2 dx + \frac{\alpha}{2}\int_\Omega v^2 dx - \int_\Omega F(v)dx \\
\geq& \frac{1}{2}\int_\Omega |\nabla v|^2 dx + \frac{\alpha}{2}\int_\Omega v^2 dx - 2\pi\alpha\left(\int_\Omega v^2 dx\right)^{\frac{1}{2}} |\Omega|^{\frac{1}{2}} - 4\pi^2\alpha|\Omega| - 2\lambda|\Omega|.
\end{aligned}$$

We see that $I_\lambda(v) \to \infty$ as $\|v\|_{H^1(\Omega)} \to \infty$. Therefore, $I_\lambda(v)$ is coercive in $H^1(\Omega)$. □

LEMMA 2.5. *$I_\lambda(v)$ satisfies the (PS) condition.*

PROOF. For any $u \in L^2(\Omega)$, define $Lu \in H^1(\Omega)$ by

$$< u, w >_{L^2(\Omega)} = < Lu, w >_{H^1(\Omega)}, \quad \forall w \in H^1(\Omega).$$

Thus, L is a bounded linear operator from $L^2(\Omega)$ to $H^1(\Omega)$. For any $v,\, w \in H^1(\Omega)$, we have

$$\begin{aligned} I'_\lambda(v)\cdot w &= \int_\Omega \nabla v\cdot\nabla w dx + \alpha\int_\Omega vw dx - \int_\Omega f(v)w dx \\ &= < v - Lf(v), w >_{H^1(\Omega)}. \end{aligned}$$

Thus,

$$I'_\lambda(v) = v - Lf(v).$$

Assume $\{v_n\}\subset H^1(\Omega)$ is a sequence satisfying $|I_\lambda(v_n)|\le M$ for some $M>0$ and $I'_\lambda(v_n)\to 0$ as $n\to\infty$. By Lemma 2.4, we have that $\{v_n\}$ is bounded in $H^1(\Omega)$. Thus $\{v_n\}$ admits a weakly convergent subsequence, still denoted by $\{v_n\}$. By Sobolev imbedding [**Ad**], $\{v_n\}$ converges in $L^2(\Omega)$. Since $f(t)$ is Lipschitz continuous and L is a bounded linear operator from $L^2(\Omega)$ to $H^1(\Omega)$, we have the convergence of $\{Lf(v_n)\}$ in $H^1(\Omega)$. Since

$$v_n = I'_\lambda(v_n) + Lf(v_n),$$

we have the convergence of $\{v_n\}$ in $H^1(\Omega)$. □

Let $\{\lambda_j \mid 0=\lambda_1<\lambda_2<\cdots<\lambda_n<\cdots, j=1,2,\ldots\}$ be the set of eigenvalues of the Neumann boundary value problem

$$(2.10)\qquad \begin{cases} -\Delta\phi_j = \lambda_j\phi_j & \text{on } \Omega,\\ \dfrac{\partial\phi_j}{\partial n} = 0 & \text{on } \partial\Omega, \end{cases}$$

where $\{\phi_j\}$ satisfying $\int_\Omega \phi_j^2(x)dx = 1$ are the corresponding eigenfunctions. By elliptic regularity [**GT**], $\phi_j\in C^\infty(\overline{\Omega})$.

THEOREM 2.6. *If $\lambda>\lambda_2$, then $I_\lambda(v)$ has at least two nonconstant critical points w in $H^1(\Omega)$. Consequently,* (2.8) *admits at least two nonconstant solutions.*

PROOF. By Lemmas 2.3 and 2.5, $I_\lambda\in C^1(H^1(\Omega),\mathbb{R})$ satisfies the (PS) condition, and 0 is a local minimum of $I_\lambda(v)$ in $H^1(\Omega)$ and $I_\lambda(0)=0$. Note that $I_\lambda(2\pi)=0$. By the Mountain Pass Lemma [**AR**], we obtain that $I_\lambda(v)$ has at least a critical point $w_0\in H^1(\Omega)$ such that

$$(2.11)\qquad I_\lambda(w_0) = \inf_{\gamma\in\Gamma}\max_{t\in[0,1]} I_\lambda(\gamma(t)) \ge \delta_0 > 0,$$

for some $\delta_0>0$ defined in Lemma 2.1, where $\Gamma=\{\gamma\in C([0,1],H^1(\Omega))\mid \gamma(0)=0, \gamma(1)=2\pi\}$. The remaining part is to prove $w_0\ne\pi$, where $I_\lambda(\pi)=2\lambda|\Omega|$.

Let ϕ_1 and ϕ_2 be defined by (2.10). If $\lambda>\lambda_2$, then for any $|t|<\dfrac{\pi}{\|\phi_2\|_\infty}$,

$$\begin{aligned} I_\lambda(\pi+t\phi_2) - I_\lambda(\pi) &= \frac{t^2}{2}\int_\Omega |\nabla\phi_2|^2 dx - \lambda\int_\Omega (1-\cos(t\phi_2))dx \\ &= \frac{\lambda_2 t^2}{2}\int_\Omega \phi_2^2\, dx - \lambda\int_\Omega \left[\frac{t^2\phi_2^2}{2!} - \frac{t^4\phi_2^4}{4!} + \frac{t^6\phi_2^6}{6!} - \cdots\right] dx \\ &= t^2\left\{\frac{(\lambda_2-\lambda)}{2}\int_\Omega \phi_2^2\, dx + t^2\lambda\int_\Omega\left[\frac{\phi_2^4}{4!} - \frac{t^2\phi_2^6}{6!} + \cdots\right] dx\right\}. \end{aligned}$$

Thus

$$< I''_\lambda(\pi)\phi_2,\ \phi_2 >_{H^1} = (\lambda_2 - \lambda) \int\limits_\Omega \phi_2^2 dx < 0.$$

Note that ϕ_1 is a constant. From the definition of I_λ, one can easily verify that

$$< I''_\lambda(\pi)\phi_1,\ \phi_1 >_{H^1} = -\lambda|\Omega|\phi_1^2 < 0.$$

Since ϕ_1 and ϕ_2 are orthogonal in $H^1(\Omega)$, the Morse index of I_λ at $w = \pi$, as defined by the dimension of the maximal subspace of $H^1(\Omega)$ where $I''_\lambda(\pi)$ is negative definite [**Ch**], is at least 2. However, since w_0 is a critical point of I_λ of the Mountain-Pass type, the Morse index of I_λ at w_0 is less than or equal to 1 [**Ch**]. Thus, $w_0 \neq \pi$. By Lemma 2.2, $0 \leq w_0 \leq 2\pi$. Let $w_1 = 2\pi - w_0$. It is easy to check that $I_\lambda(w_1) = I_\lambda(w_0)$ and $I'_\lambda(w_1) = -I'_\lambda(w_0) = 0$. Hence w_1 is also a critical point of $I_\lambda(v)$. Therefore, $I_\lambda(v)$ has two non-constant critical points, w_0 and w_1 in $H^1(\Omega)$, if $\lambda > \lambda_2$. □

By Lemma 2.2 and Theorem 2.6, the Neumann boundary value problem (2.3) (or (2.1)) of the elliptic sine-Gordon equation admits infinitely many non-constant solutions.

By using a different approach based upon the Mountain Pass Lemma in order intervals [**LW**], C. Li [**Li**] has studied a class of nonlinear elliptic equations with oscillating nonlinearities similar to the elliptic sine-Gordon equation. However, our proofs are more elementary, and provide an explicit topological geometry of the solutions, as our approach has taken advantage of the special nature of the elliptic sine-Gordon equation.

The following theorem gives us some additional information about the bounds of the solutions of (2.1).

THEOREM 2.7. (2.1) *does not admit any nonconstant solution w such that $0 < w(x) < \pi$ on Ω, for any $\lambda > 0$. Consequently,* (2.1) *does not admit any nonconstant solution w such that*

$$2n\pi < w(x) < (2n+1)\pi,$$

for any integer n.

PROOF. Assume that (2.1) admits a nonconstant solution $w(x)$ such that $0 < w(x) < \pi$. By integrating both sides of (2.1) on Ω, we have

$$\lambda \int_\Omega \sin w dx = 0,$$

a contradiction. □

3. The elliptic sine-Gordon equation (*II*): Dirichlet boundary condition

For the Dirichlet boundary value problem of the elliptic sine-Gordon equation

$$\begin{cases} -\Delta w = \lambda \sin w & \text{in } \Omega, \\ w = 0 & \text{on } \partial\Omega, \end{cases} \tag{3.1}$$

define $J_\lambda : H_0^1(\Omega) \to \mathbb{R}$ by

$$J_\lambda(v) = \frac{1}{2}\int_\Omega |\nabla v|^2 dx - \lambda \int_\Omega (1 - \cos v) dx. \tag{3.2}$$

It is straightforward to verify that $J_\lambda(0) = 0$, $J_\lambda \in C^1(H_0^1(\Omega), \mathbb{R})$ and the weak solutions of (3.1) in $H_0^1(\Omega)$ correspond exactly to critical points of J_λ in $H_0^1(\Omega)$. If $v \in H_0^1(\Omega)$ is a weak solution of (3.1), then by elliptic regularity [**GT**], $v \in W^{2,p}(\Omega)$ for any $1 \le p < \infty$. By the Sobolev imbedding theorem [**Ad**], $v \in C^{1,\nu}(\overline{\Omega})$ for any $0 < \nu \le 1$. Again by elliptic regularity [**GT**], we have $v \in C^{2,\nu}(\overline{\Omega})$ for any $0 < \nu \le 1$.

Let $\{\mu_n \mid 0 < \mu_1 < \mu_2 < \cdots < \mu_n < \cdots,\ n = 1, 2, 3, \ldots\}$ denote the set of eigenvalues of the homogeneous Dirichlet boundary value problem

$$\begin{cases} -\Delta \psi_n = \mu_n \psi_n & \text{in } \Omega, \\ \psi_n = 0 & \text{on } \partial\Omega, \end{cases} \tag{3.3}$$

where $\{\psi_n\}$ satisfying $\int_\Omega \psi_n^2(x)dx = 1$ are the corresponding eigenfunctions. By elliptic regularity [**GT**], $\psi_n \in C^\infty(\overline{\Omega})$.

THEOREM 3.1. *Let λ satisfy $0 < \lambda < \mu_1$. Then $w \equiv 0$ is the only critical point of $J_\lambda(v)$ in $H_0^1(\Omega)$.*

PROOF. Assume that $w \in H_0^1(\Omega)$ is a critical point of J_λ and $w \ne 0$. Then w is a weak solution of (3.1). Multiplying (3.1) by w and integrating over Ω, we obtain

$$0 < \int_\Omega |\nabla w|^2 dx = \lambda \int_\Omega w \sin w dx \le \lambda \int_\Omega w^2 dx.$$

Apply the Poincaré inequality,

$$0 < \int_\Omega |\nabla w|^2 dx \le \frac{\lambda}{\mu_1} \int_\Omega |\nabla w|^2 dx.$$

Thus, $\lambda \ge \mu_1$, contradicting the assumption that $0 < \lambda < \mu_1$. Therefore, $w = 0$ is the only critical point of $J_\lambda(v)$ if $0 < \lambda < \mu_1$. □

LEMMA 3.2. *For any $\lambda > 0$, there is a $v_0 \in H_0^1(\Omega)$ such that*

$$J_\lambda(v_0) = \inf_{v \in H_0^1(\Omega)} J_\lambda(v).$$

PROOF. For any $v \in H_0^1(\Omega)$, we have

$$J_\lambda(v) = \int_\Omega \left[\frac{1}{2}|\nabla v|^2 - \lambda(1 - \cos v)\right] dx \ge \frac{1}{2}\int_\Omega |\nabla v|^2 dx - 2\lambda|\Omega|. \tag{3.4}$$

Thus $J_\lambda(v)$ is bounded below in $H_0^1(\Omega)$. Let $\{v_n\} \subset H_0^1(\Omega)$ such that

$$\lim_{n\to\infty} J_\lambda(v_n) = \inf_{v \in H_0^1(\Omega)} J_\lambda(v).$$

By (3.4), $J_\lambda(v)$ is coercive in $H_0^1(\Omega)$. Then we have that $\{v_n\}$ is bounded in $H_0^1(\Omega)$. Thus $\{v_n\}$ has a weakly convergent subsequence, still denoted by $\{v_n\}$, such that

$$\lim_{n\to\infty} v_n = v_0 \quad \text{weakly in} \quad H_0^1(\Omega),$$

for some $v_0 \in H_0^1(\Omega)$. By Theorem 1.6 in [**St**], $J_\lambda(v)$ is weakly lower semi-continuous. Thus we have

$$\inf_{v \in H_0^1(\Omega)} J_\lambda(v) = \lim_{n\to\infty} J_\lambda(v_n) \ge J_\lambda(v_0) \ge \inf_{v \in H_0^1(\Omega)} J_\lambda(v).$$

Thus v_0 is a global minimizer of $J_\lambda(v)$ in $H_0^1(\Omega)$. □

By Theorem 3.1 and Lemma 3.2, $v_0 = 0$ if $0 < \lambda < \mu_1$. The following theorem indicates that $v_0 \neq 0$ if $\lambda > \mu_1$.

THEOREM 3.3. *If $\lambda > \mu_1$, then the global minimizer v_0 in Lemma 3.2 is nontrivial: $v_0 \not\equiv 0$. Therefore, $-v_0$ is also a global minimizer.*

PROOF. We need only to prove that there is a $v^* \neq 0$, $v^* \in H_0^1(\Omega)$ such that $J_\lambda(v^*) < 0 = J_\lambda(0)$.

Let $v = t\psi_1$ in $J_\lambda(v)$. Then

$$\begin{aligned} J_\lambda(t\psi_1) &= \frac{t^2}{2}\int_\Omega |\nabla\psi_1|^2 dx - \lambda\int_\Omega (1-\cos(t\psi_1))dx \\ &= \frac{\mu_1 t^2}{2}\int_\Omega |\psi_1|^2 dx - \lambda\int_\Omega (1-\cos(t\psi_1))dx. \end{aligned}$$

Using

$$\cos x = 1 - \frac{x^2}{2!} + \frac{x^4}{4!}\cdots\cdots, \qquad x \in \mathbb{R},$$

we have

$$J_\lambda(t\psi_1) = \frac{1}{2}(\mu_1 - \lambda)t^2\int_\Omega |\psi_1|^2 dx + \lambda\int_\Omega \left[\frac{t^4|\psi_1|^4}{4!} - \cdots\right]dx.$$

Since ψ_1 is C^∞ on Ω, $\|\psi_1\|_{L^\infty(\Omega)} < +\infty$, and

$$\begin{aligned} J_\lambda(t\psi_1) &= t^2\left\{-\frac{1}{2}(\lambda-\mu_1)\int_\Omega |\psi_1|^2 dx + \lambda t^2\int_\Omega \left[\frac{|\psi_1|^4}{4!} - \frac{t^2|\psi_1|^6}{6!} + \cdots\right]dx\right\} \\ &\equiv t^2\{T_1 + \lambda t^2 T_2\}, \end{aligned}$$

where T_1 is a negative constant, and T_2 depends on t such that if $|t|$ is small, then T_2 is positive and bounded.

Thus, there exists $\delta_0 > 0$ such that

$$J_\lambda(t\psi_1) < 0 \quad \text{if} \quad |t| < \delta_0.$$

The proof is complete. □

THEOREM 3.4. *Let $\lambda > \mu_1$. If v_0 is a global minimizer of J_λ on $H_0^1(\Omega)$, then $-\pi \le v_0(x) \le \pi$ in Ω.*

PROOF. If v_0 is the global minimizer of J_λ on $H_0^1(\Omega)$, then v_0 is a weak solution of (3.1) in $H_0^1(\Omega)$. Hence $v_0 \in C^{2,\nu}(\overline{\Omega})$ for any $0 < \nu \le 1$. Assume that $|v_0(x)|$ crosses above π. Then define

$$v_\pi(x) = \begin{cases} \pi, & \text{if } v_0(x) \ge \pi, \\ v_0(x), & \text{if } |v_0(x)| < \pi, \\ -\pi, & \text{if } v_0(x) \le -\pi. \end{cases}$$

We have $v_\pi \in H_0^1(\Omega)$, and

$$\begin{aligned} J_\lambda(v_0) - J_\lambda(v_\pi) &= \left[\frac{1}{2}\int_\Omega |\nabla v_0|^2 dx - \lambda \int_\Omega (1-\cos v_0)dx\right] \\ &\quad - \left[\frac{1}{2}\int_\Omega |\nabla v_\pi|^2 dx - \lambda \int_\Omega (1-\cos v_\pi)dx\right] \\ &= \frac{1}{2}\int_{\Omega^*} |\nabla v_0|^2 dx - \lambda \int_{\Omega^*} (\cos v_\pi - \cos v_0)dx, \end{aligned}$$

where $\Omega^* = \{x \in \Omega \mid |v_0(x)| \geq \pi\}$, giving

$$\begin{aligned} J_\lambda(v_0) - J_\lambda(v_\pi) &= \frac{1}{2}\int_{\Omega^*} |\nabla v_0|^2 dx - \lambda \int_{\Omega^*} (-1 - \cos v_0)dx \\ &= \frac{1}{2}\int_{\Omega^*} |\nabla v_0|^2 dx + \lambda \int_{\Omega^*} (1 + \cos v_0)dx > 0, \end{aligned}$$

if $|v_0(x)|$ crosses above π, i.e., $|\Omega^*| > 0$. This contradicts the fact that v_0 is a global minimizer of J_λ on $H_0^1(\Omega)$. Hence $-\pi \leq v_0(x) \leq \pi$. □

Theorem 3.4 indicates that any global minimizer of J_λ for $\lambda > \mu_1$ is bounded between $-\pi$ and π on Ω. Our numerical example in the next section seems to indicate that a global minimizer of J_λ is of one sign. In the following, we provide a few additional theorems on positive solutions of (3.1).

THEOREM 3.5. *Let $\lambda > \mu_1$. Then (3.1) admits a unique positive solution $w \in H_0^1(\Omega) \cap C^{2,\nu}(\overline{\Omega})$ satisfying $0 < w(x) < \pi$, where $0 < \nu \leq 1$.*

PROOF. Rewrite (3.1) as

$$(3.5) \qquad \begin{cases} (-\Delta + \alpha)w = \alpha w + \lambda \sin w, & \text{in } \Omega, \\ w > 0 & \text{in } \Omega, \\ w = 0 & \text{on } \partial\Omega, \end{cases}$$

where $\alpha > \lambda$ is a given number. Define $\mathcal{A}_\alpha = -\Delta + \alpha$ on $H_0^1(\Omega) \cap C^{2,\nu}(\overline{\Omega})$, and $f(u) = \alpha u + \lambda \sin u$. One can easily check that the following holds:

(1) $\sigma(\mathcal{A}_\alpha) = \{\alpha + \mu_n \mid n = 1, 2, 3, \ldots\}$, the spectrum of $\mathcal{A}_\alpha$;

(2) $f(u)$ is monotonically increasing, i.e., if $u < v$, $f(u) < f(v)$.

Let $\overline{w}_0 = \pi$ on Ω. Then $\overline{w}_0$ is a super-solution of (3.5). Also let $\overline{v}_0 = \varepsilon\psi_1$ for some small $\varepsilon > 0$. Then $\overline{v}_0$ is a sub-solution of (3.5). Thus, define

$$\overline{w}_{n+1} = \mathcal{A}_\alpha^{-1} f(\overline{w}_n),$$

and

$$\overline{v}_{n+1} = \mathcal{A}_\alpha^{-1} f(\overline{v}_n),$$

Then $\{\overline{v}_n\}$ and $\{\overline{w}_n\}$ are monotone sequences such that

$$0 < \overline{v}_0 \leq \overline{v}_1 \leq \cdots \leq \overline{v}_n \leq \overline{v}_{n+1} \leq \overline{w}_{n+1} \leq \overline{w}_n \leq \cdots \leq \overline{w}_0 = \pi.$$

By the Monotone Iteration Scheme [**Sa**] (see also [**Ni**]), there exist $\overline{v},\ \overline{w} \in H_0^1(\Omega) \cap C^{2,\nu}(\overline{\Omega})$ such that

$$\lim_{n\to\infty} \overline{v}_n = \overline{v}, \qquad \lim_{n\to\infty} \overline{w}_n = \overline{w},$$

and $\overline{v} \le \overline{w}$ on Ω. Thus $\overline{v}$ and $\overline{w}$ are solutions of (3.1).

If w is a positive solution of (3.1) satisfying $0 < w(x) < \pi$ on Ω, then we can choose a small $\epsilon > 0$ such that

$$0 < \overline{v}_0 = \epsilon\psi_1 < w < \overline{w}_0 = \pi.$$

The Monotone Iteration Scheme yields again

$$0 < \overline{v} \le w \le \overline{w} \le \pi.$$

Since both $\overline{v}$ and $\overline{w}$ are solutions of (3.1), we have

$$\begin{aligned} 0 &= \int_\Omega [(-\Delta\overline{v})\overline{w} - (-\Delta\overline{w})\overline{v}]dx \\ &= \int_\Omega [\lambda \sin\overline{v}\cdot\overline{w} - \lambda\sin\overline{w}\cdot\overline{v}]dx \\ &= \lambda \int_\Omega \overline{v}\overline{w}\left[\frac{\sin\overline{v}}{\overline{v}} - \frac{\sin\overline{w}}{\overline{w}}\right]dx. \end{aligned}$$

Note that $\overline{v}(x)\overline{w}(x) > 0$ on Ω, and $\dfrac{\sin x}{x}$ is monotonically decreasing on $[0,\pi]$, thus,

$$0 = \lambda \int_\Omega \overline{v}\overline{w}\left[\frac{\sin\overline{v}}{\overline{v}} - \frac{\sin\overline{w}}{\overline{w}}\right]dx$$

implies

$$\overline{v}\overline{w}\left[\frac{\sin\overline{v}}{\overline{v}} - \frac{\sin\overline{w}}{\overline{w}}\right] \equiv 0$$

on Ω and hence $\overline{v} = \overline{w} = w$ on Ω. Therefore, (3.1) admits a unique positive solution $w(x)$ satisfying $0 < w(x) \le \pi$. By a result in [**AH**, Theorem 6.2],

$$\|w\|_\infty \ne \pi.$$

Hence $0 < w(x) < \pi$. □

Theorem 3.5 asserts that (3.1) admits a unique positive solution satisfying $0 < w(x) < \pi$ on Ω. However, it is unclear whether (3.1) admits a unique positive solution. The following example indicates that if Ω is a disk in $\mathbb{R}^2$, then (3.1) admits a unique positive solution.

EXAMPLE 3.6 (Uniqueness of the positive solution on the disk). Let $\lambda > \mu_1$. Consider

$$\left\{ \begin{array}{ll} -\Delta u = \lambda \sin u & \text{in} \quad \Omega, \\ u > 0 & \text{in } \Omega, \\ u = 0 & \text{on } \partial\Omega, \end{array} \right. \tag{3.6}$$

where $\Omega = \{(x_1, x_2) \mid x_1^2 + x_2^2 < R^2\}$ for some $R > 0$. By the results in Gidas, Ni and Nirenberg [**GNN**], all solutions of (3.6) are radially symmetric. Thus, let

$$w(r) = u(r, \theta);$$

$w(r)$ satisfies

$$(3.7)\qquad \begin{cases} \dfrac{d^2w}{dr^2} + \dfrac{1}{r}\dfrac{dw}{dr} = -\lambda \sin w, & 0 \le r < R, \\ w(r) > 0, & 0 \le r < R, \\ w'(0) = 0, \quad w(R) = 0. \end{cases}$$

Multiply both sides of (3.7) by $\dfrac{dw}{dr}$, and integrate from 0 to r,

$$\frac{1}{2}\left(\frac{dw}{dr}\right)^2 + \int_0^r \frac{1}{r}\left(\frac{dw}{dr}\right)^2 dr = \lambda(\cos w - \cos w(0)), \qquad 0 \le r < R.$$

Since the left-hand side is always nonnegative, we must have $\cos w(0) = -1$ if $w(r)$ crosses above π, in order to maintain the nonnegativity of the left-hand side. If such a case happens, then

$$\frac{1}{2}\left(\frac{dw}{dr}\right)^2 + \int_0^r \frac{1}{r}\left(\frac{dw}{dr}\right)^2 dr = \lambda(1 + \cos w).$$

If $w(r)$ crosses π at r_0 for some $0 < r_0 < R$, i.e., $w(r_0) = \pi$, then

$$\frac{1}{2}\left(\frac{dw}{dr}\right)^2 (r_0) + \int_0^{r_0} \frac{1}{r}\left(\frac{dw}{dr}\right)^2 dr = \lambda(1 + \cos w(r_0)) = 0.$$

Thus $\dfrac{dw}{dr}(r) = 0$, $0 \le r \le r_0$. We deduce that $w(r) \equiv \pi$, $0 \le r \le r_0$. Therefore

$$0 < w(r) \le \pi \quad \text{if} \quad 0 \le r < R.$$

Again by a result in [**AH**], we have

$$\|w\|_\infty \not\equiv \pi.$$

Thus, for $\lambda > \mu_1$, a positive solution w satisfies

$$0 < w(r) < \pi.$$

Consequently, by Theorem 3.5, the positive solution is unique. □

The preceding theorem and Example 3.6 offer strong support to the conjecture that there is a *unique* positive solution to (3.1), if $\lambda > \mu_1$. However, a complete proof was not available when this paper was prepared. In a recent paper [**DCL**], it is proved that (3.1) admits a unique positive solution on any convex domain or domains with certain symmetries. The conjecture remains open for non-convex domains. However, the numerical examples presented here and in [**DCL**] have supported that such a conjecture also holds on non-convex domains.

Sign-changing solutions of the elliptic sine-Gordon equation are of considerable significance because they correspond to the so-called Josephson vortex and, thus, are related to physical phenomena of practical interest. So let us consider sign-changing solutions of (3.1) in the rest of this section.

THEOREM 3.7. *Let $\lambda > \mu_2$. Then (3.1) admits at least four nontrivial solutions.*

PROOF. From (3.2), $J_\lambda(v)$ is coercive on $H_0^1(\Omega)$. By repeating a similar argument as the proof of Lemma 2.5, $J_\lambda(v)$ satisfies the (PS) condition. By Theorem 3.3, if $\lambda > \mu_1$, then (3.1) admits at least two nontrivial solutions w_0 and $-w_0$, where $\pm w_0 \in H_0^1(\Omega)$ are global minimizers of $J_\lambda(v)$ on $H_0^1(\Omega)$ with $J_\lambda(w_0) = J_\lambda(-w_0)$.

Thus by applying the Mountain Pass Lemma [**AR**], we obtain that $J_\lambda(v)$ has at least one critical point $w_1 \in H_0^1(\Omega)$ such that

$$J_\lambda(w_1) = \inf_{\gamma\in\Gamma} \max_{t\in[0,1]} J_\lambda(\gamma(t)) > J_\lambda(w_0),$$

where

$$\Gamma = \{\gamma \in C([0,1], H_0^1(\Omega)) \mid \gamma(0) = w_0, \gamma(1) = -w_0\}.$$

The only thing left is to prove $w_1 \neq 0$.

By repeating the same argument in the proof of Theorem 3.3, we have

$$J_\lambda(t\psi_1) = -\frac{t^2}{2}(\lambda - \mu_1)\int_\Omega |\psi_1|^2 dx + \lambda t^4 \int_\Omega \left[\frac{|\psi_1|^4}{4!} - \frac{t^2|\psi_1|^6}{6!} + \cdots\right] dx,$$

and

$$J_\lambda(t\psi_2) = -\frac{t^2}{2}(\lambda - \mu_2)\int_\Omega |\psi_2|^2 dx + \lambda t^4 \int_\Omega \left[\frac{|\psi_2|^4}{4!} - \frac{t^2|\psi_2|^6}{6!} + \cdots\right] dx.$$

Thus if $\lambda > \mu_2$, then

$$< J_\lambda''(0)\psi_1,\, \psi_1 >_{H^1} = -(\lambda - \mu_1)\int_\Omega \psi_1^2 dx < 0,$$

and

$$< J_\lambda''(0)\psi_2,\, \psi_2 >_{H^1} = -(\lambda - \mu_2)\int_\Omega \psi_2^2 dx < 0,$$

Since ψ_1 and ψ_2 correspond to different eigenvalues, they are linearly independent in $H_0^1(\Omega)$. Thus the Morse index of $J_\lambda(v)$ at $v = 0$ is greater than or equal to 2, if $\lambda > \mu_2$. Since w_1 is a critical point of $J_\lambda(v)$ of the mountain pass type, one has that the Morse index of $J_\lambda(v)$ at $v = w_1$ is less than or equal to 1. Thus $w_1 \neq 0$ if $\lambda > \lambda_2$. Therefore, (3.1) admits four nontrivial solutions: $\pm w_0$ and $\pm w_1$. □

In fact, upon applying a general theorem (Theorem 9.6) in Rabinowitz [**Ra**], we can check that the four assumptions therein are satisfied by (3.1). Thus we have the following.

THEOREM 3.8. *If $\lambda > \mu_k$, then* (3.1) *admits at least k different pairs of nontrivial solutions.* □

We note that even though our Theorem 3.7 appears to be only a special case of Theorem 3.8, our proof contains slightly more solution information.

4. Visualization of positive solutions of the elliptic sine-Gordon equation subject to the homogeneous Dirichlet boundary condition

In this section, we compute positive solutions by minimizing J on $H_0^1(\Omega)$ for a dumbbell-shaped domain in 2D. The algorithm (based on the method of steepest descent) for finding the global minimum of J as guaranteed by Theorem 3.3 and proceed as follows (assuming $\lambda > \mu_1$):

(1) Choose an initial guess $u_0 \in H_0^1(\Omega)$ such that

$$0 < u_0(x) \leq \pi, \qquad J_\lambda(u_0) < 0;$$

(2) Find a $T > 0$ such that $J_\lambda(Tu_0) \geq 0$;

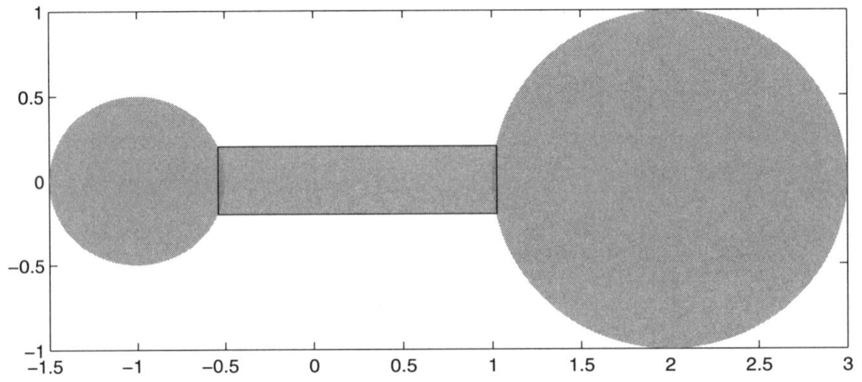

FIGURE 2. The Dumbbell-shaped domain Ω

(3) Find a $t^* \in (0,T)$ such that

$$J_\lambda(t^* u_0) = \min_{t\in[0,T]} J_\lambda(t u_0);$$

(4) Let u_0 be replaced by $t^* u_0$. Compute $v = \nabla J_\lambda(u_0)$ by

$$\begin{cases} \Delta v = \Delta u_0 - \lambda \sin u_0, & \text{in } \Omega, \\ v = 0, & \text{on } \partial\Omega; \end{cases} \tag{4.1}$$

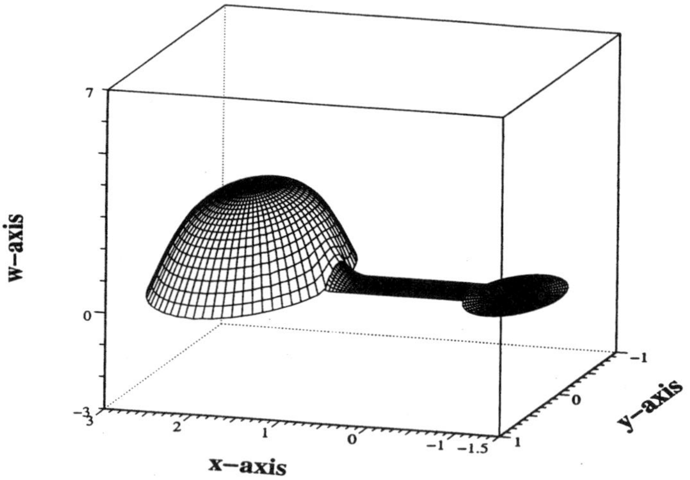

FIGURE 3. Profile of the global minimizer w for J_λ on $H_0^1(\Omega)$, $\lambda = 20$, $J_\lambda(w) = -33.47$, $\max w = 2.883$.

(5) If $\|v\| < \varepsilon$, then output u_0; otherwise, solve

$$s = \arg\min_{s>0} J_\lambda(u_0 - sv),$$

replace u_0 by $u_0 - sv$, and goto Step (3).

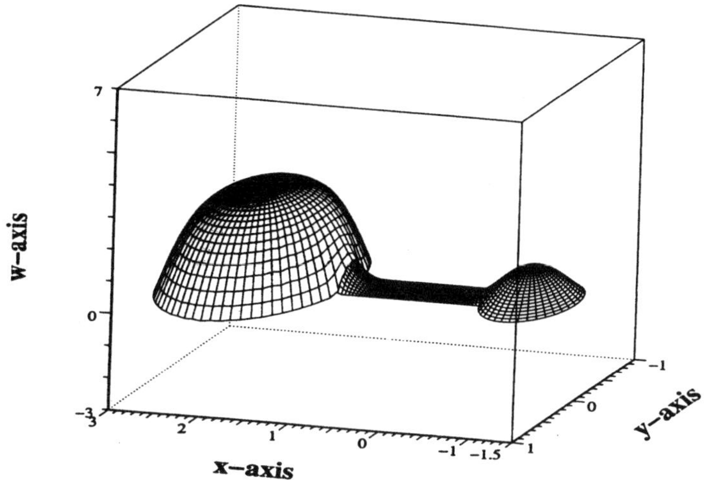

FIGURE 4. Global minimizer w, $\lambda = 25$, $J_\lambda(w) = -51.73$, $\max w = 2.984$.

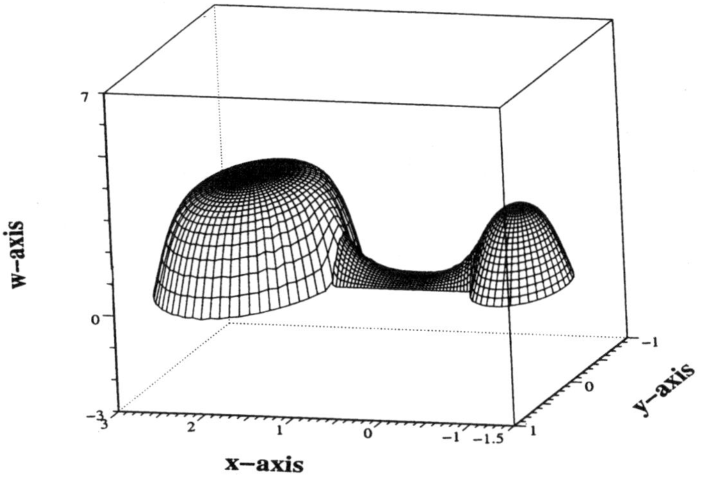

FIGURE 5. Global minimizer w, $\lambda = 50$, $J_\lambda(w) = -169.7$, $\max w = 3.119$.

The dumbbell-shaped domain Ω is given in Fig. 2. It is formed by a disk on the left, centered at $(-1, 0)$ with radius $1/2$, connected through a rectangular corridor with width $1/2$, to a disk on the right centered at (2,0) with radius 1. A boundary element method [**CZ**] is used to compute the numerical solution of the linear elliptic BVP (4.1) in Step (4). Five values of λ are used:

$$\lambda = 20, \quad 25, \quad 50, \quad 100, \quad \textit{and} \quad 500.$$

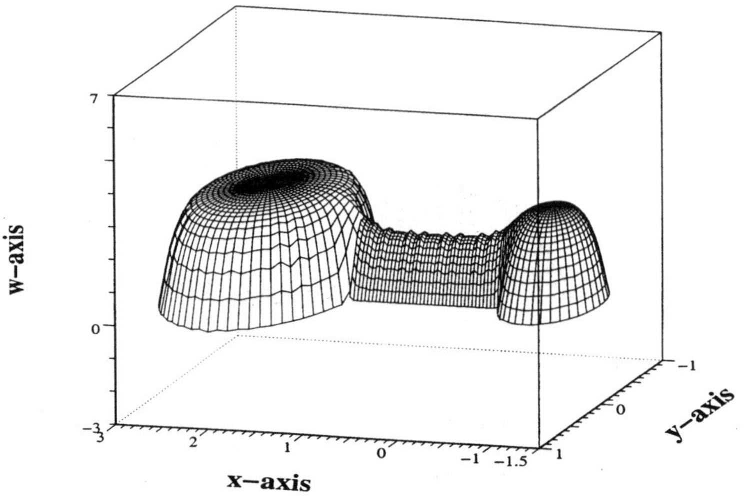

FIGURE 6. Global minimizer w, $\lambda = 100$, $J_\lambda(w) = -469.9$, $\max w = 3.14$.

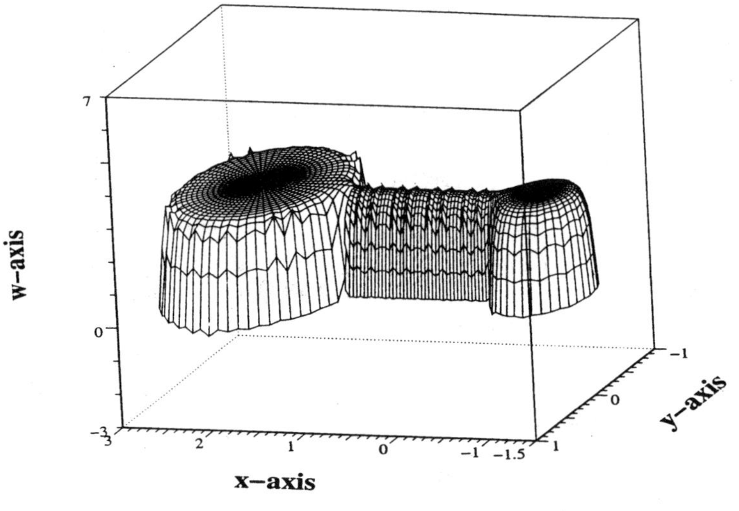

FIGURE 7. Global minimizer w, $\lambda = 500$, $J_\lambda(w) = -3506.7$, $\max w = 3.142$. Note that because of the large value of λ, numerical instability becomes noticeable.

Their graphics are given, in sequential order, in Figs. 3–7, along with the values of J and the maximum value of the function. We see that such solutions' maxima converge to a value close to $\pi = 3.1416$.

Acknowledgement. We thank Professor Zhi-Qiang Wang for helpful discussions.

References

[Ad] R. Adams, *Sobolev Spaces*, Academic Press, New York, 1978.

[AH] A. Ambrosetti and P. Hess, J. Math. Anal. Appl., **73**, 411 (1980).

[AR] A. Ambrosetti and P. H. Rabinowitz, J. Funct. Anal., **14**, 349 (1973).

[Ch] K. C. Chang, *Infinite Dimensional Morse Index Theory and Multiple Solution Problems*, Birkhäuser, Boston, 1993.

[CZ] G. Chen and J. Zhou, *B*oundary Element Methods, Academic Press, San Diego, 1992.

[DCL] Z. Ding, G. Chen and S. Li, On positive solutions of the elliptic sine-Gordon equation, preprint.

[Dr] P. G. Drazin, *Solitons*, Cambridge Univ. Press, London, 1983, p. 101.

[GNN] B. Gidas, W. M. Ni and L. Nirenberg, Comm. Math. Phys., **68**, 209 (1979).

[GT] D. Gilberg and N. S. Trudinger, *Elliptic Partial Differential Equations of Second Order*, 2nd Edition, Springer-Verlag, New York, 1983.

[Ku] I. O. Kulik, Sov. Phys. JETP **1**, 251 (1962); T. Kontos et al., Phys. Rev. Lett. **89**, 137007 (2002); Y. Blum et al., Phys. Rev. Lett. **89**, 187004 (2002).

[La] G. L. Lamb, Jr., *Elements of Soliton Theory*, John Wiley and Sons, New York, 1980, Sec. 6.4

[Li] C. Li, Nonlinear Analysis, **54**, 431 (2003).

[LW] S. J. Li and Z. Q. Wang, J. d'Anal. Math., **81**, 373 (2000).

[MK] K. Maki and P. Kumar, Phys. Rev. B **14** , 3920 (1976), and earlier references cited therein.

[Mi] H. J. Mikeska, J. Phys. C **11**, L29 (1978).

[Ni] W. M. Ni, *Some Aspects of Semilinear Elliptic Equations*, National Tsing Hua University, Taiwan, R.O. C., 1987.

[Ra] P. H. Rabinowitz, *Minimax methods in critical point theory with applications to differential equations*, CBMS 65, AMS, 1986.

[RS] C. Rogers and W. F. Shadwick, *Bäcklund Transformations and Their Applications*, Academic Press, New York, 1982.

[Sc] A. C. Scott et al., Nuovo Cimento B **69**, 241 (1970).

[Sa] D. Sattinger, *Topics in Stability and Bifurcation Theory*, Lecture Notes in Mathematics No. 309, Springer-Verlag, New York, 1973.

[SR] M. Sigrist and T. M. Rice, J. Phys. Soc. Jpn. **61**, 4283 (1992); C. C. Tsuei and J. R. Kirtley, Rev. Mod. Phys. **72**, 969(2000).

[St] M. Struwe, *Variational Methods*, Springer-Verlag, New York, 1996.

Department of Mathematics, Texas A&M University, College Station, TX 77843
E-mail address: gchen@math.tamu.edu

Department of Mathematical Sciences, University of Nevada, 4505 Maryland Parkway, Las Vegas, NV 89154-4020
E-mail address: dingz@unlv.nevada.edu

Department of Physics, Texas A&M University, College Station, TX 77843
E-mail address: hu@rainbow.physics.tamu.edu

School of Mathematics, University of Minnesota, Minneapolis, MN 55455
E-mail address: ni@math.umn.edu

Department of Mathematics, Texas A&M University, College Station, TX 77843
E-mail address: jzhou@math.tamu.edu

Contemporary Mathematics
Volume **357**, 2004

Convergence Analysis of an Optimal Scaling Algorithm for Semilinear Elliptic Boundary Value Problems

Goong Chen, Berthold G. Englert, and Jianxin Zhou

ABSTRACT. Proof of convergence for iterative schemes for finding unstable solutions of semilinear elliptic boundary value problems is an arduous task. In perspective is a special iterative algorithm using the idea of *scaling*. In the form called *Scaling Iterative Algorithm (SIA)* based on normalizing each iterate's function value to be 1 at a given interior point of the domain, it is found that SIA is computationally quite advantageous. Yet no convergence analysis is available. In this paper, we present a different idea of scaling which is an *optimal scaling* in the sense that the first integral is optimized. For this *Optimal Scaling Iterative Algorithm (OSIA)*, we prove the convergence under certain assumptions on the nonlinearity and stipulated stepsize rule.

1. Introduction

In the computation of numerical solutions of semilinear elliptic boundary value problems (BVPs) in the form

$$(1.1)\qquad \begin{cases} \Delta u(x) + f(x, u(x)) = 0, & x \in \Omega, \\ Bu|_{\partial\Omega} = 0, \end{cases}$$

where in the above, Ω is a bounded open domain in $\mathbb{R}^N$ with sufficiently smooth boundary $\partial\Omega$, and B is a linear boundary operator involving u and $\frac{\partial u}{\partial n}$ (the normal derivative of u on $\partial\Omega$), several algorithms have been found to be effective ([**CNZ**])

(1) *DIA*: the direct iteration algorithm;
(2) *MIA*: the monotone iteration algorithm;
(3) *MMA*: the minimax algorithm, including the mountain-pass (MPA), the high-linking algorithm (HLA), etc.;
(4) *SIA*: the scaling iteration algorithm;
(5) Variational derivative iteration type algorithm such as Newton's algorithm or Sobolev gradients [**Ne**],

etc. Each algorithm has its advantages and disadvantages and should be chosen according to certain a priori knowledge of the solution manifold structure, if any

---

2000 *Mathematics Subject Classification.* Primary 35J65, Secondary 65N12.

Key words and phrases. Semilinear elliptic boundary value problems, optimal scaling, convergence analysis.

Supported in part by a TITF initiative from Texas A&M University.

such knowledge is available. For example, DIA and MIA are useful in capturing the so-called *stable solutions* (Morse index=0) while MPA and SIA are useful in capturing the so-called mountain pass solutions (Morse index=1) which are *unstable*. As for algorithm complexity, DIA, MIA and SIA are one-level algorithm and easy to program; MMA is a two-level algorithm which can be used to capture saddle points of high Morse index for a more general functional.

Associated with the various algorithms, whether theoretical or numerical, there is the all important question of *convergence*. Algorithms enable the finding of a solution of a nonlinear problem through *iterations* where at each iteration often an iterate of a *linear* problem is sought. Will the sequence or a subsequence thereof of iterates converge, in a certain sense, to a definitive solution of the given nonlinear problem? The answer of this question is well understood in the case of MIA, where the sequences of sub-solutions and super-solutions will sandwich a bona fide (stable) solution in the $C^{2,\alpha}(\Omega)$-norm [**Am, Sa**], under certain relatively easily verifiable) conditions on (1.1). However, for the various algorithmic versions of MPA and HLA ([**CNZ, CM1, CM2, DCC**] e.g.), proofs of convergence to an unstable solution turn out to be much more challenging. In the original formulation of the famous Mountain Pass Lemma by Ambrosetti and Rabinowitz [**AR**], its proof is *not totally constructive*. Also compounding this difficulty is an associated variational problem which must be treated *globally*. For a theoretical method to be amenable to numerical implementation, *adaptations* are necessary. One of the main components of such adaptations is the *localization* of the global procedure in the Mountain Pass Lemma, resulting in several possible versions of MMA. Only recently, convergence analysis for a local MMA has been successfully carried out; see [**LZ1, LZ2, LZ3**].

In this paper, our main interest is in SIA, the scaling iterative algorithm. The version of SIA as presented in [**CNZ**, pp. 1573–1574] (see more details in (2.3)–(2.4) in Section 2) was found to be very effective in computing numerical solutions of a certain class of unstable solutions of semilinear elliptic BVPs, especially the Lane–Emden type equation whose nonlinearity is a *power law*. The accuracy and convergence rate of SIA in [**CNZ**] totally agree with the MPA in every example computed in [**CNZ**]. Yet SIA in [**CNZ, CNPZ**] has a major advantage over MPA in that SIA is an one-level algorithm and considerably easier to program and implement. However, the form of SIA as presented and used in [**CNZ**] has not yet been placed a solid rigorous foundation because no convergence proof was established, despite many years of serious efforts by the authors of [**CNZ**].

Recently, in a continuing attempt to analyze the convergence of SIA, we have come up with a new version of SIA which may be called an *optimal scaling iteration algorithm (OSIA)*. The SIA in [**CNZ**] was based on scaling at a single interior point of domain Ω. However, the new OSIA is based on scaling along a given direction. It turns out to work much in the same way as in a theoretical study done much earlier by Ding and Ni [**DN**]. This confluence of ideas is indeed a pleasure. Even though the work in [**DN**] is not relevant to any iterative algorithms, credits are due to the authors of [**DN**] for the priority of the intersected ideas. Here, our main interest lies in treating OSIA as an algorithm, with emphasis on its convergence analysis.

Because of the closeness and similarity, we are able to utilize the ideas of convergence analysis developed earlier by Li and Zhou [**LZ1**]–[**LZ3**] for MMA and applied them to OSIA. In Section 2, we describe the procedures of OSIA and relevant matters. In Section 3, convergence analysis of OSIA is given. Even though we must concede that convergence proof for SIA (as presented in [**CNZ**]) is still *open*, from our work in Section 3 one can begin to understand why such a proof is challenging to clinch. Hopefully, this paper will contain at least some more than tangential ideas and will also draw attention toward a convergence proof of SIA.

The various algorithms, DIA, MIA, MMA, SIA, Newton's, etc., may be implemented with finite difference (FDM), finite elements (FEM) and boundary elements (BEM). At the present time, *error estimates* for such algorithms and discretization schemes are available *only for MIA*; see [**CDNZ, DCNZ, HMW, Is, Pa**], for example. This is a wide open area in numerical analysis of nonlinear PDEs where more investigations are most welcome.

2. Scaling Iteration Algorithms

It is well-known that semilinear elliptic PDEs of the form

$$\tag{2.1} \begin{cases} \Delta u - au + bu^p = 0 & \text{on } \Omega, \\ u > 0 & \text{on } \Omega, \\ u|_{\partial\Omega} = 0, \end{cases}$$

where $a \geq 0$ and $b > 0$ are two arbitrarily given constants and $p > 1$, satisfy the "scaling property". For example, if u is a solution of (2.1), then from u we can find a solution v to the problem

$$\begin{cases} \Delta v - av + b\lambda v^p = 0 & \text{on } \Omega, \\ v > 0 & \text{on } \Omega, \\ v|_{\partial\Omega} = 0, \end{cases}$$

for any constant $\lambda > 0$ by taking $v = \lambda^{\frac{1}{1-p}} u$. This property can be exploited for computational purposes as follows. Consider solving (2.1). Define

$$v(x) = \frac{u(x)}{u(x_0)}, \quad \text{for some fixed} \quad x_0 \in \Omega.$$

Then v satisfies

$$\tag{2.2} \begin{cases} \Delta v(x) - av(x) + \lambda bv^p(x) = 0, & \text{on } \Omega, \quad \lambda \equiv [u(x_0)]^{p-1}, \\ v(x_0) = 1, \\ v > 0 \text{ on } \Omega, \\ v|_{\partial\Omega} = 0. \end{cases}$$

We now define SIA as follows

SIA

Step 1: Choose any $v_0(x) \underset{\not\equiv}{>} 0$ on $\Omega; v_0$ sufficiently smooth;

Step 2: Find $\lambda_{n+1} \geq 0$ and $v_{n+1}(\cdot)$ such that

$$\tag{2.3} \begin{cases} \Delta v_{n+1}(x) - av_{n+1}(x) = -\lambda_{n+1} bv_n^p(x), & x \in \Omega, \\ v_{n+1}(x_0) = 1, \\ v_{n+1}|_{\partial\Omega} = 0. \end{cases}$$

Step 3: If $\|v_{n+1} - v_n\|_{H_0^1(\Omega)} < \varepsilon$ output and stop. Else set $n = n + 1$ and goto Step 2.

In the above, $v_0(x) \underset{\not\equiv}{>} 0$ on Ω means $v_0(x) \geq 0$ on Ω but $v_0(x)$ is not identically to 0 on Ω. Also, note that λ_{n+1} can be easily determined by the pointwise normalization condition $v_{n+1}(x_0) = 1$.

REMARK 2.1. It is clear that if in Step 3, we have $v_{k+1} - v_k = 0$ then v_{k+1} is a solution to (2.2) and consequently $u^* = \lambda_{n+1}^{\frac{1}{1-p}} v_{n+1}$ solves (2.1). Assume that DIA, the direct iteration algorithm, were applied to (2.1) as follows: with chosen $v_0(x) \underset{\not\equiv}{>} 0$ on Ω, for $n = 0, 1, 2, \ldots$, solve

$$\begin{cases} \Delta v_{n+1} - a v_{n+1} = -b v_n^p & \text{on } \Omega, \\ v_{n+1}|_{\partial\Omega} = 0. \end{cases}$$

Then what we have found numerically is that the above iteration either converges to the trivial solution $v_\infty \equiv 0$ or diverges to infinity. Thus the motivation for setting $v_{n+1}(x_0) = 1$ in (2.3) is to prevent the convergence of v_∞ to the trivial solution 0, and to prevent v_n from running away to infinity. □

Before we present a variant of SIA, let us digress from iterative schemes for semilinear PDEs and instead look at the common *Newton's algorithm* for scalar-valued functions. This is intended to provide heuristics for the subsequent discussion. Recall Newton's method for solving a nonlinear equation $g(x) = 0$:

iteration: Take an initial guess x_0,

and then let

$$x_{n+1} = x_n - \frac{g(x_n)}{g'(x_n)}. \tag{2.4}$$

See Fig. 1.

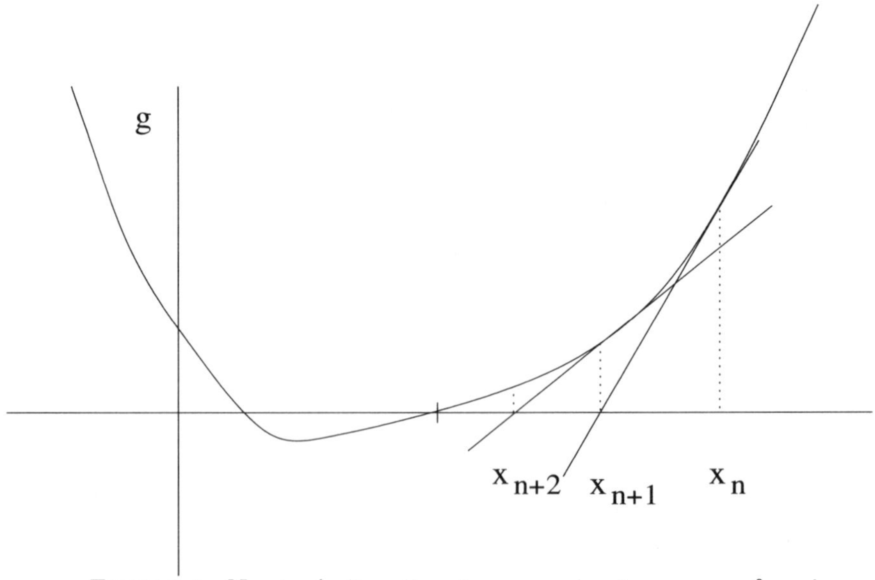

FIGURE 1. Newton's iteration to approximate a zero of a given function g.

Newton's method *does not assume a variational structure* (which actually should be utilized as much as possible if it is available). It may also fail to converge in the following situation; see Fig. 2.

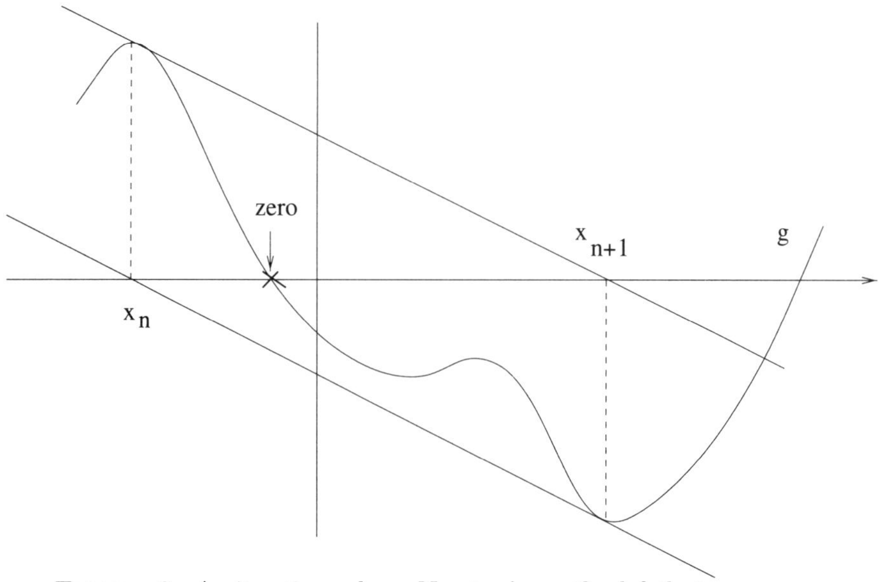

FIGURE 2. A situation where Newton's method fails to converge to a zero of g. The iteration repeats itself between x_n and x_{n+1}.

The method is known to be *unstable.* (Here, "stability" means insensitivity of convergence relative to small perturbations of the initial conditions.) Suppose that we want to find a critical point u^* of a variational problem with first integral $J(v)$, i.e.,

$$\text{find } u^* \text{ s.t. } J'(u^*) = 0. \tag{2.5}$$

The infinite dimensional analogue of Newton's method becomes

$$u_{n+1} = u_n - [J''(u_n)]^{-1} J'(u_n). \tag{2.6}$$

It involves computation of the second derivative $J''(u_n)$ of an infinite-dimensional functional J at u_n. Note that the iteration (2.6) could face serious trouble when u^* satisfying (2.5) is a *degenerate critical point*, i.e., the operator

$$J''(u^*) \quad \text{does not exist or is not invertible.} \tag{2.7}$$

In some sense, degeneracy is hard to avoid in the finding of a saddle point. Say, let u_1 be a nondegenerate local minimum critical point of J. So all eigenvalues of $J''(u_1)$ are positive. Let u_2 be a mountain pass solution (saddle point) to be found. We assume that u_2 is nondegenerate; $J''(u_2)$ has a negative eigenvalue. If J is C^2, then there will be a point u_0 between u_1 and u_2 such that $J''(u_0)$ has a zero eigenvalue. Thus $J''(u_0)$ is not invertible and around u_0, J'' is ill-conditioned. Such a case can be bypassed if an initial guess is chosen to be close to u_2.

As to enhance the stability of Newton's algorithm, there is some rescue from optimization theory. An important idea as suggested from the counterexample in Fig. 2 is to introduce a *variable stepsize*:

$$u_{n+1} = u_n - s_n[J''(u_n)]^{-1}J'(u_n), \tag{2.8}$$

where s_n satisfies certain stepsize rules (such as Armijo's, Goldstein's, or Powell's, etc.; see [**Ke**] for example). This modification facilitates convergence and enhances stability.

Now we are in a position to present an *optimal scaling algorithm with variable stepsize*, acronymed *OSIA*.

Consider the semilinear elliptic BVP

$$\begin{cases} \Delta u(x) + f(x, u(x)) = 0, & x \in \Omega, \\ u|_{\partial\Omega} = 0. \end{cases} \tag{2.9}$$

Let the corresponding first integral be

$$J(u) = \int_\Omega \left[\frac{1}{2}|\nabla u|^2 - F(x,u)\right] dx \tag{2.10}$$

where F satisfies

$$\frac{\partial}{\partial u}F(x,u) = f(x,u).$$

For the semilinear problem (2.9), let $H = H_0^1(\Omega)$ and $\|\ \|$ be the $H_0^1(\Omega)$-norm and denote $S = \{u \in H : \|u\| = 1\}$. we now define the following iterative scheme:

OSIA

Step 1: Set $k = 0$. Given $u_k \in S$.

Step 2: Find $\lambda_k = \arg_\lambda \max J(\lambda u_k)$ and set $v_k = \lambda_k u_k$, for $k = 0$.

Step 3: Solve $\bar{v}_k$ from

$$\begin{cases} -\Delta \bar{v}_k = f(\cdot, v_k), & \text{on } \Omega, \\ \bar{v}_k|_{\partial\Omega} = 0. \end{cases}$$

Step 4: If $\|\bar{v}_k - v_k\| < \varepsilon$ then stop and exit, else goto Step 5.

Step 5: Denote

$$u_k(s) = \frac{s\bar{v}_k + (1-s)v_k}{\|s\bar{v}_k + (1-s)v_k\|} \in S,\ 0 < s \le 1 \quad \text{and} \quad \lambda_k(s) = \arg\max_\lambda J(\lambda u_k(s)),$$

choose $0 < s_k \le 1$ satisfying certain stepsize rule, e.g., (2.11).

Step 6: Set $v_{k+1} = \lambda_k(s_k)u_k(s_k)$ and update $k = k+1$, then goto Step 3.

REMARK 2.2. Note that the problem settings for both SIA and OSIA assume a variational structure. But SIA does not utilize its variational structure, while Steps 2 and 5 in OSIA use its variational structure. It is clear that the problem setting (2.9) of OSIA is more general than the problem setting (2.1) of SIA. If in Step 4 of OSIA, we have $\bar{v}_k - v_k = 0$, then $\bar{v}_k$ is a solution to (2.9). The normalization in Step 5 of OSIA signifies "return to the unit sphere S". □

There are several different ways to define a stepsize rule. For example, we may define the following.

DEFINITION 2.3. Let $v_k, \bar{v}_k, \lambda_k, \lambda_k(s)$ and $u_k(s)$ be defined as in OSIA. If $\bar{v}_k \neq v_k$ then we define the stepsize rule by

$$\begin{aligned} s_k = \max\Big\{0 < s \leq 1 \;\Big|\; & J(\lambda_k(s)u_k(s)) - J(v_k) \\ & \leq -\frac{1}{2}\frac{s\lambda_k}{\|v_k + s(\bar{v}_k - v_k)\|}\|\bar{v}_k - v_k\|^2\Big\}. \end{aligned} \tag{2.11}$$

□

EXAMPLE 2.4. A model problem is the Lane–Emden equation

$$\begin{cases} \Delta u(x) + u^p(x) = 0 & \text{on } \Omega, \\ u(x) > 0 & \text{on } \Omega, \\ u|_{\partial\Omega} = 0. \end{cases} \tag{2.12}$$

The first integral is

$$J(v) = \int_\Omega \left[\frac{1}{2}|\nabla v|^2 - \frac{1}{p+1}|v|^{p+1}\right] dx.$$

The key iteration step in Steps 3 says that $\bar{v}_k$ is determined from solving

$$-\Delta \bar{v}_k = \lambda_k^p u_k^p, \tag{2.13}$$

where λ_k is chosen such that $J(\lambda u_k)$ is maximized at $\lambda = \lambda_k$. Thus

$$\frac{d}{d\lambda}\int_\Omega \left[\frac{1}{2}\lambda^2|\nabla u_k|^2 - \frac{1}{p+1}\lambda^{p+1}|u_k|^{p+1}\right] dx = 0,$$

i.e.,

$$\lambda_k = \left[\int_\Omega |\nabla u_k|^2 dx \Big/ \int_\Omega |u_k|^{p+1} dx\right]^{1/p-1}.$$

Therefore, the scaling in (2.13) is "optimal" in this sense. □

3. Convergence of OSIA

With the variational functional J in (2.10), under some standard conditions on f, a direct computation shows that (weak) solutions to (2.9) coincide with critical points of J. We now proceed to prove the convergence of OSIA under certain assumptions on the nonlinearity f and the stipulated stepsize rule (2.11).

Let $\langle\ ,\ \rangle$ denote the inner product in $H_0^1(\Omega)$ and let $d = -J'(u)$ be the negative gradient of J at u. (Because the gradient J' is utilized here, OSIA can be viewed as a Sobolev gradient method.) For each $v \in H_0^1(\Omega)$, we have

$$\langle d, v\rangle = \langle -J'(u), v\rangle \equiv -\frac{d}{dt}J(u+tv)|_{t=0}.$$

Then

$$\int_\Omega \nabla d \cdot \nabla v \; dx = \int_\Omega [-\nabla u \cdot \nabla v + f(\cdot, u)v] \; dx,$$

or

$$-\int_\Omega \Delta d \cdot v \; dx = \int_\Omega [\Delta u + f(\cdot, u)]v \; dx.$$

Thus $d = -J'(u)$ solves

$$\begin{cases} \Delta d = -[\Delta u + f(\cdot, u)] \text{ on } \Omega, \\ d|_{\partial\Omega} = 0. \end{cases} \tag{3.1}$$

Note that if

$$\begin{cases} -\Delta \bar{v}_k = f(\cdot, v_k) \quad \text{on } \Omega, \\ \bar{v}_k|_{\partial\Omega} = 0, \end{cases}$$

then

$$\begin{cases} -\Delta(\bar{v}_k - v_k) = \Delta v_k + f(\cdot, v_k) \quad \text{on } \Omega, \\ (\bar{v}_k - v_k)|_{\partial\Omega} = 0, \end{cases}$$

or $d_k \equiv \bar{v}_k - v_k = -J'(v_k)$ is the negative gradient of J at v_k. Now Step 4 in OSIA reads:

If $\|d_k\| < \varepsilon$ then stop and exit, else goto Step 3.

OSIA becomes a critical point approximation algorithm. To prove its convergence, let us rewrite $u_k(s)$ in Step 5 of OSIA as

$$u_k(s) = \frac{v_k + s(\bar{v}_k - v_k)}{\|v_k + s(\bar{v}_k - v_k)\|} = \frac{v_k + sd_k}{\|v_k + sd_k\|} = \frac{u_k + s_v d_k}{\|u_k + s_v d_k\|}$$

where $s_v = \frac{s}{\|v_k\|}$. The stepsize rule (2.11) becomes

$$s_k = \max\left\{0 < s \leq 1 \,\middle|\, J(\lambda_k(s)u_k(s)) - J(v_k) \leq -\frac{1}{2}\frac{s\lambda_k}{\|v_k + sd_k\|}\|d_k\|^2\right\}. \tag{3.2}$$

With $H = H_0^1(\Omega)$ and $S = \{u \in H : \|u\| = 1\}$, let J be given in (2.10). For each $u \in S$, define the selection of the peak along u or, briefly, *peak selection* of J by $p(u) = \lambda_u u$ where $\lambda_u = \arg\max_\lambda J(\lambda u)$. It is clear that OSIA becomes a variant of the local minimax method presented in [**LZ1**]. Thus in general, OSIA is a two-level algorithm. However, when an explicit formula for computing λ_k is available, such as in Example 2.4, OSIA remains a one-level algorithm. The relation between the negative gradient and the steepest descent direction can be seen as follows. The steepest descent direction of J at a point u is a unit vector d_u such that

$$d_u = \arg \max_{d \in H, \|d\|=1} \langle -J'(u), d\rangle. \tag{3.3}$$

When H is a Hilbert space, the right hand side of (3.3) is equal to $\| - J'(u)\|$. Thus $d_u = \frac{-J'(u)}{\|J'(u)\|}$ is the steepest descent direction which differs from the negative gradient only by a positive factor. Note that we are looking for a critical point u^*, i.e., $J'(u^*) = 0$. Dividing by a term going to zero usually will introduce extra error in numerical computation. But there is a stepsize before the steepest descent direction in the algorithm; this stepsize can absorb a positive factor anyway. We can simply use the negative gradient to replace the steepest descent direction in the algorithm.

Using similar arguments as in [**LZ1, LZ2**], we can prove the following.

THEOREM 3.1 (Theorem 4.3 in [**LZ1**]). *Assume that Conditions (h1) – (h5′) in* [**LZ1**] *are satisfied and that there exist* $a_5 > 0$ *and* $a_6 > 0$ *s.t. for* s *as specified in (h2),*

$$|f_\xi(x, \xi)| \leq a_5 + a_6|\xi|^{s-1}. \tag{3.4}$$

Then the peak selection p *of* J *is* C^1. □

Now we assume that Conditions (h1) – (h5′) in [**LZ1**] and (3.4) hold.

THEOREM 3.2 ([**LZ1, LZ2**]). *Let u^k and $v^k = p(u^k), k = 0, 1, 2, \ldots$ be the sequence generated by OSIA, then*

(a) *$J(v_{k+1}) < J(v_k)$, i.e., OSIA is a strictly decreasing algorithm and is thus stable;*
(b) *$\{v^k\}_{k=0}^{\infty}$ has a subsequence $\{v^{k_i} = p(u^{k_i})\}$ that converges to a critical point;*
(c) *any limit point of $\{v^k\}$ is a critical point of J. In particular, any convergent subsequence $\{v^{k_i}\}$ of $\{v^k\}$ converges to a critical point.* □

If a critical point is not isolated, we can only expect some point-to-set convergence as similar to Theorem 3.3 in [**LZ2**]. When a critical point is isolated, by a similar proof of Theorem 3.4 in [**LZ2**], we can establish the following point-to-point convergence.

THEOREM 3.3. *If $J(p(\bar{u})) = loc\min_{u\in S} J(p(u))$ and $\bar{v} = p(\bar{u}) \neq 0$ is an isolated critical point, then there exists an open set U in H, such that $\bar{u} \in U \cap S$ and by starting from any $u_0 \in U \cap S$, the sequence $\{v^k\}_{k=0}^{\infty}$ generated by OSIA converges to $\bar{v} = p(\bar{u})$.* □

THEOREM 3.4 (Theorem 3.3 in [**LZ3**]). *Assume that Conditions (h1) – (h5′) in* [**LZ1**] *and (3.4) are satisfied, then a critical point found by OSIA has Morse index 1.* □

4. Numerical Examples

We present computational results in this paper. An irregular shaped domain Ω is chosen as in Fig. 3. A disk with radius 2 appears in the upper left corner of Ω with a cavity of radius 1. The center of that cavity is located at $(-3.5, 3)$. The lower left corner is a disk with radius 1/2, while the right side has two disks with radius 1. A unit square occupies the upper right corner, which is connected to the various disks by rectangular corridors of widths 1/2 and 1/4. The domain Ω is discretized by 10321 triangular C^0 Lagrangian finite elements.

We compute solutions of

$$(4.1) \qquad \begin{cases} \Delta u(x) + u^3(x) = 0, & x \in \Omega, \\ u > 0 & \text{on } \Omega, \\ u = 0 & \text{on } \partial\Omega \end{cases}$$

by OSIA, such as Example 2.4, with $p = 3$ therein. Note that the usual finite element method is used in solving the linear elliptic Dirichlet BVP in Step 3 of OSIA. Multiple single-peak solutions have been obtained, see their profiles in Fig. 4-13 arranged in ascending magnitudes of J. Also displayed on their right are the contour lines. These solution profiles are consistent with the numerical solutions in [**CNZ**].

For the comparison purpose, we have also computed equation (4.1) using SIA as in [**CNZ**, pp. 1573–1574]. The numerical efficiency, as evaluated by the required number of iterations, is tabulated in Table 1. The reader may find that the efficiencies of OSIA and SIA are comparable.

Conclusion

(1) Even though the data in Table 1 shows that OSIA is comparable with SIA in the required number of iterations in order to achieve a prescribed accuracy, computationally we have found that for OSIA, the numerical error

	number of iterations	
Solution in	OSIA	SIA
Fig. 4	14	12
Fig. 5	14	15
Fig. 6	18	20
Fig. 7	16	15

TABLE 1. Comparison of numerical efficiency between OSIA and SIA according to the required number of iterations for numerical convergence.

decreases *monotonically*. This is an added advantage of OSIA. Best of all, the convergence of OSIA is proven here, while that of SIA still remains open. (Nevertheless, SIA is easier to code for the programmer. We still hope that a rigorous proof or a disproof for SIA could be established)

(2) When the nonlinearity f in (1.1) has two different powers, say $f(x,u) = f(u) = -c_1 u^{p_1} + c_2 u^{p_2}$, when c_1, c_2 and p_1, p_2 are positive and $p_2 > p_1 > 1$, we have not yet fully understood how well OSIA works because the two powers scale differently. This remains to be investigated. In addition, we wish to know more about the theoretical properties of solutions corresponding to such a nonlinearity for the purpose of confirmation through visualization.

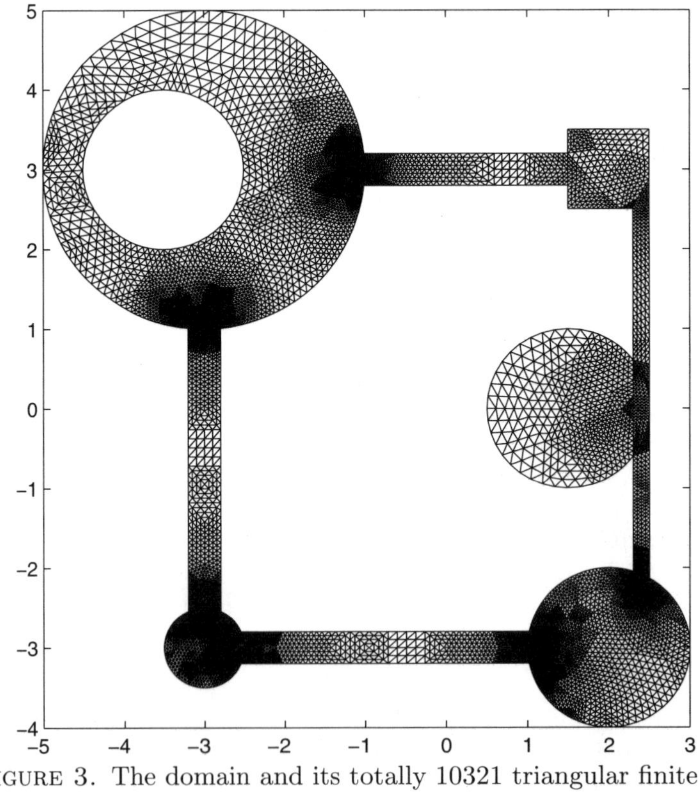

FIGURE 3. The domain and its totally 10321 triangular finite elements.

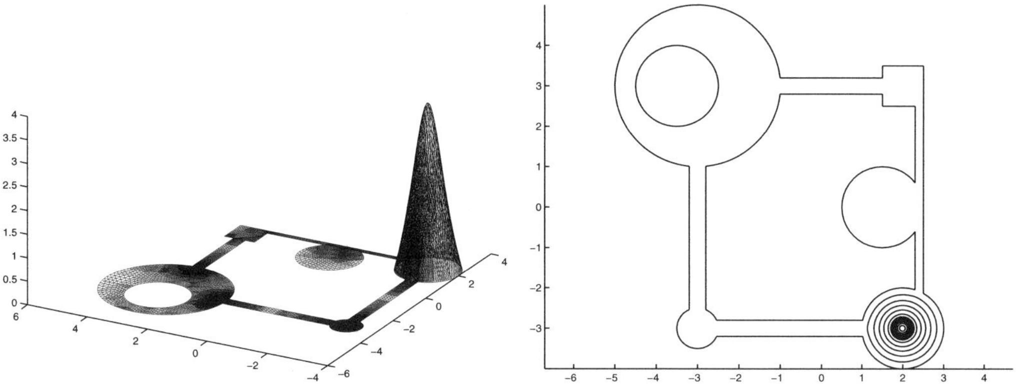

FIGURE 4. A solution to (4.1) with $J = 10.9576$

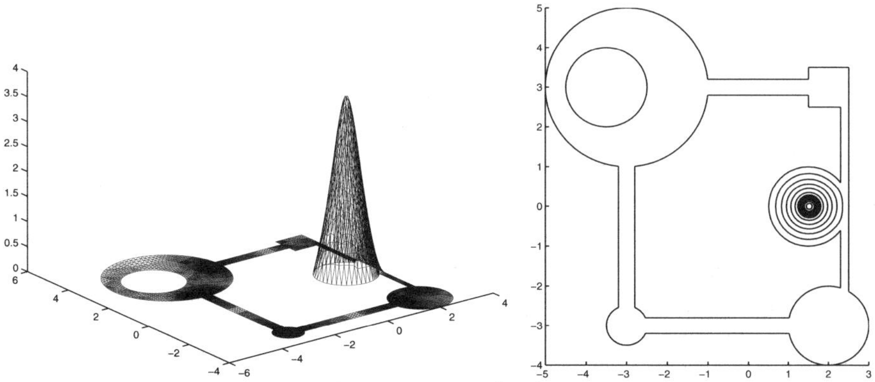

FIGURE 5. A solution to (4.1) with $J = 11.0232$

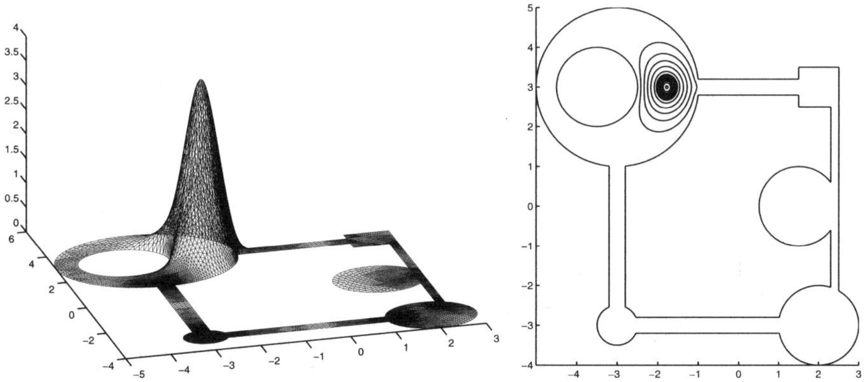

FIGURE 6. A solution to (4.1) with $J = 11.6854$

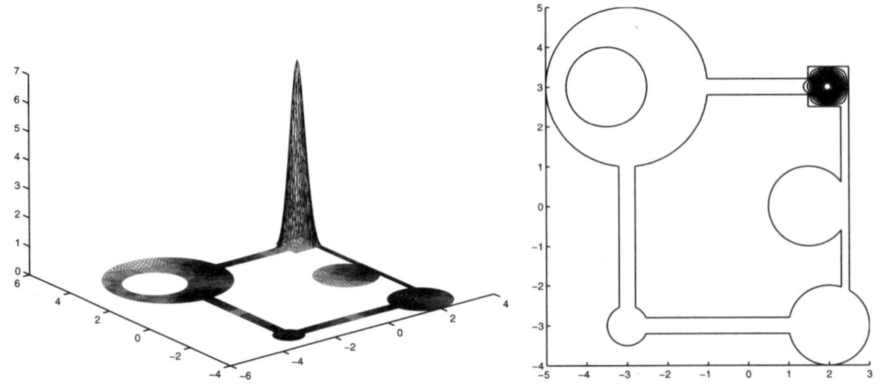

FIGURE 7. A solution to (4.1) with $J = 37.07$

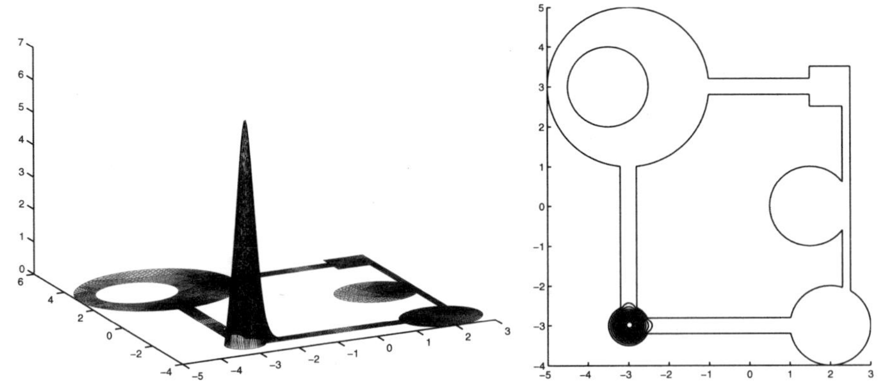

FIGURE 8. A solution to (4.1) with $J = 41.4318$

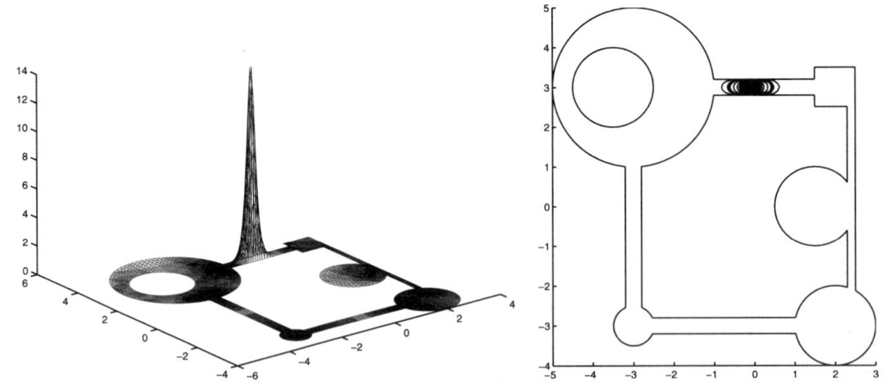

FIGURE 9. A solution to (4.1) with $J = 179.9271$

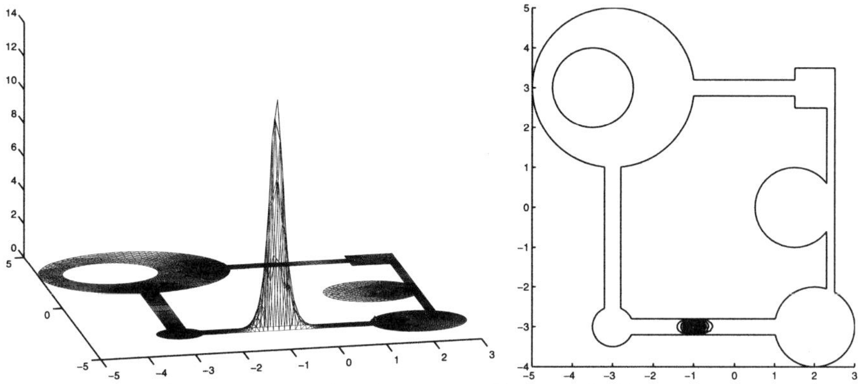

FIGURE 10. A solution to (4.1) with $J = 195.9461$

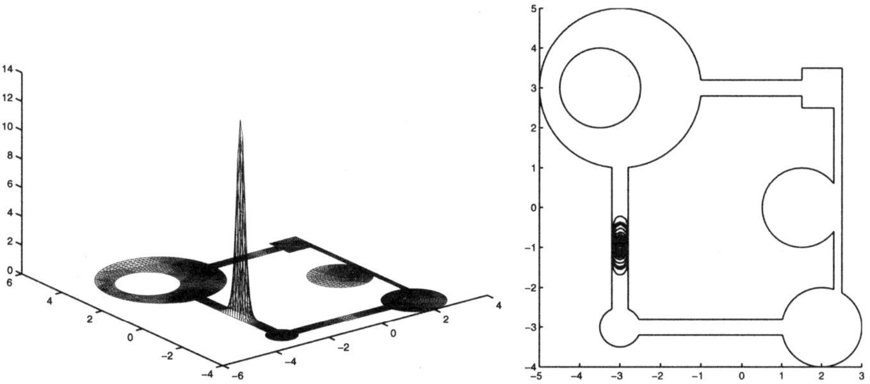

FIGURE 11. A solution to (4.1) with $J = 195.954$

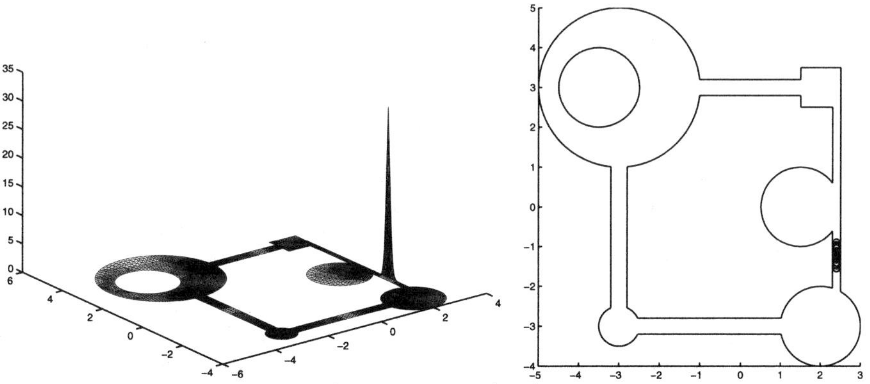

FIGURE 12. A solution to (4.1) with $J = 865.4277$

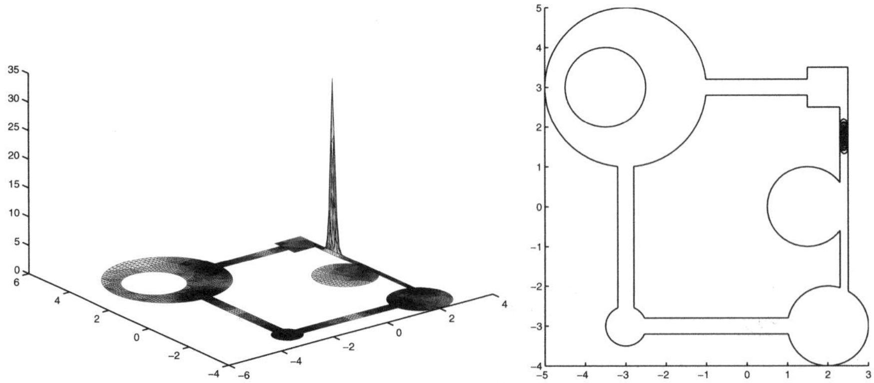

FIGURE 13. A solution to (4.1) with $J = 971.1523$

References

[Am] H. Amann, Supersolution, monotone iteration and stability, *J. Diff. Eq* **21** (1976), 367–377.

[AR] A. Ambrosetti and P.H. Rabinowitz, Dual variational methods in critical point theory and applications, *J. Funct. Anal.* **14** (1973), 349–381.

[CDNZ] G. Chen, Y. Deng, W.M. Ni and J. Zhou, Boundary element monotone iteration scheme for semilinear elliptic partial differential equations, Part II: Quasimonotone iteration for coupled 2×2 systems, *Math. Comp.* **69** (1999), 629–652.

[CNZ] G. Chen, W.M. Ni and J. Zhou, Algorithms and visualization for solutions of nonlinear elliptic equations, *Int. J. Bifurc. & Chaos* **7** (2000), 1565–1612.

[CNPZ] G. Chen, W.M. Ni, A. Perronnet and J. Zhou, ibid, Part II: Dirichlet, Neumann and Robin boundary conditions and problems in 3D, *Int. J. Bifurc. & Chaos* **11** (2001), 1781–1799.

[CM1] Y.S. Choi and P.J. McKenna, A mountain pass method for the numerical solution of semilinear elliptic problems, *Nonlin. Anal.* **20** (1993), 417–437.

[CM2] Y.S. Choi and P.J. McKenna, Numerical mountain pass methods for nonlinear boundary value problems, in "*Nonlinear Problems in Applied Mathematics*", T.S. Angell, et al., eds., SIAM, Philadelphia, 1996, 86–95.

[DCNZ] Y. Deng, G. Chen, W.M. Ni and J. Zhou, Boundary element monotone iteration scheme for semilinear elliptic partial differential equations, *Math. Comp.* **65** (1996), 943–982.

[DN] W.Y. Ding and W.M. Ni, On the existence of positive entire solutions of a semilinear elliptic equation, *Arch. Rat. Mech. Anal.* **91** (1986), 283–308.

[DCC] Z. Ding, D. Costa and G. Chen, A high linking method for sign changing solutions of semilinear elliptic equations, *Nonlin. Anal.* **38** (1999), 151–172.

[HMW] C.U. Huy, P.J. McKenna and W. Walter, Finite difference approximations to the Dirichlet problem for elliptic systems, *Numer. Math.* **49** (1986), 227–237.

[Is] K. Ishihara, Monotone explicit iterations of the finite element approximations for the nonlinear boundary value problems, *Numer. Math.* **45** (1984), 419–437.

[Ke] C. T. Kelley, *Iterative Methods for Linear and Nonlinear Equations*, SIAM, Philadelphia, 1995.

[LZ1] Y. Li and J. Zhou, Local characterizations of saddle points and their Morse indices, in "*Control Nonlinear Distributed Parameter Systems*", G. Chen, I. Lasiecka and J. Zhou, eds., Chap. 11, Lecture Notes in Pure & Appl. Math., Vol. 218, Marcel Dekker, New York, 2001, 233–251.

[LZ2] Y. Li and J. Zhou, A minimax method for finding multiple critical points and its applications to semilinear PDE, *SIAM J. Sci. Comp.* **23** (2001), 840–865.

[LZ3] Y. Li and J. Zhou, Convergence results of a local minimax method for finding multiple critical points, *SIAM J. Sci. Comp.* **24** (2002), 865–885.

[Ne] J.W. Neuberger, *Sobolev Gradients and Differential Equations*, Lecture Notes in Mathematics, Vol. 1670, Springer, Berlin–Heidelberg–New York, 1997.

[Pa] C.V. Pao, Block monotone iterative methods for numerical solutions of nonlinear elliptic equations, *Numer. Math.* **72** (1995), 239–262.

[Sa] D. Sattinger, Monotone methods in nonlinear elliptic and parabolic boundary value problems, *Indiana Univ. Math. J.* **21** (1971/72), 979–1000.

Department of Mathematics, Texas A&M University, College Station, TX 77843
E-mail address: gchen@math.tamu.edu

Department of Physics, National University of Singapore, Singapore
E-mail address: phyebg@nus.edu.sg

Department of Mathematics, Texas A&M University, College Station, TX 77843
E-mail address: jzhou@math.tamu.edu

Contemporary Mathematics
Volume **357**, 2004

Sobolev Gradients: Introduction, Applications, Problems

J. W. Neuberger and R. J. Renka

ABSTRACT. Sobolev gradient is defined and a simple example is given. A number of applications are described. A number of problems are stated, some of which are open problems for research.

1. Introduction to Sobolev Gradients

Sobolev gradients provide an efficient method of calculating solutions to a wide variety of systems of partial differential equations. Successful applications have been made to problems in transonic flow, Ginzburg-Landau equations for superconductivity, elasticity, minimal surfaces, and oil-water separation problems.

A basic reference is [**13**]. This reference contains motivation and background for Sobolev gradients as well as applications up to the time of publication in 1997. The present paper will summarize more recent applications and will present a number of problems. Some of the problems are exercises, some are very open ended, and some may be impossible.

Suppose X is a Banach space with the property that if f is a continuous linear function from X to R, then, given $c > 0$, there is a unique element $h \in X$ so that

$$\sup_{g \in X, \|g\|_X = c} fg = fh. \tag{1}$$

Such spaces X include all Hilbert spaces as well as many Sobolev spaces which are not Hilbert spaces (see [**21**]). If ϕ is a C^1 functional on the Sobolev space X with the above property, then the Sobolev gradient of ϕ at $x \in X$ is the element $(\nabla\phi)(x)$ so that

$$\sup_{g \in X, \|g\|_X = |\phi'(x)|} \phi'(x)g = \phi'(x)(\nabla\phi(x)).$$

Our general reference for Sobolev spaces is [**1**].

Given a function ϕ as above, one seeks a critical point of ϕ by either continuous or discrete steepest descent. Discrete steepest descent is the classical process of constructing a sequence $\{x_k\}_{k=0}^{\infty}$ so that x_0 is given and

$$x_k = x_{k-1} - \delta_k(\nabla\phi)(x_{k-1})), \; k = 1, 2, \ldots, \tag{2}$$

where for each k, δ_k is chosen so that

$$\phi(x_{k-1} - \delta_k(\nabla\phi)(x_{k-1}))$$

is minimal in some appropriate sense. A critical point of ϕ is sought as

$$u = \lim_{k\to\infty} x_k.$$

On the other hand, continuous steepest descent consists of constructing a function $z : [0, \infty) \to X$ so that

$$z(0) = z_0 \in X, \; z'(t) = -(\nabla\phi)(z(t)), \; t \geq 0 \tag{3}$$

and then seeking $u \in X$ so that

$$u = \lim_{t\to\infty} z(t)$$

so that u is a critical point of ϕ.

Continuous steepest descent provides a theoretical starting point for proving convergence of discrete steepest descent.

Functions ϕ of interest fall into two groups. The first group consists of problems in which ϕ comes from a classical variational principle, for example, ϕ is an energy functional. A critical point of such a functional ϕ is frequently a solution to a system of partial differential equations under some boundary conditions.

The second group of problems is given by a function ϕ for which there is a function F from X to a second Banach space Y so that

$$\phi(x) = \|F(x)\|_Y^2/2, \; x \in X. \tag{4}$$

We think of F as being so that a zero u of F is a solution to a system of partial differential equations. It is quite common that a critical point of ϕ is actually a zero of F, something not so common for general optimization problems. This point is discussed in some detail in [**13**]. Functionals ϕ of the second type are generally used only when one has a system of partial differential equations for which a conventional variational principle is not available.

For numerical computation, finite dimensional versions of functions ϕ as above are considered. The main practical point in the use of Sobolev gradients in applications is the following: If ϕ either represents a conventional variational principle or is given by (4), for ϕ to be C^1, it must be that the norm on X is chosen properly with respect to any derivatives contained in the definition of ϕ (we are thinking here that X is a Sobolev space of functions). In finite dimensional approximations to ϕ by, say a function ϕ_n,

n being the number of grid points used in an approximation, the norm on a corresponding finite dimensional space should simulate the required norm on X. An ordinary gradient of ϕ_n could be calculated (in the usual way, as a list of partial derivatives), but such a gradient is quite different from the corresponding Sobolev gradient (and has vastly inferior numerical properties). The reader might consult [**13**] for a detailed discussion of this point. To illustrate these comments we use the following very simple example.

Suppose X is the Sobolev space $H^{1,2}([0,1])$ and $K = L_2([0,1])$. Define $F : X \to K$ by

(5) $$F(u) = u' - u, \; u \in X.$$

With ϕ defined as in (4) in relation to this F, we see that zeros of ϕ are precisely the solutions u on $[0,1]$ to

(6) $$u' - u = 0,$$

these being, of course, just constants times the exponential function.

PROBLEM 1. *Show that for F as in* (5) *and ϕ as in* (4)*, any critical point of ϕ is also a zero of F.*

The next few problems lead to the construction of the Sobolev gradient $\nabla_S\phi$ of ϕ as defined in (4) where F is defined in (5). We denote the Sobolev gradient of ϕ by

$$\nabla_S\phi$$

to be so that

$$(\phi'(u))h = \langle h, (\nabla_S\phi)(u)\rangle_X, \; u, h \in X.$$

PROBLEM 2. *Show that*

$$(\phi'(u))h = \int_0^1 (h' - h)(u' - u), \; u, h \in X.$$

PROBLEM 3. *Show that*

$$(\phi'(u))h = \langle \begin{pmatrix} h \\ h' \end{pmatrix}, P\begin{pmatrix} u - u' \\ u' - u \end{pmatrix}\rangle_{L_2([0,1])^2} \; u, h \in X,$$

where P is the orthogonal projection of $L_2([0,1])^2$ onto

$$W = \{\begin{pmatrix} y \\ y' \end{pmatrix} : y \in X\}.$$

PROBLEM 4. *Show that*

$$W^\perp = \{\begin{pmatrix} -y' \\ y \end{pmatrix} : y \in X, \; y(0) = 0 = y(1)\}.$$

Use this fact to obtain an expression for the orthogonal projection P and hence for $\nabla_S\phi$.

We now begin some numerics. Divide the interval $[0,1]$ into n pieces of equal length, define ϕ_n by (for $y = (y_0, y_1, \ldots, y_n)$)

$$\phi_n(y) = \sum_{k=1}^{n}((y_k - y_{k-1})/\alpha - (y_k + y_{k-1})/2)^2/2$$

where $\alpha = 1/n$. Now ϕ_n has an ordinary gradient. Discrete steepest descent as described above may be attempted using this ordinary gradient. Even though it is possible to *prove* that this process converges to a zero of ϕ_n, numerical performance is extremely poor, as documented in [**13**]. When using conventional gradients for problems in differential equations, one finds such poor performance universal. We now contrast this situation with a corresponding Sobolev gradient. Note first that if $\nabla\phi_n$ denotes the ordinary gradient function, then

$$(\phi'(x))h = \langle h, (\nabla\phi_n)(x)\rangle_{R^{n+1}}, \; x, h \in R^{n+1}.$$

The Sobolev gradient, $\nabla_S\phi_n$ is defined to be the function on R^{n+1} so that

$$(\phi'(x))h = \langle h, (\nabla_S\phi_n(x)\rangle_{S,n}, \; x, h \in R^{n+1}$$

where

$$\langle *, * \rangle_{S,n} \tag{7}$$

denotes the inner product associated with the following second norm on R^{n+1}:

$$\|y\|_{S,n}^2 = \sum_{k=1}^{n}((y_k - y_{k-1})/\alpha)^2 + ((y_k + y_{k-1})/2)^2,$$

$y = (y_0, y_1, \ldots, y_n)$, $\alpha = 1/n$.

Note that the inner product (7) emulates the corresponding inner product in $H^{1,2}([0,1])$. Such emulation is used throughout the theory of Sobolev gradients. The motivating idea is that ϕ does not have an ordinary gradient, at least not one which is continuous and everywhere defined. The gradients $\nabla\phi_n$ have, in a sense, nothing to which to converge. By contrast, the gradients $\nabla_S\phi_n$ converge in an orderly way to the Sobolev gradient of ϕ. The poor numerical performance of ordinary gradients and the good performance of corresponding Sobolev gradients give an instance of the first law of numerical analysis: "Analytical Difficulties and Numerical Difficulties Always Come in Pairs".

To see a form for the above Sobolev gradient, define

$$D : R^{n+1} \to R^{2n}$$

so that if $y = (y_0, y_1, \ldots, y_n)$, then Dy is the member of R^{2n} whose component k is

$$(y_k + y_{k-1})/2$$

and whose $n+k$ component is

$$(y_k - y_{k-1})/\alpha, \; k = 1, 2, \ldots, n.$$

PROBLEM 5. *Show that*

$$(\nabla_S \phi_n)(x) = (D^t D)^{-1} (\nabla \phi)_n, \; x \in R^{n+1}.$$

The above simple example serves as a prototype for creating Sobolev gradients for a wide variety of finite dimensional functionals of interest in numerical approximation of solutions to partial differential equations. Perhaps with the aid of [**13**] the reader may begin to write code using steepest descent with Sobolev gradients. The following is always the same in cases in which the relevant spaces are inner product spaces: First calculate an ordinary gradient; then the Sobolev gradient is a smoothed (preconditioned) version of this ordinary gradient. The smoother $(D^t D)^{-1}$ is a positive definite, symmetric matrix which depends on the relationship between a Euclidean and a (finite dimensional) Sobolev metric. This relationship ultimately depends on how the Sobolev space in question (here $H^{1,2}([0,1])$) is embedded in the underlying space (here $L_2([0,1])$). Again, see [**13**] for details.

PROBLEM 6. *An exercise which has often been used to gain insight into Sobolev gradients is the following. Denote* $H^{1,2}([0,1]), L_2([0,1])$ *by* H, K *respectively. Here denote by* D *the differentiation operator on* H

$$Df = f', \; f \in H.$$

Find an expression for D^**, the adjoint of* D *so that*

$$\langle Df, g \rangle_K = \langle f, D^* g \rangle_H, \; f \in H, g \in K.$$

What some have found helpful is the use of P*, the orthogonal projection of* $K \times K$ *onto the range of* D *(see* [**20**]*). A solution to this problem is in* [**13**].

Boundary conditions may be incorporated in a Sobolev gradient. This is described in considerable detail in [**13**]. Here we illustrate this point for our above example. Suppose we wish to enforce the boundary condition $u(0) = 1$ to (6). In this case, We let H_0 be the subspace of H consisting of those members of H which are 0 at 0. Instead of the orthogonal projection P as above, use the orthogonal projection P_0 of $L_2([0,1])^2$ onto

$$\{ \begin{pmatrix} y \\ y' \end{pmatrix} : y \in X, \; y(0) = 1 \}.$$

PROBLEM 7. *Construct* P_0.

A similar consideration holds in the numerical case.

Several applications are now indicated. Related graphical output may be found in [**12**],[**13**],[**15**],[**17**],[**14**],[**19**].

2. Ginzburg-Landau Functional and Superconductivity

Suppose that Ω is a bounded region in R^d, for $d = 2$ or 3. Denote by J the functional on $H = H^{1,2}(\Omega, C) \times H^{1,2}(\Omega, R^d)$ defined by

$$(8)\ J(u, A) = \int_\Omega \|\nabla u - iuA\|^2/2 + \|\nabla \times A\|^2/2 + \frac{\kappa^2}{4}(|u|^2 - 1)^2,\ (u, A) \in H.$$

The function u is called an order parameter, A is called a vector potential. H_0 specifies an imposed magnetic field(it need not be constant on Ω), κ specifies material properties (it also need not be constant: variable κ corresponds to an inhomogeneous material). Once a critical point (u, A) of J is determined, $|u|^2$ gives the density of superconducting electrons; $\nabla \times A$ gives the corresponding induced magnetic field. The corresponding superconducting current can be determined from (u, A). There is a vast literature on this functional, but we won't attempt to give a guide to this literature here except to mention [**18**],[**5**].

We seek numerical critical points of J by means of Sobolev gradients as indicated in Section 1. Results with graphical output have been reported in [**15**],[**14**] as well as in [**13**]. We have observed many critical points of J for various regions Ω, some simply connected and some with various configurations of holes.

A significant problem is the appearance computationally of multiple critical points. There are several questions of interest in this connection:

- Are all apparent critical points actual critical points?
- Are 'metastable', i.e., 'near' critical points of physical interest?
- If there are multiple critical points, are they of equal scientific interest?
- Is there a preferred selection from among multiple critical points?

Following are some comments on these four questions.

The first questions might be rephrased: Are there metastable critical points? Steepest descent with Sobolev gradients often stays at an energy level for many iterations, only to fall to a lower level (these levels are dictated by number of flux quanta). It seems likely that the answer to the first question is in the negative. Such metastable critical points seem likely to be of physical interest; 'near' critical points may indicate superconducting states which last days as opposed to actual critical points which indicate states lasting for years. A speculation is that if only actual critical points are found (for example solutions to the Euler equations for (8)), then one has the possibility of missing meaningful metastable critical points. It is a challenging problem to articulate workable definitions for metastable states. It seems likely that the answer to the second question is in the affirmative relative to some interesting definition of the term metastable.

Concerning the third question, our experience is that in many circumstances there are multiple critical points, particularly if there are at least two holes in Ω. If a descent computation is started with a simulated vortex situated at one of the holes, it seems to develop into a true vortex pinned to that hole. In short, it appears that the answer to the third question is in the affirmative.

Concerning the fourth question, it seems likely that the answer is in the affirmative. Recently we have developed a code which simulates cooling of a superconducting device [**16**]. We attempt to describe background for this code and then raise a rather definite mathematical question related to this code. We go into some detail on the last question in the above list.

Given a region Ω of a certain shape, possibly containing holes, and a number $c > 0$ denote by $c\Omega$ the region scaled in linear dimensions by a factor c. Once Ω is fixed, scaling Ω by a factor c seems of comparable strength to changing κ or H_0. So if a shape is chosen and one agrees to consider the case of constant κ and H_0, one essentially has a three-parameter system to consider. Many problems of current scientific interest involve choosing parameters c, κ, H_0 so that critical points of the corresponding J have desired properties.

Fix now a bounded region Ω and also fix κ, H_0. For each $c > 0$ define

(9)

$$J_c(u, A) = \int_{c\Omega} \|\nabla u - iuA\|^2/2 + \|\nabla \times A - H_c\|^2/2 + \frac{\kappa^2}{4}(|u|^2 - 1)^2, \ (u, A) \in S_c$$

where S_c denotes the same as H except the relevant spaces are based on $c\Omega$. H_c denotes a scaling of H_0 so that

$$\int_{\Omega} H_0 = \int_{c\Omega} H_c, \ c \geq 0.$$

Without going into the physics of the situation, we now discuss the idea behind our recent 'cooling' code. Pick an interval $[c_s, c_f]$ of positive numbers. Pick a positive integer k and divide $[c_s, c_f]$ into k pieces of equal length using $c_0, c_1, \ldots, c_k$. Find a critical point of (u_0, A_0) of J_{c_0}. Scale this critical point linearly to the region $c_1\Omega$ and use the up scaled version of (u_0, A_0) to start a steepest descent (always using Sobolev gradients) using J_{c_1}, taking care to scale H_0 to H_{c_1}. Continue this process for $k-1$ additional steps, at each step scaling the imposed magnetic field so that total magnetic flux is not changed. Question: Is there a limiting value as $k \to \infty$ and $c_f \to \infty$, at least if c_s is positive but sufficiently small? For graphical and mathematical purposes keep account of the computation by plotting results on Ω itself using the indicated back scaling. Recent computational evidence (not yet published) points in the direction of this question having an affirmative answer. We realize that the above question leaves some latitude in interpretation, particularly that of the relative rates of $k \to \infty$ and $c_f \to \infty$, but numerical simulations of the above have been positive. For example if one starts with Ω a rectangular region with a large hole to the left on the horizontal centerline; a smaller hole to the right on the horizontal centerline and a simulated vortex on the horizontal centerline but to the right of the small hole, the computed 'limiting' arrangement has been observed to be a vortex captured in the large hole. On a fixed region if one starts with such an initial configuration, the vortex is likely to be captured by the small hole. Somehow the 'cooling' process helps the process find the physically most appealing solution (i.e., the minimal energy one). We are led to believe that actual practice in superconductivity experiments follows this pattern: Start with a device at the end of a rod (say at room temperature), put the rod into a tank of liquid helium (all the time maintaining an imposed magnetic field), then make measurements.

PROBLEM 8. *Formulate and prove a precise conjecture suggested by the above description.*

PROBLEM 9. *A search for critical points of functionals J as in* (8) *represents an attempt to augment experiments. A well functioning code has the possibility of guiding expensive and time consuming experiments which involve constructing and testing superconducting devices. This is such an important problem that it is certainly desirable to have a number of independently written codes that deal with this simulation.*

PROBLEM 10. *In* **[6]**, *for example, there is a description of various Yang-Mills functionals. Someone examining* **[6]** *is certain to be struck by the similarity in form of various Yang-Mills functionals and Ginzburg-Landau functionals of superconductivity. The reader is challenged to write a code using Sobolev gradients for some form of the Yang-Mills equations. A main difference between these two problems is that in our present Ginzburg-Landau problems, the order parameter u is complex valued whereas for Yang-Mills, it may be two-by-two unitary matrix valued, for example. Coding for Yang-Mills should not be vastly more difficult than for Ginzburg-Landau for superconductivity.*

PROBLEM 11. *A final problem in this section deals with the relationship with Sobolev gradient steepest descent with J in* (8) *and the corresponding time dependent Ginzburg-Landau equations (TDGL), see* **[2]**. *In our experience, TDGL evolves states (u, A) in time toward a state of lower energy. Sobolev gradient steepest descent does likewise. It is an appealing thought that somehow these two processes go through quite similar paths. It seems a worthwhile problem to articulate a clear mathematical formulation of this comparison problem. Similar problems occur with other energy functionals.*

PROBLEM 12. *In* **[12]** *there are results on multiple critical points for some energy-like functionals. Can techniques of this reference be used in the case of Ginzburg-Landau functionals?*

3. Problems in Transonic Flow

A full potential functional for transonic flow may be expressed as

$$\mathbf{J}(\mathbf{u}) = \int_{\boldsymbol{\Omega}} [\mathbf{1} + \frac{\gamma - \mathbf{1}}{\mathbf{2}} \|\nabla \mathbf{u}\|^{\mathbf{2}}]^{\gamma/(\gamma-\mathbf{1})}, \tag{10}$$

where Ω is a bounded region in R^d for $d = 1$,2 or 3 and the domain of J is $H = H^{1,7}(\Omega)$. We make this choice of H since for air, $\gamma = 1.4$ In [**13**] there are pictures of critical points of such a J in case $d = 2$. It is a fact that there are uncountably many critical points of such a J given appropriate boundary conditions, but generally only one physically viable solution. This situation is clear even in one dimension: Shocks, that is, jumps from subsonic to supersonic, do not occur in the direction of flow, but the opposite kind of jump is observed to occur. Even in one dimension, there are uncountably many critical points which exhibit jumps from subsonic to supersonic in the flow direction, but only one which shows the opposite jump. How does one pick this physically viable critical point? For details we refer to Chapter 13 of [**13**] which deals with transonic flow, but we indicate that one starts by finding a critical point of a functional similar to (10) but containing a rather heavy viscosity term. The viscosity used is considerably greater than is physically realistic. With this heavy viscosity, no sharp shocks appear. We use this critical point as a starting estimate for a Sobolev gradient descent procedure for a functional similar to (10) but with a lighter viscosity term. After several repetitions of this process we arrive at a critical point which demonstrates sharp shocks which are physically consistent. The process is similar in spirit to the cooling process described in the previous section.

In [**13**] there are results of two-dimensional runs of flow around a standard airfoil, one ambient subsonic with supersonic shocks above and below the airfoil and one ambient supersonic with a subsonic stagnation pocket in front of the airfoil.

Now (10) requires $H^{1,7}(\Omega)$ for a proper definition. For a C^1 function J from $H^{1,7}(\Omega) \to R$ and $u \in H^{1,7}(\Omega)$, the Sobolev gradient $(\nabla J)(u)$ is the element $h \in H^{1,7}(\Omega)$ which, following (1), maximizes

$$J'(u)h \text{ subject to } \|h\|_H^{1,7}(\Omega) = |J'(u)|.$$

See [**21**] and [**13**] for computational details.

PROBLEM 13. *Code a three dimensional version of* (10). *In computational fluid dynamics, there are many instances of 'shock fitting' (computational intervention in places where exceptional effects are anticipated) vs. 'shock capturing' (start with a grid or finite element setting and compute without special intervention). Shock fitting is not so difficult in one dimension; there are many who have learned to do it in two dimensions; few seem to have ambitions for trying it in three dimensions. The pictures for transonic flow in* [**13**] *were done using only shock capture and hence it might be suspected that the method has good prospects for three dimensions.*

4. Some Speculation on Boundary Conditions

The above process shows a feature of Sobolev gradient that has been informally recognized from the start of calculations with Sobolev gradients (in 1976-77 or so but the term 'Sobolev gradient' had not yet been invented): Continuous steepest descent (3) with Sobolev gradients (in the presence of boundary conditions which are not sufficient to dictate a single solution) tends to converge to the nearest solution in the appropriate Sobolev metric. As indicated in [**13**] this is proved for linear problems, but in nonlinear problems this remains at the observational level. It is suspected that at least for continuous steepest descent with Sobolev gradients, a convergent process converges to the nearest solution in some yet to be articulated sense, perhaps along some sort of geodesic. Chapter 14 of [**13**] gives a start of a theory of this sense. Briefly, it is shown that under appropriate hypotheses, if continuous steepest descent converges for every starting value in a function space, then equivalence classes (two starting elements are equivalent if they lead via continuous steepest descent to the same solution) give a way to tag solutions. One hopes to eventually understand sufficiently for significant problems corresponding foliations into equivalence classes in order to provide a solution to the problem of the presentation of all solutions to a given PDE system on a given region. This remains one of the most challenging problems in all of mathematics. In the spirt of the present volume, this challenge is offered to the reader. Following are some specific problems for which the solution might give ideas for a theory on the subject.

PROBLEM 14. *Suppose* $H = H^{2,2}(\Omega)$ *where* $\Omega = [0,1] \times [-1,1]$ *and* $F : H \to K = L_2(\Omega)$ *is defined by*

$$(F(u))(x,y) = yu_{11}(x,y) + u_{22}(x,y), \ (x,y) \in \Omega.$$

Define $\phi : H \to R$ *by*

$$\phi(u) = \frac{1}{2}\int_\Omega F(u)^2, \ u \in H.$$

Find a useful characterization of

$$Q = \{u \in H : \ \textit{if } z(0) = u, z \textit{ satisfies } (3), \ \textit{then } \lim_{t\to\infty} z(t) = 0\}.$$

Observe that $Q^\perp$ *is the set of all zeros of* F, *that is the set of all solutions* $u \in H$ *to the Tricomi equation*

$$yu_{11}(x,y) + u_{22}(x,y) = 0, \ (x,y) \in \Omega. \tag{11}$$

It seems that despite considerable effort, boundary conditions for (11) have never been articulated (see [**10**] for a recent reference on Tricomi's equation). The question is raised as to whether some other mechanism than boundary conditions is in order to specify the set of all solutions. In the present linear problem,

$$H \setminus Q$$

is a foliation of H, each member of which contains exactly one solution to (11). A sufficient understanding of this foliation might lead to the comprehension of the set of all solutions to (11). Computations with Sobolev gradients as outlined in the introduction should serve as a laboratory for the understanding of (11). See Chapter 14 of [**13**] for more discussion on this matter.

PROBLEM 15. *Write a Sobolev gradient code for* (11). *Try to discern properties common to all members of a given equivalence class.*

PROBLEM 16. *For this problem, suppose that* $\Omega = [0,1]^2$ *and* $H = C^1(\Omega)$. *Consider Burger's equation (without viscosity) on* Ω*: Find all members* $u \in H$ *such that*

$$u_1 + uu_2 = 0. \tag{12}$$

Observe that if $u \in H$ *satisfies* (12), *then* Ω *may be filled with intervals so that* u *is constant on each of these intervals and that moreover, the slope of each interval is that constant value. Reflect on the difficulties of determining the set of all solutions in* H *to* (12) *by specifying boundary conditions on some portion of the boundary of* Ω. *Now define*

$$F(u) \;=\; u_1 + uu_2,\; u \in H^{1,2}(\Omega)$$

and define ϕ *in terms of* F *as in the previous problem. Construct a Sobolev gradient for* ϕ, *both for the function space* $H^{1,2}(\Omega)$ *and for finite difference approximations. Use resulting numerics to gain insight on how limiting values of* (3) *correspond to initial estimates. Try to characterize the set of all solutions on* Ω *to this version of Burger's equation.*

This computational problem was the first PDE problem done by the first author (around 1976-77).

PROBLEM 17. *Find a new paradigm to replace boundary conditions in the study of PDE.*

5. Minimal Surfaces and Variable Metrics

Suppose that $\Omega = [0,1]^2$ and $H = H^{1,2}(\Omega, R^3)$. Denote by H_0 those members of H which are zero on the boundary of Ω. Define $\phi : H \to R$ so that if $u_1 \times u_2 \neq 0$ on all of Ω, then

$$\phi(u) \;=\; \int_\Omega \|u_1 \times u_2\|,\; u \in H,$$

the area function for members of H. Taking a Fréchet derivative,

$$(\phi'(u))h \;=\; \int_\Omega \langle u_1 \times h_2 + h_1 \times u_2, u_1 \times u_2\rangle / \|u_1 \times u_2\|,\; u \in H, h \in H_0.$$

For u as above, consider the following inner product on H_0:

$$\langle h,k\rangle_u \;=\; \int_\Omega \langle u_1 \times h_2 + h_1 \times u_2, u_1 \times k_2 + k_1 \times u_2\rangle / \|u_1 \times u_2\|,\; h \in H_0. \tag{13}$$

Representing the functional $\phi'(u)$ in the inner product (13), we define a Sobolev gradient by means of the defining relationship

$$(\phi'(u))h = \langle h, (\nabla\phi)(u)\rangle_u.$$

This gives us something like a Hilbert manifold: For each u as above, we attach inner products to the tangent spaces at each element of H. For each $u \in H$ a Sobolev gradient of ϕ at u is defined. As the descent process runs, the metric constantly changes; this is an example of a variable metric where at each step, the metric is adapted to the current estimate.

For $u \in H$ and $\|u_1 \times u_2\|$ bounded away from 0 on Ω define the norm:

$$\|w\|^2_{L_{2,u}} = \int_\Omega \|w\|^2 \, \|u_1 \times u_2\|, \; w \in H_0.$$

For this fixed u and for $w \in L_{2,u}$, consider the linear functional $T :\rightarrow R$ defined by

$$T_w z = \int_\Omega zw\|u_1 \times u_2\|, \; z \in H_0.$$

The transformation T is also continuous with respect to the norm $\| \cdot \|_{L_{2,u}}$ and so it may be represented in the corresponding inner product;

$$T_w z = \langle z, Mw\rangle_u, \; z \in H_0.$$

The indicated transformation M is the inverse of the laplacian associated with the pair $L_{2,u}, H_0$ (see [**13**], Chapter 5, [**3**]). A finite dimensional version of such a transformation M is used as a basis for computation in [**17**]. Such an M is a surface based laplacian which is distinct from the Laplace-Beltrami operator acting on functions on the range of u. Much remains to be worked out about such transformations M.

PROBLEM 18. *Make a systematic study of transformations M and the corresponding laplacian M^{-1}.*

6. An elasticity Problem

Suppose $H = H^{1,2}(\Omega, R^2)$, $\Omega = [0,1]^2$ and

$$\phi(u,v) = \int_\Omega E(u,v)^4 - E(u,v)^2 + \|(\nabla E)(u,v)\|^2 + \frac{1}{2}(F(u,v)^2 + G(u,v)^2)$$

for all $(u,v) \in H$ (interpret $(u(x), v(x))$ as displacement at $x \in \Omega$) where

- $2 * E(u,v) = u_1 + v_2$, (deviatoric strain)
- $2 * F(u,v) = u_1 - v_2$, (compression strain)
- $2 * G(u,v) = u_2 + v_1$, (shear strain)

For boundary conditions take $v = 0$ on $\partial\Omega$, $u = 0$, bottom,right, u linear, top,left. In [**19**] critical points of a numerical version of this function are found.

7. An Oil-water Separation Problem

Take $\Omega = [0,1]^2, H = H^{1,2}(\Omega, R^2)$

$$\phi(u) = \int_\Omega [\frac{\alpha}{4}(1-u^2)-Tu+\frac{T}{2}((1+u)\ln(1+u)+(1-u)\ln(1-u))+\frac{\kappa}{2}(\|\nabla u\|]^2]$$

for $u \in H$. For boundary conditions take

$$u = 0 \text{ top, left }, u = 0.5 \text{ bottom }, u = -0.5 \text{ right.}$$

The non-local condition

$$\int_\Omega u = 0.2$$

is also imposed. Sobolev gradients frequently can be made to accommodate such conditions. This is work in progress by the authors of [**19**]. Excellent identification of oil-water boundary has been obtained without any special computational devices.

8. Monge-Ampere Problems

For this section denote $[0,1]^2$ by Ω. Suppose $g \in L_2(\Omega)$ is given. Take $H = H^{1,2}(\Omega)$ and denote by H_0 those members of H which are zero on $\partial\Omega$. Consider the problem of finding $u \in H$ so that

$$F(z) = g, \; z \in H_0. \tag{14}$$

where

$$F(z) = z_{11}z_{22} - z_{12}z_{21}, \; u \in H.$$

Define $Q : H^{1,2}(\Omega)^2 \to L_2(\Omega)$ by

$$Q(u,v) = u_1v_2 - u_2v_1, \; u,v \in H^{1,2}(\Omega).$$

Reformulate (14) as the problem of finding $(u,v) \in H^{1,2}(\Omega)^2$ so that $u = 0$ on the top and bottom of Ω, $v = 0$ on the left and right of Ω and (u,v) is a critical point of ϕ where

$$\phi(u,v) = \frac{1}{2}\int_\Omega (Q(u,v) - g)^2, \; u,v \text{ as above,} \tag{15}$$

and (u,v) are related by the condition $u_2 = v_1$. Consequently z can be obtained from (u,v) by integration.

PROBLEM 19. *Write a code to find critical points of a numerical version of* (15).

9. Convergence Problems

We give here a theorem dealing with convergence of descent processes. A number of others are contained in [**13**].

THEOREM 1. *Suppose that H is a Hilbert space, $\Omega \subset H$ and $\phi : H \to [0,\infty)$ is a C^1 function with locally lipschitzian derivative. Suppose also that $x \in \Omega$ and $z : [0,\infty) \to \Omega$ such that*

$$z(0) = x,\ z'(t) \ = \ -(\nabla\phi)(z(t)),\ t \geq 0. \tag{16}$$

Suppose finally that there is $c > 0$, $\theta \in (0,1)$ so that if $x \in \Omega$,

$$\|(\nabla\phi)(x)\| \ \geq c\phi(x)^{\theta}.$$

Then

$$u \ = \ \lim_{t\to\infty} z(t) \text{ exists and } \phi(u) = 0.$$

This is a slight generalization of Theorem 4.8 of [**13**] in which it was required that $\theta = 1/2$. Now (16) is called a Łojasiewicz inequality. In finite dimensional cases, such an inequality is a consequence of analyticity (see [**9**]) in a neighborhood of a zero of ϕ. An infinite dimensional analogue does not always hold but there are several recent results in which infinite dimensional cases have been proved under additional hypothesis (see [**4**] for a very recent account of this matter). See also [**5**] for application of these ideas to Ginzburg-Landau functionals.

PROBLEM 20. *Read [**4**] and attempt to apply ideas to variational problems in this note, in [**13**], [**7**] and problems in the present volume.*

10. Recomended Reading

In [**8**] there is an account of the use of Sobolev gradients (Knowles calls Sobolev gradients by another name) in ill-posed problems. This interesting work is related to such problems as numerical differentiation and a number of other topics. This paper is highly recommended as a source of new ideas on the traditionally difficult subject of ill-posed problems.

In [**11**] there is an account of computational advantages of constructing Sobolev gradients using weighted Sobolev spaces. The interest here is in choosing weights which are appropriate to singular problems. In a sense, choice of metric for minimal surface problems in this note gives a weighted Sobolev space. These two instances are just the tip of an iceberg on this topic.

It has been a privilege to view a prepublication copy of a new book by Faragó and Karátson [**7**]. This book contains a great number of interesting problems to which Sobolev gradients apply. The book also contains detailed discussions of these gradients as well as a wealth of information on computations.

References

[1] R.A. Adams, *Sobolev Spaces*, Academic Press, 1978.

[2] A. Ardelea, G. F. Carey, A. Pardhanani and W. B. Richardson, *Simulation of Macroscopic Superconductivity for Microelectronics*, Physica C: Superconductivity, V. 341-348, Part 4, (2000), 2649–2650.

[3] A. Beurling and J. Deny, Dirichlet spaces, Proc. Nat. Acad. Sci. 45 (1959), 208–215.

[4] R. Chill, *On the Łojasiewicz-Simon Gradient Inequality* (preprint).

[5] S.-Z. Huang and P. Takač, Convergence in gradient-like systems which are asymptotically autonomous and analytic, Nonlinear Anal. 46 (2001), 675–698.

[6] A. Jaffe and C. Taubcs, *Vortices and Monopoles. Structure of Static Gauge Theories*, Progress in Physics 2, 1980.

[7] Faragó, I., Karátson, J., *Numerical solution of nonlinear elliptic problems via preconditioning operators: Theory and applications.* Advances in Computation, Volume 11, NOVA Science Publishers, New York, 2002.

[8] I. Knowles, *Variational methods for ill-posed problems*, http://www.math.uab.edu/knowles/pubs.html .

[9] S. Łojasiewicz, Une propriété topologique des sous-ensembles analytiques rés, Colloques internationaux du C.N.R.S.: Les équations aux dérivées partielles, Editions du C.N.R.S., Paris (1963), 87–89.

[10] D. Lupo and K.R. Payne, *On the maximum principle for generalized solutions to the Tricomi problem*, Commun. Contemp. Math. 2 (2000), 535–557.

[11] W. T. Mahavier, *A convergence result for discrete steepest descent in weighted Sobolev spaces*, Abstract Appl. Analy. 2 (1997), 67–72.

[12] John M. Neuberger, *A numerical method for finding sign-changing solutions of superlinear Dirichlet problems*, Nonlinear World 4 (1997), 73–83.

[13] J.W. Neuberger, *Sobolev Gradients and Differential Equations*, Springer Lecture Notes in Mathematics 1670, 1997.

[14] J. W. Neuberger and R. J. Renka, *Sobolev gradients and the Ginzburg-Landau equations*, Siam J. Sci. Comp. 20 (1998), 582–590.

[15] J. W. Neuberger and R. J. Renka, *Critical points of the Ginzburg-Landau functional on multiply connected domains*, J. Experimental Mathematics 9 (2000), 523–533.

[16] J. W. Neuberger and R. J. Renka, *Numerical determination of vortices in superconductors: simulation of cooling*, Superconductor Science and Technology, 16 (2003), 1413–1416.

[17] R. J. Renka and J. W. Neuberger, *Minimal surfaces and Sobolev gradients*, Siam J. Sci. Comp. 16 (1995), 1412–1427.

[18] J. Rubinstein, *Six lectures on superconductivity. Interfaces and Transitions.*, CRM Proc. Lec. Notes 13, Amer. Math. Soc., 1998.

[19] S. Sial, J. Neuberger, T. Lookman, A. Saxena, *Energy minimization using Sobolev gradients: applications to phase separation and ordering.*, J. Comp. Physics, 189 (2003), 88–97.

[20] J. von Neumann, *Functional Operators II*, Annls. Math. Stud. 22, (1940).

[21] M. Zahran, *Steepest descent on a uniformly convex space*, Rocky Mountain J. Math 33 (2004).

Contemporary Mathematics
Volume **357**, 2004

Unbounded Perturbations of Resonant Schrodinger Equations

David G. Costa and Hossein Tehrani

ABSTRACT. We consider the question of existence of solution for resonant Schrodinger equations of the form

$$-\Delta u + V(x)u = \lambda u + g(x,u), \quad x \in \mathbb{R}^N,$$

where the potential $V(x)$ vanishes at infinity and $\lambda < 0$ is an eigenvalue of the corresponding Schrodinger operator $L_V := \Delta + V(x)$. The (possibly unbounded) perturbation $g(x,u)$ is assumed to have at most a sublinear growth and satisfy a variant of the Ahmad-Lazer-Paul condition.

Introduction

In this note we consider the question of existence of solution for equations of the form

$$(P_\lambda) \qquad -\Delta u + V(x)u = \lambda u + g(x,u)\ , \quad x \in \Omega = \mathbb{R}^N\ ,$$

when the parameter $\lambda = \widehat{\lambda} < 0$ is an eigenvalue of the Schrodinger operator $L_V := -\Delta + V(x)$ in $\mathbb{R}^N$. Here, the potential $V \in C(\mathbb{R}^N, \mathbb{R})$ is such that $\lim_{|x|\to\infty} V(x) = 0$ and L_V has negative eigenvalues, while the perturbation $g \in C(\mathbb{R}^N \times \mathbb{R}, \mathbb{R})$ has a sub-linear growth in the sense that

$$(G_1)_\alpha \qquad |g(x,s)| \leq b_1(x)|s|^\alpha + b_2(x)$$

for some $\alpha \in [0,1)$ and a class of integrable functions $b_1(x), b_2(x)$ (precise conditions will be stated later on). Under these conditions (P_{λ_0}) is a *resonant* problem. In case this equation is considered in a bounded domain $\Omega \subset \mathbb{R}^N$ (with, say, Dirichlet boundary condition) and $g(x,s)$ is a bounded function ($\alpha = 0$, $b_1(x), b_2(x) \in L^\infty$), there is a large literature on such resonant problems. We refer the reader to the landmark papers [**20, 1, 22, 3**] and to [**2, 6, 9, 10, 11, 13-18, 21, 23, 25-29**] and references therein. Of course, since Ω is bounded, the spectrum $\sigma(L_V)$ consists of a *countable* set of eigenvalues with finite multiplicities and in the "rare" case of *resonance* a solution may not exist if no further assumption is made on the nonlinearity (cf. Fredholm alternative). In the case of a bounded domain $\Omega \subset \mathbb{R}^N$

1991 *Mathematics Subject Classification.* Primary 35J20, 35J60; Secondary 35J10.
Key words and phrases. Schrodinger equations, unbounded perturbations, resonant.

(with Dirichlet boundary condition), Ahmad-Lazer-Paul [**1**] and Rabinowitz [**22**] considered the situation $V(x) = 0$ and showed that (P) has a solution provided $g(x,s)$ is a bounded nonlinearity and the primitive $G(x,s)$ is either *coercive* or *anti-coercive* in the sense that one of the conditions $(G_2)^+$ or $(G_2)^-$ below is satisfied,

$$(G_2)^{\pm} \qquad \lim_{\|u_0\|\to\infty,\ u_0\in E_0} \int_\Omega G(x,u_0(x))\,dx = \pm\infty\ ,$$

where $E_0 = N(-\Delta - \widehat{\lambda})$ is the eigenspace corresponding to the eigenvalue $\widehat{\lambda}$ of $-\Delta$ on $H_0^1(\Omega)$.

On the other hand, to our knowledge, very little has been done when $\Omega = \mathbb{R}^N$, i.e., in the case of problem (P_λ). In fact, in the non-resonant situation we are only aware of the papers [**30, 19, 12**] and, aside from [**19**], we believe no results have been proved in the resonant situation. One of the difficulties in this case is the fact that the spectrum of the operator L includes an *essential* part, namely $[0,\infty)$, so that one has to deal with a much more complicated *resonant set.* The other difficulty in dealing with such problems in $\mathbb{R}^N$ is the *lack of compactness* exhibited by the corresponding energy functional, say, as measured by the well-known Palais-Smale condition. In this paper, we allow unbounded nonlinearities satisfying condition $(G_1)_\alpha$ stated earlier, provided that the primitive $G(x,s)$ satisfies either one of the conditions $(G_2)^+_\alpha$ or $(G_2)^-_\alpha$ below,

$$(G_2)^{\pm}_\alpha \qquad \lim_{\|u_0\|\to\infty,\ u_0\in E_0} \frac{1}{\|u_0\|^{2\alpha}} \int_{\mathbb{R}^N} G(x,u_0(x))\,dx = \pm\infty\ ,$$

where again $E_0 = N(-\Delta - \widehat{\lambda})$ denotes the $\widehat{\lambda}$-eigenspace, this time of the operator $-\Delta + V(x)$. In the case of a bounded domain similar conditions were considered in [**27, 8, 24**] for resonance at an eigenvalue of $-\Delta$ on $H_0^1(\Omega)$.

1. Preliminary Results

In addition to conditions $(G_1)_\alpha$ and (one of) $(G_2)^{\pm}_\alpha$, let us also state the conditions on the potential $V(x)$ under which problem (P_λ) will be considered:

(V_1) $V \in C^\beta(\mathbb{R}^N,\mathbb{R})$ $(0<\beta<1)$ satisfies $\lim_{|x|\to\infty} V(x) = 0$;

(V_2) $\displaystyle\int_{\mathbb{R}^N} [|\nabla\varphi|^2 + V(x)\varphi^2]\,dx < 0$ for some $\varphi \in C_c^\infty(\mathbb{R}^N)$.

We observe that (V_1), (V_2) imply that the Schrodinger operator $L_V = -\Delta + V(x)$ is a self-adjoint operator defined on $H^2(\mathbb{R}^N) \subset L^2(\mathbb{R}^N)$ and has an *essential spectrum,* $\sigma_{ess}(L_V) = [0,\infty)$, as well as a *point spectrum* $\sigma_p(L_V) = \{\lambda_1 < \lambda_2 < \ldots\} \subset (-\infty, 0)$ consisting of isolated eigenvalues of finite multiplicity, possibly accumulating at $\lambda = 0$ ([**4**]). Of course, there may also exist other eigenvalues embedded in the essential spectrum. Here we are considering problem $(P_{\widehat{\lambda}})$ with $\widehat{\lambda} = \lambda_k$ being one of the isolated eigenvalues outside the *essential spectrum.* Next we consider the shifted operator $L := L_V - \lambda_k$. Using the spectral theorem we define the unbounded selfadjoint operator

$$A = |L|^{\frac{1}{2}} = \int_{-\infty}^{\infty} |\lambda|^{\frac{1}{2}}\ dE(\lambda),$$

that is,

$$D(A) = \{u \in L^2 \mid \int_{-\infty}^{\infty} |\lambda| \; d(E(\lambda)u, u)_2 < \infty \;\}, \tag{1.1}$$

$$(Au, v)_2 = \int_{-\infty}^{\infty} |\lambda|^{\frac{1}{2}} \; d(E(\lambda)u, v)_2, \qquad \forall u \in D(A), \; v \in L^2, \tag{1.2}$$

where $\{E(\lambda) | \lambda \in \mathbb{R}\}$ is a *spectral family for* L. As is well-known, the closedness of the operator A implies that $E := D(A)$ is a Hilbert space with the *graph-inner product*

$$(u, v)_E := (u, v)_2 + (Au, Av)_2,$$

and corresponding *graph-norm*

$$||u||_E = (|u|_2^2 + |Au|_2^2)^{\frac{1}{2}} = \left(\int_{-\infty}^{\infty} (1 + |\lambda|) \; d(E(\lambda)u, u)_2 \right)^{\frac{1}{2}}.$$

Let us write $\mathbb{R} = \mathbb{R}_+ \cup \{0\} \cup \mathbb{R}_-$ and denote

$$P_+ = \chi_{\mathbb{R}_+}(L), \quad P_0 = \chi_{\{0\}}(L), \quad P_- = \chi_{\mathbb{R}_-}(L), \tag{1.3}$$

where χ_S denotes the characteristic function of a set $S \subset \mathbb{R}$. Then, P_+ , P_0 and P_- are orthogonal projections such that $P_+ \oplus P_0 \oplus P_- = I$. Correspondingly, we have the orthogonal decomposition

$$E = E_+ \oplus E_0 \oplus E_- \;, \tag{1.4}$$

where $E_+ = P_+(E)$, $E_0 = P_0(E)$, and $E_- = P_-(E)$. Since by assumption $\lambda = 0$ is an isolated point of $\sigma(L)$, it is clear that

$$||u|| := \left(|AP_+u|_2^2 + |AP_-u|_2^2 + |P_0u|_2^2\right)^{\frac{1}{2}} \qquad u \in E, \tag{1.5}$$

defines an equivalent norm to $||\cdot||_E$. Throughout this paper, unless stated otherwise, all integrals are understood as taken over all of $\mathbb{R}^N$ and we will consider the space E equipped with the norm $||\cdot||$ above. Also, we will freely use the following result:

LEMMA 1.1. *The space* E *coincides with* $H^1(\mathbb{R}^N)$ *and the norm* $||\cdot||$ *is equivalent to the standard norm* $||\cdot||_1$ *of* $H^1(\mathbb{R}^N)$, *given by* $||u||_1 = (|u|_2^2 + |\nabla u|_2^2)^{\frac{1}{2}}$. *In addition we have*

$$||P_+u||^2 - ||P_-u||^2 = \int \left(|\nabla u|^2 + V(x)u^2 - \lambda_k u^2\right) \; dx \qquad \forall u \in E = H^1(\mathbb{R}^N) \;. \tag{1.6}$$

PROOF. Let $-M = \inf_{x \in \mathbb{R}^N} [V(x) - \lambda_k]$. We first define an equivalent norm $||\cdot||_M$ on E through the inner product

$$(u, v)_M := \int_{\sigma(L)} (\lambda + 2M) d(E(\lambda)u, v) + (u, v)_2, \qquad u, v \in E.$$

Thus,

$$||u||_M^2 = |u|_2^2 + \int_{\sigma(L)} |\lambda| d(E(\lambda)u, u)$$

$$+2 \int_{\lambda \in \sigma(L), \lambda < 0} (\lambda + M) d(E(\lambda)u, u) + 2 \int_{\lambda \in \sigma(L), \lambda > 0} M d(E(\lambda)u, u).$$

Since we have $-M \leq \lambda$ for $\lambda \in \sigma(L), \lambda < 0$, it follows that

$$||u||_E^2 \leq ||u||_M^2 \leq (2M + 1)||u||_E^2 \qquad \text{for all } u \in E. \tag{1.7}$$

Now, for $\phi \in C_0^\infty(\mathbb{R}^N)$, we have

$$\begin{aligned} ||\phi||_M^2 &= \int \lambda d(E(\lambda)\phi,\phi) + (2M+1)|\phi|_2^2 = (L\phi,\phi) + (2M+1)|\phi|_2^2 \\ &= \int |\nabla\phi|^2 + (V(x) - \lambda_k + 2M + 1)\phi^2. \end{aligned}$$

Therefore, there exists $C_1, C_2 > 0$ such that

$$C_1||\phi||_1^2 \le ||\phi||_M^2 \le C_2||\phi||_1^2, \qquad \text{for all } \phi \in C_0^\infty(\mathbb{R}^N),$$

that is, $||\cdot||_M$ defines a norm equivalent to $||\cdot||_1$ on the class $C_0^\infty(\mathbb{R}^N)$. So, denoting the completion of $C_0^\infty(\mathbb{R}^N)$ under the norm $||\cdot||_M$ by H, we conclude that $H = H^1(\mathbb{R}^N)$. Next, in order to show that $E = H^1(\mathbb{R}^N)$, we observe that (1.7) implies $H \subset E$ and we argue by contradiction assuming that $H \neq E$. Then, there would exist $u_0 \in E$, $u_0 \neq 0$ such that

$$(\phi, u_0)_M = 0 \qquad \text{for all } \phi \in C_0^\infty(\mathbb{R}^N)\ ,$$

that is,

$$\int_{\sigma(L)} \lambda d(E(\lambda)\phi, u_0) + (2M+1)(\phi,u_0)_2 = (L\phi,u_0)_2 + (2M+1)(\phi,u_0)_2\ ,$$

or,

$$((L + (2M+1)I)\,\phi, u_0)_2 = 0 \quad \text{ for all } \phi \in C_0^\infty(\mathbb{R}^N).$$

Since $C_0^\infty(\mathbb{R}^N)$ is dense in $H^2(\mathbb{R}^N)$, we conclude that $u_0 \in D(L^*) = D(L)$ and

$$(L + (2M+1)I)u_0 = 0\ .$$

But this says that $-(2M+1) \in \sigma(L)$, which contradicts the fact that $\sigma(L) \subset [-M,\infty)$. Therefore the proof of that $E = H^1(\mathbb{R}^N)$ is established.

Finally, for $\phi \in C_0^\infty(\mathbb{R}^N)$, we can write

$$\begin{aligned} ||P_+\phi||^2 - ||P_-\phi||^2 &= \int_{\lambda>0} |\lambda| d(E(\lambda)\phi,\phi) - \int_{\lambda<0} |\lambda| d(E(\lambda)\phi,\phi) \\ &= \int_{\sigma(L)} \lambda d(E(\lambda)\phi,\phi) = (L\phi,\phi) \\ &= \int \left(|\nabla\phi|^2 + (V(x)-\lambda_k)\phi^2\right) dx\ , \end{aligned}$$

so that a density argument completes the proof of Lemma 1.1 . □

We are now ready to state the following basic lemma.

LEMMA 1.2. *Assume* $(V_1), (V_2)$, $(G_1)_\alpha$ *with* $\alpha \in [0,1)$, $b_2(x) \in L^2(\mathbb{R}^N)$ *and* $b_1(x) \in L^p(\mathbb{R}^N)$ *for some* p *in the interval* $[\frac{2^*}{2^*-(1+\alpha)}, \frac{2}{1-\alpha}]$. *Then the functional given by*

$$\begin{aligned} I(u) &:= \frac{1}{2}\left(||P_+u||^2 - ||P_-u||^2\right) - \int G(x,u(x))dx \\ &= \frac{1}{2}\int \left(|\nabla u|^2 + V(x)u^2 - \lambda_k u^2\right) - N(u) \end{aligned}$$

is well-defined and of class C^1 *on* $H^1(\mathbb{R}^N)$, *with its critical points being precisely the solutions* $u \in C^{2,\beta}(\mathbb{R}^N) \cap H^1(\mathbb{R}^N)$ *of problem* (P_{λ_k}) *(Here, as usual,* $G(x,s)$ *denotes the primitive* $\int_0^s g(x,t)\,dt$*).*

PROOF. Let us denote the norms in $H^1(\mathbb{R}^N)$ and $L^r(\mathbb{R}^N)$ by $\|\cdot\|$ and $|\cdot|_r$, respectively. In order to prove the proposition, it suffices to verify that our conditions render the functional

$$N(u) := \int G(x, u(x))\, dx$$

well-defined and of class C^1 on $H^1(\mathbb{R}^N)$, with

$$\langle \nabla N(u), h \rangle = \int g(x, u(x))h\, dx$$

for all $u, h \in H^1(\mathbb{R}^N)$. This follows from a simple application of Holder and Sobolev inequalities.

In fact consider the function $\Psi(r, s) = 1 - \frac{1}{r} - \frac{\alpha}{s}$ defined on the rectangle $[2, 2^*] \times [2, 2^*]$. Note that the image of this continuous function is the closed interval $[\frac{1-\alpha}{2}, \frac{2^*-(1+\alpha)}{2^*}]$. Therefore, for $p \in [\frac{2^*}{2^*-(1+\alpha)}, \frac{2}{1-\alpha}]$, there exist $2 \le r, s \le 2^*$ such that $\frac{1}{p} + \frac{\alpha}{s} + \frac{1}{r} = 1$. Keeping $(G_1)_\alpha$ in mind we have

$$\int |G(x, u)|\, dx \le \tfrac{1}{\alpha+1} |b_1|_p |u|_s^\alpha |u|_r + |b_2|_2 |u|_2 \le C \left(|b_1|_p \|u\|^\alpha + |b_2|_2 \right) \|u\| ,$$

and, similarly,

$$\int |g(x, u)h|\, dx \le |b_1|_p |u|_s^\alpha |h|_r + |b_2|_2 |h|_2 \le C \left(|b_1|_p \|u\|^\alpha + |b_2|_2 \right) \|h\| .$$

□

LEMMA 1.3. *Assume the conditions of Lemma 1.2 hold true. Then the gradient* $\nabla N : H^1(\mathbb{R}^N) \longrightarrow H^1(\mathbb{R}^N)$ *is a compact compact.*

PROOF. Let $u_n \rightharpoonup u$ *weakly* in $H^1(\mathbb{R}^N)$. We must show that $\nabla N(u_n) \to \nabla N(u)$ *strongly* in $H^1(\mathbb{R}^N)$. It suffices to verify that

$$\delta(n, h) := \int |(g(x, u_n) - g(x, u))h|\, dx \longrightarrow 0$$

as $n \to \infty$, uniformly for $h \in H^1(\mathbb{R}^N)$ with $\|h\| = 1$. Setting $B_R^c := \mathbb{R}^N \setminus B_R$, where B_R denotes the ball of radius $R > 0$ in $\mathbb{R}^N$ centered at the origin, we can write

$$\begin{aligned} \delta(n, h) &= \int |(g(x, u_n) - g(x, u))h| = \delta_1(n, h, R) + \delta_2(n, h, R) \\ &= \int_{B_R} |(g(x, u_n) - g(x, u))h| + \int_{B_R^c} |(g(x, u_n) - g(x, u))h| \\ &= \delta_1(n, h, R) + \delta_2(n, h, R) \end{aligned}$$

where the second integral $\delta_2(n, h, R)$ can be estimated as

$$\begin{aligned} \delta_2(n, h, R) &\le \left(\int_{B_R^c} |b_1(x)|^p \right)^{\frac{1}{p}} \left(\int_{B_R^c} (|u_n|^s + |u|^s) \right)^{\frac{\alpha}{s}} \left(\int_{B_R^c} |h|^r \right)^{\frac{1}{r}} \\ &\quad + \left(\int_{B_R^c} |b_2(x)|^2 \right)^{\frac{1}{2}} \left(\int_{B_R^c} |h|^2 \right)^{\frac{1}{2}} \\ &\le C \left[\left(\int_{B_R^c} |b_1(x)|^p \right)^{\frac{1}{p}} + \left(\int_{B_R^c} |b_2(x)|^2 \right)^{\frac{1}{2}} \right] \end{aligned}$$

uniformly for $\|h\| = 1$, where $\frac{1}{r} + \frac{\alpha}{s} + \frac{1}{2^*} = 1$. Note that this last expression can be made arbitrarily small by taking $R > 0$ large. Next we show that for given R, the first integral $\delta_1(n, h, R)$ conveges to zero as n goes to infinity uniformly for $\|h\| = 1$. In fact, since $b_1(x) \in L^p$ and $b_2(x) \in L^2$, for any given $\epsilon > 0$ there exists $\gamma > 0$ such that

$$\left(\int_K |b_1(x)|^p\right)^{\frac{1}{p}} + \left(\int_K |b_2(x)|^2\right)^{\frac{1}{2}} < \epsilon$$

for all $K \subset B_R$ with $\mu(K) < \gamma$, where $\mu(K)$ represents the Lebesque measure of K. Therefore

$$\begin{aligned}\int_K |(g(x, u_n) - g(x, u))h| &\leq \left(\int_K |b_1(x)|^p\right)^{\frac{1}{p}} \left(\int_K (|u_n|^s + |u|^s)\right)^{\frac{\alpha}{s}} \left(\int_K |h|^r\right)^{\frac{1}{r}} \\ &\quad + \left(\int_K |b_2(x)|^2\right)^{\frac{1}{2}} \left(\int_K |h|^2\right)^{\frac{1}{2}} \\ &\leq C\left((\int_K |b_1(x)|^p)^{\frac{1}{p}} + (\int_K |b_2(x)|^2)^{\frac{1}{2}}\right) \\ &\leq C\epsilon\end{aligned}$$

where C is independent of n and h. Hence, in view of Vitali's convergence theorem, $\delta_1(n, h, R) \to 0$ as $n \to \infty$ uniformly for $\|h\| = 1$. We conclude that $\nabla N(u_n) \to \nabla N(u)$ *strongly* in $H^1(\mathbb{R}^N)$. □

LEMMA 1.4. *In addition to conditions of Lemma 1.2, assume also* $(G_2)^+_\alpha$. *Then the restrictions* $I|E_+$ *and* $(-I)|(E_- \oplus E_0)$ *are coercive, that is*

(a) $I(u) \longrightarrow +\infty$ *as* $\|u\| \to \infty$, $u \in E_+$,
(b) $I(u) \longrightarrow -\infty$ *as* $\|u\| \to \infty$, $u \in E_- \oplus E_0$.

PROOF. Using $(G_1)_\alpha$ and Sobolev's Theorem, we obtain for $u \in E_+$ that

$$I(u) \geq \frac{1}{2}\|u\|^2 - \int G(x, u(x))dx \geq \frac{1}{2}\|u\|^2 - \gamma\|u\|^{\alpha+1} - \beta\|u\| ,$$

where $\beta = |b_2|_2$ and $\gamma = |b_1|_p$. Hence, since $\alpha + 1 < 2$, it follows that $I(u) \longrightarrow +\infty$ as $\|u\| \to \infty$, $u \in E_+$, i.e., (a) holds true.

On the other hand, for $u = u_- + u_0 \in E_- \oplus E_0$, using $(G_1)_\alpha$, and the mean value theorem, we obtain

$$\begin{aligned}I(u) &\leq -\frac{1}{2}\|u\|^2 - \int [G(x, u) - G(x, u_0)]\, dx - \int G(x, u_0)\, dx \\ &\leq -\frac{1}{2}\|u_-\|^2 - \int g(x, u_0 + \tau(x)u_-)u_-\, dx - \int G(x, u_0)\, dx \\ &\leq -\frac{1}{2}\|u_-\|^2 + \gamma\left(\|u_0\|^\alpha + \|u_-\|^\alpha\right)\|u_-\| + \beta\|u_-\| - \int G(x, u_0)\, dx .\end{aligned}$$

Hence, using the elementary estimate $cd \leq \epsilon c^2 + \frac{1}{4\epsilon}d^2$ with $0 < \epsilon < \frac{1}{2}$ on the term $\|u_-\|\gamma\|u_0\|^\alpha$, it follows that

$$I(u) \leq -(\frac{1}{2} - \epsilon)\|u_-\|^2 + \|u_-\|(\gamma\|u_-\|^\alpha + \beta) + \frac{\gamma^2}{4\epsilon}\|u_0\|^{2\alpha} - \textstyle\int G(x, u_0)\, dx$$

$$= -(\frac{1}{2} - \epsilon)\|u_-\|^2 + \|u_-\|(\gamma\|u_-\|^\alpha + \beta) + \|u_0\|^{2\alpha}\left[\frac{\gamma^2}{4\epsilon} - \frac{1}{\|u_0\|^{2\alpha}}\textstyle\int G(x, u_0)\, dx\right] ,$$

where the expression in square brackets tends to $-\infty$ as $\|u_0\| \to \infty$, in view of $(G_2)^+_\alpha$. Therefore, since $\alpha + 1 < 2$, the above estimate shows that $I(u) \longrightarrow -\infty$ as $\|u\|^2 = \|u_-\|^2 + \|u_0\|^2 \to \infty$, $u \in E_- \oplus E_0$, i.e., (b) also holds true. $\square$

2. Proof of Main Result

We first show that the functional $I(u)$ satisfies the Palais-Smale condition (PS).

PROPOSITION 2.1. *Assume the conditions of Lemma 1.3 are satisfied. If $(u_n) \subset E$ is a sequence such that $I(u_n)$ is bounded and $\nabla I(u_n) \longrightarrow 0$ then (u_n) possesses a convergent subsequence.*

PROOF. Since $\nabla I = P_+ - P_- - \nabla N$ (where P_- and P_- are the projection operators defined in (1.3)), E_0 is finite dimensional, and ∇N is a compact mapping by Lemma 1.3, it suffices to show that any sequence $(u_n) \subset E$ such that $|I(u_n)| \leq C$, $\nabla I(u_n) \longrightarrow 0$ is bounded. We argue by contradiction.

Suppose, on the contrary, that $\|u_n\| \to \infty$. We write $u_n = u_n^- + u_n^0 + u_n^+$, where $u_n^\pm \in E_\pm$, $u_n^0 \in E_0$, and we first show that

$$\|u_n^\pm\| \leq c\|u_n\|^\alpha + d \tag{2.1}$$

for some constants $c,\ d > 0$ independent of n. Indeed, since

$$|\langle \nabla I(u_n), h\rangle| = |\langle (P_+ - P_-)u_n, h\rangle - \int g(x, u_n)h\, dx| \leq \|h\| \quad \forall h \in E\ , \tag{2.2}$$

we obtain from $(G_1)_\alpha$ that

$$\|u_n^+\| \geq |\langle \nabla I(u_n), u_n^+\rangle| \geq 2\delta\|u_n^+\|^2 - (\gamma\|u_n\|^\alpha + \beta)\|u_n^+\|$$

for all n sufficiently large, hence

$$\|u_n^+\| \leq c\|u_n\|^\alpha + d$$

for some $c, d > 0$. Similarly, letting $h = u_n^-$ in (2.2) and using (1.2), $(G_1)_\alpha$, we obtain

$$-\|u_n^-\| \leq |\langle \nabla I(u_n), u_n^-\rangle| \leq -2\delta\|u_n^-\|^2 + (\gamma\|u_n\|^\alpha + \beta)\|u_n^-\|$$

for all n sufficiently large, hence

$$\|u_n^-\| \leq c\|u_n\|^\alpha + d\ .$$

Having shown (2.1), it follows that

$$\|u_n - u_n^0\| = \|u_n^+ + u_n^-\| \leq C\|u_n\|^\alpha + D \tag{2.3}$$

and

$$\lim_{n\to\infty} \frac{\|u_n^0\|}{\|u_n\|} = 1\ . \tag{2.4}$$

On the other hand, denoting $A := P_+ - P_-$, we can write $I(u_n)$ as

$$I(u_n) = \frac{1}{2}\langle A(u_n - u_n^0), u_n - u_n^0\rangle - \int [G(x, u_n) - G(x, u_n^0)]\, dx - \int G(x, u_n^0)\, dx$$

$$= I_{n1} - I_{n2} - I_{n3}\ , \tag{2.5}$$

and we shall now obtain estimates for the terms I_{n1}, I_{n2}. Indeed, (2.3) implies

$$|I_{n1}| \leq C_0(\|u_n\|^{2\alpha} + 1)\ , \tag{2.6}$$

and from the mean value theorem, $(G_1)_\alpha$ and (2.3) we obtain

$$\begin{aligned} |I_{n2}| &\leq \left[A_1(\|u_n\| + \|u_n^0\|)^\alpha + B_1\right] \|u_n - u_n^0\| \\ &\leq \left[A_2(\|u_n\| + \|u_n^0\|)^\alpha + B_2\right] (\|u_n\|^\alpha + 1) \ , \end{aligned}$$

hence

$$|I_{n2}| \leq B_2 \left(\|u_n\|^\alpha + 1\right) + \frac{A_2}{2} \left(\|u_n\|^\alpha + 1\right)^2 + \frac{A_2}{2} \left(\|u_n\| + \|u_n^0\|\right)^{2\alpha}$$

$$\leq C_1 \left(\|u_n\|^{2\alpha} + \|u_n\|^\alpha + 1\right) + C_2 \|u_n^0\|^{2\alpha} \ . \tag{2.7}$$

Therefore, using estimates (2.6), (2.7) in (2.5) yields

$$I(u_n) \leq \|u_n^0\|^{2\alpha} \left[-\frac{1}{\|u_n^0\|^{2\alpha}} \int G(x, u_n^0)\, dx + C_2 \right.$$

$$\left. + C_1 \left(\frac{\|u_n\|^{2\alpha}}{\|u_n^0\|^{2\alpha}} + \frac{\|u_n\|^\alpha}{\|u_n^0\|^{2\alpha}} + \frac{1}{\|u_n^0\|^{2\alpha}} \right) + C_0 \left(\frac{\|u_n\|^{2\alpha}}{\|u_n^0\|^{2\alpha}} + \frac{1}{\|u_n^0\|^{2\alpha}} \right) \right] ,$$

where the expression in square brackets goes to $-\infty$ in view of $(G_2)_\alpha^+$, (2.4), and the fact we are assuming that $\|u_n\| \to \infty$. Therefore, we reach the contradiction $I(u_n) \longrightarrow -\infty$ and we conclude that the sequence (u_n) must be bounded. The proof is complete. □

We can now state and prove our main result.

THEOREM 2.2. *Assume conditions* $(V_1), (V_2)$, $(G_1)_\alpha, (G_2)_\alpha^+$ *where* $\alpha \in [0,1)$, $b_2(x) \in L^2(\mathbb{R}^N)$ *and* $b_1(x) \in L^p(\mathbb{R}^N)$, *for some* $\frac{2^*}{2^*-(1+\alpha)} \leq p \leq \frac{2}{1-\alpha}$. *Let* $\lambda = \lambda_k < 0$ *be an isolated eigenvalue of the Schrodinger operator* $L = -\Delta + V(x)$ *on* $H^2(\mathbb{R}^N) \subset L^2(\mathbb{R}^N)$. *Then problem* (P_λ) *has at least one solution* $u \in C^{2,\beta}(\mathbb{R}^N) \cap H^1(\mathbb{R}^N)$

PROOF. First note that since $I|E_+$ is *coercive* by Lemma 1.4 (a), and I maps bounded sets of $H^1(\mathbb{R}^N)$ into bounded subsets of $\mathbb{R}$, we conclude that $m_+ = \inf_{E_+} I > -\infty$. And, for any given $m_0 < m_+$, Lemma 1.4 (b) implies the existence of some $R > 0$ such that $I(u) \leq m_0$ for all $u = u_- + u_0 \in E_- \oplus E_0$ with $\|u\| \geq R$. Therefore, in view of Proposition 2.1, the Saddle-Point Theorem of Rabinowitz [**22, 23**] can be used to yield a critical point $u \in E$ of I with $I(u) \geq m_+$, hence a solution of (P_{λ_k}). By the regularity assumptions made on $V(x)$ and $g(x,u)$ we also have that $u \in C^{2,\beta}(\mathbb{R}^N)$. □

REMARK 1. *Theorem 2.2 also holds true with* $(G_2)_\alpha^-$ *replacing* $(G_2)_\alpha^+$. *In this case, Proposition 2.1 gives that* $I|(E_+ \oplus E_0)$ *and* $(-I)|E_-$ *are* coercive.

REMARK 2. *Using similar arguments as in* [**24**], *by assuming an additional* sign condition *on* $G(x,s)$, *it can be shown that conditions* $(G_2)_\alpha^\pm$ *are implied respectively by*

$(\widehat{G}_2)_\alpha^\pm$
$$\lim_{|s|\to\infty} \frac{G(x,s)}{|s|^{2\alpha}} = \pm\infty \ , \quad a.e.\ x \in \mathbb{R}^N .$$

COROLLARY 2.3. *Under the same conditions as in Theorem 2.2, problem*

$$(P_{\lambda_k,h}) \qquad -\Delta u + V(x)u = \lambda_k u + g(x,u) + h \ , \quad x \in \mathbb{R}^N \ ,$$

has a solution $u \in H^1(\mathbb{R}^N) \cap C^{2,\beta}(\mathbb{R}^N)$ *for any* $h \in L^2(\mathbb{R}^N) \cap C^{\beta}(\mathbb{R}^N)$ *orthogonal to the* λ_k*-eigenspace* $N_0 = N(L_V - \lambda_k)$.

PROOF. It suffices to observe that the nonlinearity $\widehat{g}(x,s) := g(x,s) + h(x)$ satisfies conditions $(G_1)_\alpha$ and $(G_2)_\alpha^+$. $\square$

COROLLARY 2.4. *Under the conditions of Corollary 2.3, assume that* $k = 1$ *and* $0 \le \alpha \le \frac{1}{2}$. *If*

$$(G_3) \qquad \lim_{s \to \pm\infty} g(x,s) = \pm\infty \ ,$$

for a.e. $x \in \mathbb{R}^N$, *then problem* $(P_{\lambda_1,h})$ *has a solution* $u \in H^1(\mathbb{R}^N) \cap C^{2,\beta}(\mathbb{R}^N)$ *for any* $h \in L^2(\mathbb{R}^N) \cap C^{\beta}(\mathbb{R}^N)$.

PROOF. The nonlinearity $\widehat{g}(x,s) := g(x,s) + h(x)$ clearly satisfies $(G_1)_\alpha$. As to condition $(G_2)_\alpha$, we note that, since $k = 1$, the eigenspace E_0 is one-dimensional and spanned by an eigenfunction $\varphi_1 > 0$, say $\|\varphi_1\| = 1$. Therefore,

$$\frac{1}{\|t\varphi_1\|^{2\alpha}} \int \widehat{G}(x, t\varphi_1)\, dx = \frac{1}{|t|^{2\alpha}} \int \left[\frac{G(x,t\varphi_1)}{t\varphi_1} + h \right] t\varphi_1 \, dx$$

$$= |t|^{1-2\alpha} sign(t) \int A(x,t)\varphi_1(x)\, dx \ ,$$

where $A(x,t) \longrightarrow \pm\infty$ as $t \to \pm\infty$, for *a.e.* $x \in \mathbb{R}^N$, in view of (G_3). So, it follows that condition $(G_2)_\alpha^+$ is also satisfied. $\square$

REMARK 3. *The above corollary extends a well-known result due to Landesman-Lazer* [**20**]. *Also, a similar result for the wave equation was obtained by Brezis in* [**5**].

REMARK 4. *A natural class of nonlinearities satisfying* $(G_1)_\alpha$ *and* $(G_2)_\alpha^+$ *is*

$$g(x,s) = b(x)|s|^{\alpha-1}s \ ,$$

where $b \in L^p(\mathbb{R}^N)$ *(p as in Theorem 2.2),* $b(x) > 0$ *a.e., and* $\alpha \in [0,1)$. *In case* $\alpha \in (0, \frac{1}{2}]$, *such* $g(x,s)$*'s also satisfy* (G_3).

References

[1] S. Ahmad, A.C. Lazer, J.L. Paul, *Elementary Critical Point Theory and Perturbations of Elliptic Boundary Value Problems at Resonance*, Indiana Univ. Math. J. **25** (1976), 933–944.

[2] D. Arcoya, D.G. Costa, *Nontrivial Solutions for a Strongly Resonant Problem*, Diff. Int. Eqs. **8** (1995), 151–159.

[3] P. Bartolo, V. Benci, D. Fortunato, *Abstract Critical Point Theorems and Applications to Some Nonlinear Problems with Strong Resonance at Infinity*, Nonl. Anal. TMA **7** (1983), 981–1012.

[4] F.A. Berezin, M.A. Shubin, The Schrodinger Equation, Kluwer, Dordrecht, 1991.

[5] H. Brézis, *Periodic Solutions of nonlinear Vibrating Strings and Duality Principles*, Bull. Amer. Math. Soc. **8** (1983), 409–426.

[6] A. Capozzi, D. Lupo, S. Solimini, *On the Existence of a Nontrivial Solution to Nonlinear Problems at Resonance*, Nonl. Anal. TMA **13** (1989), 151–163.

[7] G. Cerami, *Un criterio de esistenza per i punti critici su varietà ilimitate*, Rc. Ist. Lomb. Sci. Lett. **112** (1978), 332–336.

[8] D.G. Costa, *A Note on Unbounded Perturbations of Linear Resonant Problems*, Preprint Univ. Brasilia, Trabalhos de Matematica #245 (1989).

[9] D. G. Costa and C. A. Magalhães, *A Unified Approach to a Class of Strongly Indefinite Functionals*, J. Diff. Eqs. **125** (1996), 521–547.

[10] D.G. Costa, E.A. Silva, *Existence of Solution for a Class of Resonant Elliptic Problems*, J. Math. Anal. Appl. **175** (1993), 411–424.

[11] D.G. Costa, E.A. Silva, *On a Class of Resonant Problems at Higher Eigenvalues*, Diff. Int. Eqs. **8** (1995), 663–671.

[12] D.G. Costa, H. Tehrani, *On a Class of Asymptotically Linear Elliptic Problems in* $\mathbb{R}^N$, J. Diff. Eqs. **173** (2001), 470–494.

[13] D.G. de Figueiredo, J.-P. Gossez, *Condition de Non-resonance pour Certains Problèmes Elliptiques Non-linéaires*, C. R. Acad. Sci. Paris **302** (1986), 543–545.

[14] D.G. de Figueiredo, J.-P. Gossez, *Nonresonance below the First Eigenvalue for a Semilinear Elliptic Problem*, Math. Ann. **281** (1988), 589–610.

[15] D.G. de Figueiredo, I. Massabó, *Semilinear Elliptic Equations with Primitive of the Nonlinearity Interacting with the First Eigenvalue*, J. Math. Anal. Appl. **156** (1991), 381–394.

[16] D.G. de Figueiredo, O.O. Miyagaki, *Semilinear Elliptic Equations with the Primitive of the Nonlinearity away from the Spectrum*, Nonl. Anal. TMA **17** (1991), 1201–1219.

[17] A. Fonda, J.-P. Gossez, *Semicoercive Variational Problems at Resonance: An Abstract Approach*, Diff. Int. Eqs. **3** (1990), 695–708.

[18] J. V. Gonçalves and O. H. Miyagaki, *Existence of nontrivial solutions for semilinear elliptic equations at resonance*, Houston J. of Math. **16** (1990), 583–595.

[19] L. Jeanjean, *On the Existence of Bounded Palais-Smale Sequences and Application to a Landesman-Lazer type Problem set on* $\mathbb{R}^N$, Proc. Roy. Soc. Edinburgh **129A** (1999), 787–809.

[20] E.M. Landesman, A.C. Lazer, *Nonlinear Perturbations of Linear Elliptic Boundary Value Problems at Resonance*, J. Math. Mech. **19** (1970), 609–623.

[21] J. Mawhin, M. Willem, *Critical Points of Convex Perturbations of Some Indefinite Quadratic Forms and Semilinear Boundary Value Problems at Resonance*, Ann. Inst. H. Poincaré, Anal. Non Linéaire **3** (1986), 431–453.

[22] P.H. Rabinowitz, *Some Minimax Theorems and Applications to Nonlinear Partial Differential Equations*, in *Nonlinear Analysis, Ed. Cesari, Kannan and Weinberger*, Academic Press, 1978, 161–177.

[23] P.H. Rabinowitz, *Minimax Methods in Critical Point Theory with Applications to Differential Equations*, CBMS Regional Conf. Ser. in Math. **65**, AMS, Providence, RI, 1986.

[24] M. Ramos, *Remarks on Resonance Problems with Unbounded Nonlinearities*, Diff. Int. Eqs. **6** (1993), 215–223.

[25] M. Schechter, *The Saddle Point Alternative*, Amer. J. Math. **117** (1995), 1603–1626.

[26] M. Schechter, *Resonance Problems which Intersect Many Eigenvalues*, Math. Nachr. **183** (1997), 210–210.

[27] E.A.B. Silva, *Critical Point Theorems and Applications to Differential Equations*, Ph.D. Thesis, University of Wisconsin-Madison, 1988.

[28] E.A. Silva, *Linking Theorems and Applications to Semilinear Elliptic Problems at Resonance*, Nonl. Anal. TMA **16** (1991), 455–477.

[29] S. Solimini, *On the Solvability of some Elliptic Partial Differential Equations with the Linear Part at Resonance*, J. Math. Anal. Appl. **117** (1986), 138–152.

[30] C.A. Stuart, H.S. Zhou, *Applying the Mountain-Pass Theorem to an Asymptotically Linear Elliptic Equation on* $\mathbb{R}^N$, Comm. Part. Diff. Eqs. **24** (1999), 1731–1758.

Department of Mathematical Sciences - UNLV, Las Vegas, NV 89154
E-mail address: `costa@unlv.nevada.edu`

Department of Mathematical Sciences - UNLV, Las Vegas, NV 89154
E-mail address: `tehranih@unlv.nevada.edu`

Contemporary Mathematics
Volume **357**, 2004

Quasilinear Boundary Value Problems: Existence and Multiplicity Results

Jan Čepička, Pavel Drábek, and Petr Girg

1. Historical survey.

Let us consider the following problem for the Dirichlet p-Laplacian Δ_p in a bounded domain $\Omega \subset \mathbb{R}^N$:

$$(1.1)\qquad \begin{cases} -\Delta_p u = g(x,u) + f(x) & \text{in } \Omega\,; \\ u = 0 & \text{on } \partial\Omega\,. \end{cases}$$

Here, $\Delta_p u \stackrel{\text{def}}{=} \operatorname{div}(|\nabla u|^{p-2}\nabla u)$, where $p \in (1,\infty)$ is a fixed number, g and f satisfy some additional assumptions to be specified later. The existence results for (1.1) are closely related to the asymptotic properties of nonlinear function g and to the properties of the eigenvalue problem

$$(1.2)\qquad \begin{cases} -\Delta_p u = \lambda |u|^{p-2}u & \text{in } \Omega\,; \\ u = 0 & \text{on } \partial\Omega \end{cases}$$

which is nonlinear if $p \neq 2$ and linear for $p = 2$.

We say that $\lambda \in \mathbb{R}$ is an *eigenvalue* of (1.2) if there is a nonzero function u which solves (1.2) in whatever sense (usually weak solutions are considered). Without specifying precise assumptions on the data in (1.1) we will try to explain in the historical context the study of solvability of problem (1.1).

The first results in this direction go back to works of NEČAS [**30**] and POHOZAEV [**31**]. These results were generalized later by many authors (see FUČÍK, NEČAS, SOUČEK and SOUČEK [**22**] for further references). These results can be formulated as follows.

Let $\lambda \in \mathbb{R}$ *be not an eigenvalue of* (1.2) *and*

$$(1.3)\qquad g(x,u) = \lambda |u|^{p-2}u + \tilde{g}(x,u)\,, \quad \widetilde{g}(x,u) = o(|u|^{p-1}),\, |u| \to \infty.$$

Then (1.1) *has at least one solution for any* f.

1991 *Mathematics Subject Classification.* Primary 54C40, 14E20; Secondary 46E25, 20C20.

The authors acknowledge support of the Ministry of Education of the Czech Republic, # MSM 235200001.

This result can be generalized if one knows "how far" the parameter λ is from the eigenvalues of (1.2). However, in higher dimensions ($N \geq 2$) complete description of the set of all eigenvalues of (1.2) is an open problem and so satisfactory results are available in the case of one dimensional p-Laplacian only. The pioneering work in this direction was published by BOCCARDO, DRÁBEK, GIACHETTI and KUČERA [**6**]. Their results read as follows.

Let $0 < \lambda_1 < \lambda_2 < \cdots \to +\infty$ be the set of all eigenvalues of

$$\begin{cases} -(|u'|^{p-2}u')' = \lambda |u|^{p-2}u & \text{in } (0,\pi)\,; \\ u(0) = u(\pi) = 0\,. \end{cases}$$

Assume that g satisfies one of the following estimates

$$\limsup_{s\to\pm\infty} \frac{g(t,s)}{|s|^{p-2}s} \leq \lambda_1 - \delta, \tag{1.4}$$

$$\lambda_i + \delta \leq \liminf_{s\to\pm\infty} \frac{g(t,s)}{|s|^{p-2}s} \leq \limsup_{s\to\pm\infty} \frac{g(t,s)}{|s|^{p-2}s} \leq \lambda_{i+1} - \delta \tag{1.5}$$

for a.e. $t \in (0,\pi)$ with some $\delta > 0$ and $i \geq 1$. Then the boundary value problem

$$\begin{cases} -(|u'(t)|^{p-2}u'(t))' = g(t,u(t)) + f(t), & t \in (0,\pi)\,; \\ u(0) = u(\pi) = 0 \end{cases}$$

has for each f at least one solution.

In higher dimensions the properties of the first eigenvalue (principal eigenvalue) are well known due to work of ANANE [**1**] and the variational characterization of the second eigenvalue was given by ANANE and TSOULI [**3**]. More precisely, for general $N \geq 1$, there exists the least eigenvalue of (1.2), $\lambda_1 > 0$, which is isolated and simple in the sense that there is precisely one normalized eigenfunction φ_1 associated with λ_1 which satisfies $\varphi_1 > 0$ in Ω. If $\partial\Omega$ is smooth then we also have $\frac{\partial \varphi_1}{\partial \nu} < 0$ on $\partial\Omega$ (here $\frac{\partial}{\partial \nu}$ is the derivative with respect to the external normal of $\partial\Omega$). There is $\lambda_2 > \lambda_1$ which is the next eigenvalue of (1.2) in the sense that there are no other eigenvalues lying in (λ_1, λ_2). The above result of [**6**] for partial differential equations case then holds with (1.4) or with (1.5) substituted by

$$\lambda_1 + \delta \leq \liminf_{s\to\pm\infty} \frac{g(t,s)}{|s|^{p-2}s} \leq \limsup_{s\to\pm\infty} \frac{g(t,s)}{|s|^{p-2}s} \leq \lambda_2 - \delta\,. \tag{1.6}$$

Note that in semilinear case ($p = 2$) the result of [**6**] holds true for any $N \geq 1$, because the structure of the set of all eigenvalues of (1.2) for $p = 2$ is well known and it is formed by a sequence $0 < \lambda_1 < \lambda_2 < \cdots \to +\infty$ (if we do not take into the account multiplicities of λ_i). There is a vaste literature dealing with semilinear case and the reader is refered to [**6**] to find some of the relevant items.

All results mentioned above are of *non resonant type* because conditions (1.3) - (1.6) avoid the interaction of nonlinearity g with the spectrum of homogeneous part of the problem (i.e. with eigenvalues of (1.2)). The first attempt to deal with the *resonant* type problems (*Landesman–Lazer* type problems) was made by ANANE and GOSSEZ [**2**] and BOCCARDO, DRÁBEK and KUČERA [**7**]. Both papers deal with the boundary value problems of the type

$$\begin{cases} -\Delta_p u &= \lambda_1 |u|^{p-2}u + g(u) + f(x) & \text{in } \Omega\,; \\ u &= 0 & \text{on } \Omega \end{cases} \tag{1.7}$$

and prove the result which can be roughly formulated as follows:

Let $g(\pm\infty) = \lim_{u\to\pm\infty} g(u)$ *and*

$$g(+\infty)\int_\Omega \varphi_1(x)dx < \int_\Omega f(x)\varphi_1(x)dx < g(-\infty)\int_\Omega \varphi_1(x)dx. \tag{1.8}$$

Then the boundary value problem (1.7) *has at least one solution.*

While [2] used variational approach, [7] relied on the topological methods. However, none of these papers allowed to treat the problem under the condition which is dual to (1.8):

$$g(-\infty)\int_\Omega \varphi_1(x)dx < \int_\Omega f(x)\varphi_1(x)dx < g(+\infty)\int_\Omega \varphi_1(x)dx. \tag{1.9}$$

Using refined variational approach based on the Saddle Point Theorem, ARCOYA and ORSINA [4] proved that both (1.8) and (1.9) are sufficient conditions for solvability of problem (1.7). Finally, DRÁBEK and ROBINSON [20], using also the variational approach, proved an analogue of the classical result of LANDESMAN and LAZER [26]:

Let us consider

$$\begin{cases} -\Delta_p u &= \lambda|u|^{p-2}u + g(u) + f(x) & \text{in } \Omega\,; \\ u &= 0 & \text{on } \partial\Omega \end{cases} \tag{1.10}$$

and either

$$g(+\infty)\int_\Omega \varphi^+(x)dx - g(-\infty)\int_\Omega \varphi^-(x)dx < \int_\Omega f(x)\varphi(x)dx <$$
$$< g(-\infty)\int_\Omega \varphi^+(x)dx - g(+\infty)\int_\Omega \varphi^-(x)dx \tag{1.11}$$

or

$$g(-\infty)\int_\Omega \varphi^+(x)dx - g(+\infty)\int_\Omega \varphi^-(x)dx < \int_\Omega f(x)\varphi(x)dx <$$
$$< g(+\infty)\int_\Omega \varphi^+(x)dx - g(-\infty)\int_\Omega \varphi^-(x)dx \tag{1.12}$$

hold for all eigenfunctions φ *associated with* λ. *(Here* φ^+ *and* φ^- *stand for the positive and negative part of* φ*, respectively.) Then* (1.10) *has at least one solution.*

Note that above we use the following convention - if λ is not an eigenvalue of (1.2) then there is an empty set of eigenfunctions associated with λ; so both (1.11) and (1.12) are trivially satisfied and (1.10) is solvable for any f!

The results of [2], [7], [4] and [20] are slightly more general than we present here, but in principle they can be formulated as above. One more thing is worth to be emphasized here. The results of [20] deal with any λ (an eigenvalue or not) without having precise information about the structure of the set of all eignevalues of (1.2) if $N \geq 2, p \neq 2$. Common feature of all these results consists in the fact that conditions (1.8), (1.9), (1.11) and (1.12) guarantee certain compactness (a priori bounds or PS-condition) which finally helps to solve the problem. The situation is very different if we do not have conditions of this type.

As a typical example let us consider the Dirichlet problem for the p-Laplacian at the first eigenvalue

$$\left\{\begin{array}{rcll} -\Delta_p u & = & \lambda_1 |u|^{p-2}u + f(x) & \text{in } \Omega\,; \\ u & = & 0 & \text{on } \partial\Omega\,. \end{array}\right. \tag{1.13}$$

For $p \neq 2$ this problem is much harder to understand than problem (1.7) (if g satisfies one of the conditions mentioned above). On the other hand for $p = 2$ there is very nice and transparent necessary and sufficient condition for solvability of (1.13) expressed by

$$\int_\Omega f(x)\varphi_1(x)dx = 0\,. \tag{1.14}$$

The first attemps to understand problem (1.13) and to prove nonlinear analogue of the Fredholm alternative in the resonant case were made in papers by BINDING, DRÁBEK and HUANG [**8**], DRÁBEK and TAKÁČ [**21**]. In these papers the authors made clear that (1.14) *is not necessary condition* for solvability of (1.13) if $p \neq 2$. However, the structure of the set of all f's for which (1.13) is solvable was still far from being understood. The essential step in this direction was paper by DELPINO, DRÁBEK and MANÁSEVICH [**16**] where several surprising results were proved. A special asymptotics for the solution of the initial value problem was used and combined with the topological degree arguments, variational methods and upper and lower solutions technique. Unfortunately, only one dimensional case could be treated there. Let us mention the main results from [**16**]. The authors studied the problem

$$\left\{\begin{array}{l} -(|u'|^{p-2}u')' = \lambda_1 |u|^{p-2}u + f \quad \text{in } (0,T)\,; \\ \qquad u(0) = u(T) = 0\,. \end{array}\right. \tag{1.15}$$

They proved the following main results for $p \neq 2$:

(i) *Condition*

$$\int_0^T f(t)\varphi_1(t)dt = 0 \tag{1.14'}$$

is sufficient for solvability of (1.15).

(ii) *If* (1.14′) *is satisfied and* $f \neq 0$ *then the set of all solutions of* (1.15) *is apriori bounded in a certain sense.*

(iii) *There exists an open cone* $\mathcal{C}$ *such that for all* $f \in \mathcal{C}$ *problem* (1.15) *has at least two distinct solutions and*

$$\int_0^T f(t)\varphi_1(t)dt \neq 0$$

for all $f \in \mathcal{C}$.

(iv) *For* $f \neq 0$ *consider the energy functional associated with* (1.15), *i.e.*

$$E(u) = \frac{1}{p}\int_0^T |u'|^p dt - \frac{\lambda_1}{p}\int_0^T |u|^p dt - \int_0^T f\,u\,dt\,.$$

Let f *satisfy* (1.14′). *Then*

(a) *for $1 < p < 2$ the functional E is unbounded from below; the set of its critical points is nonempty and bounded;*
(b) *for $p > 2$ the functional E is bounded from below and has a global minimizer, the set of its critical points is bounded.*

It was also shown that E does not satisfy PS condition at the level 0 for $p > 2$.

These results were completed in the paper by DRÁBEK, GIRG and MANÁSEVICH [**17**] where the characterization of the set of all right hand sides f was given. The approach was a combination of variational method and the method of upper and lower solutions. Another interesting generalization of [**16**] appeared in paper MANÁSEVICH and TAKÁČ [**29**] where generalized Prüfer transformation was used to deal not only with the first but also with the higher eigenvalues. However, all papers above could treat the one-dimensional case $N = 1$ only!

The higher dimensional case seemed to be qualitatively the same but none of the approaches used in the one-dimensional case could be applied directly. The first but partial results in this direction were obtained in paper by DRÁBEK and HOLUBOVÁ [**19**]. However, the authors were able to deal only with the case $1 < p < 2$ and they did not prove any multiplicity results. On the other hand, paper [**19**] was first to show that the energy functional

$$E(u) = \frac{1}{p}\int_\Omega |\nabla u|^p dx - \frac{\lambda_1}{p}\int_\Omega |u|^p dx - \int_\Omega f\, u \tag{1.16}$$

associated with problem (1.13) has a saddle point structure and is unbounded from below if $N \geq 2, 1 < p < 2$ and f satisfies (1.14)! The case $p > 2, N \geq 2$, and the global minimizer structure of functional (1.16) under the condition (1.14) was proved soon after by FLECKINGER and TAKÁČ [**23**]. Several papers then followed where existence and multiplicity results, characterization of the set of all f's for which (1.13) is solvable as well as the description of the geometry of the energy functional (1.16) associated with (1.13) is successfully studied. Let us mention the papers or preprints by DRÁBEK [**14, 15**] and TAKÁČ [**34, 35, 36**]. These results, roughly speaking, recovered and generalized results obtained in [**16**] and [**17**] in the case of one dimension.

One-dimensional case has also been extensively studied by means of numerical experiments carried out by ČEPIČKA [**11**]. Motivated by these experiments and by paper ARCOYA and GÁMEZ [**5**], GÁMEZ and GIRG [**24**] decided to employ bifurcation theory to study the problem

$$\begin{cases} -(|u'|^{p-2}u')' = \lambda |u|^{p-2}u + f & \text{in } (0,T)\,; \\ u(0) = u(T) = 0 \end{cases} \tag{1.17}$$

for λ "close" to λ_1. The main ingredients of [**24**] were asymptotic bifurcation from infinity for the one-dimensional p-Laplacian (from the principal eigenvalue) and refinement of the asymptotic estimate derived in [**16**].

Finally, paper by DRÁBEK, GIRG, TAKÁČ and ULM [**18**] deals with the problem

$$\begin{cases} -\Delta_p u = \lambda |u|^{p-2}u + f & \text{in } \Omega\,; \\ u = 0 & \text{on } \partial\Omega \end{cases} \tag{1.18}$$

for λ "close" to λ_1 on the bounded domain $\Omega \subset \mathbb{R}^N$, for arbitrary dimension $N \geq 1$. Again, asymptotic bifurcation from infinity from the principal eigenvalue for the

p-Laplacian is combined with the asymptotic estimate which generalizes that one used in [**24**]. It should be pointed out that this asypmtotic estimate allows to determine local behavior of bifurcation branches for λ close to λ_1 and explains some of unexpected phenomena which occur if we pass from $p = 2$ to $p \neq 2$:

- *a priori* bounds for solutions to (1.13) with f satisfying (1.14);
- existence of multiple solutions to (1.13) for some f violating (1.14);
- existence of multiple solutions to (1.18) for f satisfying (1.14) and $\lambda \neq \lambda_1, \lambda$ close to λ_1.

To be more precise, let us formulate the main results of [**18**]. To this end we split $f = f^\top + a\varphi_1$ and rewrite (1.18) as

$$\left\{ \begin{array}{rcll} -\Delta_p u - \lambda |u|^{p-2} & = & f^\top + a\varphi_1 & \text{in } \Omega\,; \\ u & = & 0 & \text{on } \partial\Omega\,. \end{array} \right. \tag{1.18'}$$

Here $f^\top \in L^\infty(\Omega)$ is a given function, $\int_\Omega f^\top \varphi_1 dx = 0$ and $f^\top \neq 0; \lambda, a \neq \mathbb{R}$ are parameters. By a solution to (1.18) (or (1.18′)) we always mean the weak solution in the usual sense. Then [**18**] contains the following existence and multiplicity results:

(E1) *For $\lambda = \lambda_1$, $a = 0$, problem* (1.18) *has at least one solution; all possible solutions of* (1.18) *are a priori bounded in $C^{1,\beta}(\overline{\Omega})$, $0 < \beta < 1$, by a constant which depends on $f^\top$*
(see [**18**, Thm. 6.1]*).*

(E2) *There exist $a_0 = a_0\left(f^\top\right) > 0$ and $\delta = \delta\left(f^\top\right) > 0$ such that*
- *if either $\lambda \in (\lambda_1 - \delta, \lambda_1)$ and $a \geq a_0$, or else $\lambda \in (\lambda_1, \lambda_1 + \delta)$ and $a \leq -a_0$, then problem* (1.18) *can have only positive solutions;*
- *if either $\lambda \in (\lambda_1 - \delta, \lambda_1)$ and $a \leq -a_0$, or else $\lambda \in (\lambda_1, \lambda_1 + \delta)$ and $a \geq a_0$, then problem* (1.18) *can have only negative solutions*

(see [**18**, Thm. 6.9]*).*

(M1) *There exists $\eta = \eta\left(f^\top\right) > 0$ such that for $a = 0$ problem* (1.18) *has at least three distinct solutions (among them at least one positive and one negative) provided either $1 < p < 2$ and $\lambda \in (\lambda_1 - \eta, \lambda_1)$, or $p > 2$ and $\lambda \in (\lambda_1, \lambda_1 + \eta)$*
(see [**18**, Thms. 6.3 and 6.15]*).*

(M2) *There exists $\varepsilon > 0$ with the following properties:*
- *for every $\varepsilon' \in (0, \varepsilon)$, there is $\eta = \eta\left(f^\top, \varepsilon, \varepsilon'\right) > 0$ such that $\varepsilon' < |a| < \varepsilon$ and $\lambda \in (\lambda_1 - \eta, \lambda_1) \cup (\lambda_1, \lambda_1 + \eta)$ imply that problem* (1.18) *has at least three distinct solutions, of which at least one is positive and at least one is negative;*
- *$\lambda = \lambda_1$ and $0 < |a| < \varepsilon$ imply that problem* (1.18) *has at least two distinct solutions, of which at least one is negative if $(p-2)a < 0$, and at least one is positive if $(p-2)a > 0$*

(see [**18**, Thm. 6.17]*).*

2. Global behaviour of solution set for $\lambda < \lambda_2$

2.1. Operator formulation. Under a *solution* of (1.18) we understand a pair (λ, u) of $\lambda \in \mathbb{R}$ and $u \in W_0^{1,p}(\Omega)$ satisfying the integral equality

$$\int_\Omega |\nabla u|^{p-2} \langle \nabla u, \nabla \phi \rangle \,\mathrm{d}x - \lambda \int_\Omega |u|^{p-2} u \phi \,\mathrm{d}x = \int_\Omega f \phi \,\mathrm{d}x \tag{2.1}$$

for every $\phi \in W_0^{1,p}(\Omega)$.

Let $X = W_0^{1,p}(\Omega)$ and let X' stand for its dual space, i.e., $X' = W^{-1,p'}(\Omega)$. Then (2.1) is equivalent to the abstract operator equation

$$\mathcal{I}(u) - \lambda \mathcal{S}(u) = F\,, \tag{2.2}$$

where $\mathcal{I}, \mathcal{S} : X \to X'$ and $F \in X'$ are defined as follows, for any $u, \phi \in X$:

$$\begin{aligned}(\mathcal{I}(u), \phi)_X &= \int_\Omega |\nabla u|^{p-2} \langle \nabla u, \nabla \phi \rangle \, dx\,,\\ (\mathcal{S}(u), \phi)_X &= \int_\Omega |u|^{p-2} u \phi \, dx\,,\\ (F, \phi)_X &= \int_\Omega f \phi \, dx\,.\end{aligned}$$

Here, $(\cdot, \cdot)_X$ denotes the duality pairing between X and X' and $\langle \cdot, \cdot \rangle$ stands for the scalar product in $\mathbb{R}^N$.

It is proved in [**13**, Chapter 5] that the operator $\mathcal{I} - \lambda \mathcal{S}$ satisfies condition $\alpha(X)$ from [**33**] (which is nothing else but condition (S_+) from [**9**]) and so its (topological) degree can be defined. Namely, if $B_r(0) \subset X$ is an open ball centred at the origin with radius $r > 0$, then the degree of the mapping $u \mapsto \mathcal{I}(u) - \lambda \mathcal{S}(u) - F$ on $B_r(0)$, denoted by

$$\operatorname{Deg}[\mathcal{I} - \lambda \mathcal{S} - F, B_r(0), 0]\,,$$

is well-defined provided the equation $\mathcal{I}(u) - \lambda \mathcal{S}(u) = F$ has no solution $u \in \partial B_r(0)$. Let us note that the basic properties of the degree "Deg" are the same as those of the well-known Leray-Schauder degree "deg" for mappings from X into itself. Here, $\mathcal{I}(\cdot)$ replaces the identity mapping on X, whereas $\lambda \mathcal{S}(\cdot) + F$ plays the role of a compact perturbation.

Let $\mathcal{S}$ denote the set of all solutions to (1.18), i.e.,

$$\mathcal{S} \stackrel{\text{def}}{=} \{(\lambda, u) \in \mathbb{R} \times X : \mathcal{I}(u) - \lambda \mathcal{S}(u) = F\}\,. \tag{2.3}$$

The following results (see *e.g.* [**13**]) play indispensable role in many proofs:

- (B1) *For any* $p > 1$, $f \in L^\infty(\Omega)$ *and* $\lambda < \lambda_1$ *there exists a constant* M *such that if* $(\lambda, u) \in \mathcal{S}$ *then* $\|u\|_{W_0^{1,p}(\Omega)} < M$;
- (B2) *for any* $\varepsilon > 0$ *there exists* M' *such that for all* $(\lambda, u) \in \mathcal{S}$ *with* $\lambda \le \lambda_1 - \varepsilon$ *we have* $\|u\|_{W_0^{1,p}(\Omega)} < M'$;
- (U) *for any* $\lambda \le 0$ *there is exactly one* u_λ *such that* $(\lambda, u_\lambda) \in \mathcal{S}$;
- (D1) *there exists* R *large enough such that for any* $\lambda < \lambda_1$, $\operatorname{Deg}[\mathcal{I} - \lambda \mathcal{S} - F, B_R(0), 0]$ *is well defined and* $\operatorname{Deg}[\mathcal{I} - \lambda \mathcal{S} - F, B_R(0), 0] = 1$;
- (D2) *for any* $\varrho > 0$ *and any* $\lambda \le 0$, $\operatorname{Deg}[\mathcal{I} - \lambda \mathcal{S} - F, B_\varrho(u_\lambda), 0]$ *is well defined and* $\operatorname{Deg}[\mathcal{I} - \lambda \mathcal{S} - F, B_\varrho(u_\lambda), 0] = 1$.

REMARK 2.1. Property (U) follows from the coercivity and the strict monotony of $\mathcal{I} - \lambda \mathcal{S}$ for $\lambda \le 0$, see e.g. [**13**]. Note that we easily infer (D2) from (D1) using (U) and the *additivity property* of the degree.

2.2. One continuation result. It is well known by the strict monotony of $\mathcal{I} - \lambda \mathcal{S}$ for $\lambda < 0$ (see e.g. [**13**]) that for each $\lambda < 0$ there exists exactly one $u_\lambda \in W_0^{1,p}(\Omega)$ such that (λ, u_λ) is a solution to (1.18). A natural question arise then, *whether the map* U *defined as* $U : \mathbb{R}^- \cup \{0\} \to W_0^{1,p}(\Omega)$; $\lambda \mapsto u_\lambda$ *is continuous.* If so, there is a connected set of solutions "arriving" from $-\infty$ in the λ-direction, which invokes another question in turn: *What happens with this set when we cross $\lambda = 0$ and increase λ?*

For the convenience of the formulation and proof of our result, we shall define

$$\mathcal{E}^0_\lambda \stackrel{\text{def}}{=} \{(\nu, u) \in \mathbb{R} \times X : \nu = \lambda\}\,;$$
$$\mathcal{E}^-_\lambda \stackrel{\text{def}}{=} \{(\nu, u) \in \mathbb{R} \times X : \nu < \lambda\}\,;$$
$$\mathcal{E}^+_\lambda \stackrel{\text{def}}{=} \{(\nu, u) \in \mathbb{R} \times X : \nu > \lambda\}\,.$$

THEOREM 2.2. *The set $\mathcal{S}$ contains continuum $\mathcal{T}$ (connected set that is maximal in the sense of inclusions) which satisfies*

(i) $(\mathcal{S} \cap \mathcal{E}^-_0) \subset (\mathcal{T} \cap \mathcal{E}^-_0)$, *i.e.,* $(\mathcal{S} \cap \mathcal{E}^-_0) = (\mathcal{T} \cap \mathcal{E}^-_0)$;

(ii) $(\mathcal{T} \cap \mathcal{E}^+_0)$ *is either unbounded in λ-direction or there exists $\lambda_e \in \mathbb{R}$ such that for all $\varepsilon > 0$ the set $\mathcal{T} \cap \mathcal{E}^+_{\lambda_e - \varepsilon} \cap \mathcal{E}^-_{\lambda_e + \varepsilon}$ is unbounded in $\mathbb{R} \times X$; λ_e being an eigenvalue of $-\Delta_p$.*

REMARK 2.3. $\mathcal{T} \cap \mathcal{E}^+_{\lambda_e - \varepsilon} \cap \mathcal{E}^-_{\lambda_e + \varepsilon}$ being unbounded in $\mathbb{R} \times X$ for any $\varepsilon > 0$ means that there is an asymptotic bifurcation at λ_e, because the set $\mathcal{T} \cap \mathcal{E}^+_{\lambda_e - \varepsilon} \cap \mathcal{E}^-_{\lambda_e + \varepsilon}$ is bounded in λ-direction.

As a consequence of previous theorem, we get continuity of the map $U : \mathbb{R}^- \cup \{0\} \to W^{1,p}_0(\Omega), \lambda \mapsto u$ such that (λ, u) is a solution of BVP (1.1). Answer to the second question is directly provided in (ii) of the previous theorem.

Main ingredients of our proof of Theorem 2.2 are (B2), (U), (D2) and the following lemma (cf. [**32**]):

LEMMA 2.4. *Let $\lambda_0 \in \mathbb{R}$, $u_{\lambda_0} \in W^{1,p}_0(\Omega)$ and there does not exist a connected set $\mathcal{T} \subset \mathcal{E}^-_{\lambda_0} \cup \mathcal{E}^0_{\lambda_0}$ such that $(\lambda_0, u_{\lambda_0}) \in \mathcal{T}$ and either*

(i) *is unbounded in $\mathcal{E}^-_{\lambda_0} \cup \mathcal{E}^0_{\lambda_0}$, or*

(ii) $\mathcal{T} \cap \mathcal{E}^0_{\lambda_0} \neq \{(\lambda_0, u_{\lambda_0})\}$.

Then there exists a bounded open set $\mathcal{O} \subset \mathbb{R} \times X$ such that $(\lambda_0, u_{\lambda_0}) \in \mathcal{O}$, $\partial\mathcal{O} \cap \mathcal{S} \cap (\mathcal{E}^-_0 \cup \mathcal{E}^0_0) = \emptyset$, and $\mathcal{O}$ contains no solutions (λ_0, u) other than $(\lambda_0, u_{\lambda_0})$.

Now the proof of Theorem 2.2 (i) follows.

PROOF. Let $\mathcal{T} \subset \mathcal{S}$ be the maximal connected set (maximal in the sense of inclusions) which satisfy $(0, u_0) \in \mathcal{T}$. Note that $\mathcal{T} \neq \emptyset$ because $(0, u_0) \in \mathcal{T}$ (it may degenerate to $\{(0, u_0)\}$, however). To the contrary of assertion of Theorem 2.2 (i), assume that $(\mathcal{S} \cap \mathcal{E}^-_0) \not\subset (\mathcal{T} \cap \mathcal{E}^-_0)$. By this assumption and properties (U) and (B2), we find that $\mathcal{T} \cap (\mathcal{E}^-_0 \cup \mathcal{E}^0_0)$ is bounded. As $\mathcal{T} \subset \mathcal{S}$ is the maximal connected set containing $(0, u_0)$, all connected subsets of $\mathcal{S} \cap (\mathcal{E}^-_0 \cup \mathcal{E}^0_0)$ containing $(0, u_0)$ are bounded. Thus we are lead to (i) of Lemma 2.4. Since $\mathcal{S} \cap \mathcal{E}^0_0 = \{(0, u_0)\}$ (property (U) applied for $\lambda = 0$), we are lead to (ii) of Lemma 2.4. Therefore, we have verified assumptions (i) and (ii) of Lemma 2.4 and so there exists a bounded open set $\mathcal{O}$ with properties indicated in Lemma 2.4. Since $\partial\mathcal{O} \cap \mathcal{S} \cap (\mathcal{E}^-_0 \cup \mathcal{E}^0_0) = \emptyset$ and since $\mathcal{O}$ is bounded, we can consider the following generalized homotopy along λ (see [**27**]):

$$\begin{aligned}\operatorname{Deg}[\mathcal{I} - \lambda\mathcal{S} - F, \mathcal{O} \cap \mathcal{E}^0_\lambda, 0] = {} & \operatorname{Deg}[\mathcal{I} - \lambda\mathcal{S} - F, \mathcal{O} \cap \mathcal{E}^0_0, 0] = \\ & = \operatorname{Deg}[\mathcal{I} - \lambda\mathcal{S} - F, B_\varrho(u_0), 0] = 1\,,\end{aligned}$$

for any $\lambda \leq 0$. However, using boundedness of $\mathcal{O}$ we infer that there exists $\lambda' < 0$ such that $\mathcal{O} \cap \mathcal{E}^0_{\lambda'} = \emptyset$ and, therefore,

$$\operatorname{Deg}[\mathcal{I} - \lambda\mathcal{S} - F, \mathcal{O} \cap \mathcal{E}^0_{\lambda'}, 0] = \operatorname{Deg}[\mathcal{I} - \lambda\mathcal{S} - F, \emptyset, 0] = 0\,,$$

a contradiction. So $\mathcal{S} \cap \mathcal{E}^-_0 = \mathcal{T} \cap \mathcal{E}^-_0$.

For $\mathcal{T} \cap \mathcal{E}^+_0$ we proceed analogously; the only difference is that we have to modify Lemma 2.4 and we do *not* have *a priori* bounds which are uniform in λ in the neigbourhood of eigenvalues λ_e of $-\Delta_p$. Thus $\mathcal{T} \cap \mathcal{E}^+_0$ may blow up there. □

2.3. Global bifurcation at $\lambda = \lambda_1$.

DEFINITION 2.5. Let $\mu_0 \in \mathbb{R}$. We say that (μ_0, ∞) is an *asymptotic bifurcation point* of (2.2) if there exists a sequence of pairs $\{(\mu_n, u_n)\}_{n=1}^{\infty} \subset \mathbb{R} \times X$ such that (2.2) holds with $(\lambda, u) = (\mu_n, u_n)$; $n = 1, 2, 3, \ldots$, and $(\mu_n, \|u_n\|_X) \to (\mu_0, \infty)$.

The following standard trick transforms bifurcations from infinity to bifurcations from the continuum of trivial solutions: For $u \in X$, $u \neq 0$, set $v = u/\|u\|^2_X$. Then (2.2) is equivalent to

$$\mathcal{I}(v) - \lambda\mathcal{S}(v) = \|v\|_X^{2(p-1)} F,$$

and so the term

$$\mathcal{G}(\lambda, v) \stackrel{\text{def}}{=} \begin{cases} \|v\|_X^{2(p-1)} F & \text{if} \quad v \neq 0\,; \\ 0 & \text{if} \quad v = 0\,, \end{cases}$$

for $\lambda \in \mathbb{R}$, represents a compact perturbation "of higher order" in the variable v in the equation

$$\mathcal{I}(v) - \lambda\mathcal{S}(v) = \mathcal{G}(\lambda, v)\,. \tag{2.4}$$

It follows immediately from this transformation that the pair (μ_0, ∞) is an asymptotic bifurcation point for (2.2) if and only if $(\mu_0, 0)$ is a bifurcation point (from the set of trivial solutions) for (2.4). For $\mathcal{C} \subset \mathbb{R} \times X$ we define (the set) $\widetilde{\mathcal{C}}$ to be the closure in $\mathbb{R} \times X$ of the set of all pairs $(\mu, v) \in \mathbb{R} \times X$ such that $v \neq 0$ and $(\mu, v/\|v\|^2_X) \in \mathcal{C}$.

In [**13**, Theorem 14.18], it was proved that $(\lambda_1, 0)$ is a bifurcation point for (2.4). Let us reformulate this result in terms of problem (2.2).

THEOREM 2.6. [**18**, Prop. 3.2] *Let $F \in X'$, $F \neq 0$. Then the pair (λ_1, ∞) is an asymptotic bifurcation point for* (2.2). *Moreover, there exists a maximal (in the ordering by set inclusion) closed set $\mathcal{C} \subset \mathbb{R} \times X$, such that $\widetilde{\mathcal{C}}$ is connected in $\mathbb{R} \times X$ and the following properties hold:*

(i) *there exists a sequence $\{(\mu_n, u_n)\}_{n=1}^{\infty} \subset \mathcal{C}$ such that $(\mu_n, \|u_n\|_X) \to (\lambda_1, \infty)$;*
(ii) *either $\mathcal{C}$ is unbounded in the λ-direction, or else there exists an eigenvalue λ_0 of $-\Delta_p$ such that $\lambda_0 > \lambda_1$ and there is a sequence $\{(\mu_n, u_n)\}_{m=1}^{\infty} \subset \mathcal{C}$ satisfying $(\mu_n, \|u_n\|_X) \to (\lambda_0, \infty)$.*

REMARK 2.7. The assumption $F \neq 0$ (which corresponds to $f \not\equiv 0$ in (1.18)) implies that (2.2) cannot have the trivial solution $u = 0$. Consequently, $\mathcal{C}$ contains no sequence of pairs (μ_k, u_k) with $(\mu_k, \|u_k\|_X) \to (\hat{\mu}, 0)$. Hence, the statement of Theorem 2.6 follows directly from [**13**, Theorem 14.18] using the transformation $u \mapsto v = u/\|u\|^2_X$.

3. Asymptotic estimate and refined bifurcation result

In this section we will establish an asymptotic estimate that plays the key role in the study of the structure of the solution set to (1.18). We assume $1 < p < \infty$, $p \neq 2$, throughout the entire section. From now on, we denote by λ_2 ($\lambda_2 > \lambda_1$) the second eigenvalue of the positive Dirichlet p-Laplacian $-\Delta_p$. We use only the well-known fact from [**3**] that there is no eigenvalue of $-\Delta_p$ in the open interval (λ_1, λ_2), by a variational characterization of λ_2.

All results which we present in this section were proved and reported in work [**18**]. Therefore, we omit proofs referring interested readers to [**18**].

3.1. Asymptotic estimate. For the behaviour of solutions u with large norm, the following a priori estimate plays the key role. First we introduce some notation. We introduce a new norm on $W_0^{1,p}(\Omega)$ by

$$\|v\|_{\mathcal{D}_{\varphi_1}} \overset{\text{def}}{=} \left(\int_\Omega |\nabla\varphi_1|^{p-2}|\nabla v|^2 \,\mathrm{d}x\right)^{1/2} \quad \text{for } v \in W_0^{1,p}(\Omega), \tag{3.1}$$

and denote by $\mathcal{D}_{\varphi_1}$ the completion of $W_0^{1,p}(\Omega)$ with respect to this norm. The Hilbert space $\mathcal{D}_{\varphi_1}$ is compactly imbedded in the Lebesgue space $L^2(\Omega)$; see [**35**, Lemma 4.2]. It is also shown there that the seminorm (3.1) is in fact a norm on $W_0^{1,p}(\Omega)$. For the case $1 < p < 2$ the space $\mathcal{D}_{\varphi_1}$ needs to be redefined. We do not need it for the formulation of any theorem here therefore its definition is omitted; though it plays key role in the proofs (see [**18, 34, 35**] for details).

For the sake of brevity, we also define

$$\mathcal{A}_{\varphi_1} \overset{\text{def}}{=} |\nabla\varphi_1|^{p-2}\left(\mathbf{I} + (p-2)\,\frac{\nabla\varphi_1 \otimes \nabla\varphi_1}{|\nabla\varphi_1|^2}\right) \quad \text{for } \nabla\varphi_1 \in \mathbb{R}^N,\ \nabla\varphi_1 \neq \mathbf{0} \in \mathbb{R}^N,$$

with $\mathbf{I}$ being the $n \times n$ identity matrix and $\otimes$ being the tensor product.

THEOREM 3.1. ([**18**, Thm. 4.1]) *Let $\{\mu_n\}_{n=1}^\infty \subset \mathbb{R}$, $\{f_n\}_{n=1}^\infty \subset L^\infty(\Omega)$, $\{u_n\}_{n=1}^\infty \subset W_0^{1,p}(\Omega)$ be sequences, and let $\delta > 0$ be such that*

(i) *$\lambda_1 + \mu_n < \lambda_2 - \delta$ for all $n \in \mathbb{N}$;*
(ii) *$f_n \overset{*}{\rightharpoonup} f$ weakly-star in $L^\infty(\Omega)$;*
(iii) *$\|u_n\|_{W_0^{1,p}(\Omega)} \to \infty$ as $n \to \infty$;*
(iv) *in addition, assume that for all $n \in \mathbb{N}$ and $\phi \in W_0^{1,p}(\Omega)$,*

$$\int_\Omega |\nabla u_n|^{p-2}\langle\nabla u_n, \nabla\phi\rangle \,\mathrm{d}x = (\lambda_1 + \mu_n)\int_\Omega |u_n|^{p-2}u_n\,\phi\,\mathrm{d}x + \int_\Omega f_n\,\phi\,\mathrm{d}x\,. \tag{3.2}$$

Then $\mu_n \to 0$ and, writing $u_n = t_n^{-1}(\varphi_1 + v_n^\top)$ with $t_n \in \mathbb{R}$, $t_n \neq 0$, and $v_n^\top \in W_0^{1,p}(\Omega)^\top$, we have $t_n \to 0$, $|t_n|^{-p}t_n\,v_n^\top \to V^\top$ strongly in $\mathcal{D}_{\varphi_1}$ if $p > 2$ and in $W_0^{1,2}(\Omega)$ if $1 < p < 2$, and

$$\begin{aligned} \mu_n = &- |t_n|^{p-2}t_n\int_\Omega f_n\,\varphi_1\,\mathrm{d}x + (p-2)|t_n|^{2(p-1)}\,\mathcal{Q}_0(V^\top, V^\top) \\ &+ (p-1)|t_n|^{2(p-1)}\left(\int_\Omega f\,\varphi_1\,\mathrm{d}x\right)\left(\int_\Omega \varphi_1^{p-1}\,V^\top\,\mathrm{d}x\right) + o\left(|t_n|^{2(p-1)}\right). \end{aligned} \tag{3.3}$$

In particular, if $\int_\Omega f_n\,\varphi_1\,\mathrm{d}x = 0$ for all $n \in \mathbb{N}$, then

$$\mu_n = (p-2)|t_n|^{2(p-1)}\,\mathcal{Q}_0(V^\top, V^\top) + o\left(|t_n|^{2(p-1)}\right).$$

Moreover, $V^\top \in \mathcal{D}_{\varphi_1} \cap \{\varphi_1\}^{\perp, L^2}$ *is the (unique) solution to*

$$2 \cdot \mathcal{Q}_0(V^\top, \phi) = \int_\Omega f^\dagger \, \phi \, \mathrm{d}x \quad \text{for all } \phi \in \mathcal{D}_{\varphi_1}, \tag{3.4}$$

where we have denoted

$$2 \cdot \mathcal{Q}_0(V^\top, \phi) = \int_\Omega \langle \mathbf{A}_{\varphi_1} \nabla V^\top, \nabla \phi \rangle \, \mathrm{d}x - \lambda_1 (p-1) \int_\Omega \varphi_1^{p-2} V^\top \phi \, \mathrm{d}x$$

and $f^\dagger = f - \left(\int_\Omega f \, \varphi_1 \, \mathrm{d}x \right) \varphi_1^{p-1}$.

REMARK 3.2. The linear equation (3.4) represents the weak form of the "limiting" Dirichlet boundary value problem for the limit function $|t_n|^{-p} t_n \, v_n^\top \to V^\top$ in the approximation scheme with $u_n = t_n^{-1}(\varphi_1 + v_n^\top)$. This is a resonant problem to which a standard version of the Fredholm alternative for a selfadjoint linear operator in a Hilbert space applies. More precisely, given a function $f \in L^2(\Omega)$, a weak solution $V \in \mathcal{D}_{\varphi_1}$ to the equation

$$2 \cdot \mathcal{Q}_0(V, \phi) = \textstyle\int_\Omega f \, \phi \, \mathrm{d}x \quad \text{for all } \phi \in \mathcal{D}_{\varphi_1}, \tag{3.5}$$

exists in $\mathcal{D}_{\varphi_1}$ if and only if $\int_\Omega f \, \varphi_1 \, \mathrm{d}x = 0$. Such a solution is always unique under the orthogonality condition $\int_\Omega V \, \varphi_1 \, \mathrm{d}x = 0$.

Note that (3.5) written in divergent form reads as follows (see e.g. [**34, 35, 18**])

$$\begin{aligned} \operatorname{div}(\mathbf{A}_{\varphi_1} \nabla V^\top) - \lambda_1 \varphi_1^{p-2} V^\top &= f \quad \text{in } \Omega; \\ V^\top &= 0 \quad \text{on } \partial\Omega; \\ \int_\Omega V^\top \varphi_1 &= 0. \end{aligned}$$

REMARK 3.3. In fact we also use a variant of Theorem 3.1 (see [**18**, Cor. 4.4] for details) in order to prove some results.
Let K be a closed bounded ball in $L^\infty(\Omega)$. Assume that $f_n \equiv f$ $(n = 1, 2, \dots)$ and $t_n \to 0$ as $n \to \infty$. Then there exists a sequence $\{\eta_n\}_{n=1}^\infty \subset (0,1)$, $\eta_n \to 0$ as $n \to \infty$, such that for all $f \in K$ and for all $n = 1, 2, \dots$ we have

$$\begin{aligned} \Big| |t_n|^{-2(p-1)} \left(\mu_n - |t_n|^{p-2} t_n \int_\Omega f \, \varphi_1 \, \mathrm{d}x \right) - (p-2) \cdot \mathcal{Q}_0(V^\top, V^\top) \\ - (p-1) \left(\int_\Omega f \, \varphi_1 \, \mathrm{d}x \right) \left(\int_\Omega \varphi_1^{p-1} \, V^\top \, \mathrm{d}x \right) \Big| \le \eta_n. \end{aligned} \tag{3.6}$$

The following nonexistence result is a consequence of Theorem 3.1. Note that this result is the key point for obtaining refined bifurcation result (an analogue of Dancer's result on bifurcation from simple eigenvalues).

THEOREM 3.4. ([**18**, Prop. 5.1]) *Let $f \in L^\infty(\Omega)$, $f \not\equiv 0$. Then there exists a constant $R > 0$ such that every weak solution $u \in W_0^{1,p}(\Omega)$ of the problem*

$$\begin{cases} -\Delta_p u - \lambda_1 |u|^{p-2} u = f(x) & \text{in } \Omega; \\ u = 0 & \text{on } \partial\Omega, \end{cases} \tag{3.7}$$

satisfies the a priori bound $\|u\|_{W_0^{1,p}(\Omega)} \le R$.

3.2. Refined bifurcation result. Using the *a priori* bound from previous theorem, we can apply [**13**, Thm. 14.20] to the problem transformed by $v_n = u_n/\|u_n\|_X^2$ and we obtain following refined bifurcation result. (Recall that X and X' together with $\mathcal{I}(u)$, $\mathcal{S}(u)$ and F have been specified in Section 2.)

THEOREM 3.5. ([**18**, Thm. 5.2]) *Let $F \in X'$, $F \neq 0$. Then there is a pair of maximal closed sets $\mathcal{C}^+, \mathcal{C}^- \subset \mathbb{R} \times X$ of solutions of (2.2) such that both sets $\widetilde{\mathcal{C}}^+$ and $\widetilde{\mathcal{C}}^-$ are connected in $\mathbb{R} \times X$, where $\widetilde{\mathcal{C}}^\pm$ are the closures in $\mathbb{R} \times X$ of the respective sets of all pairs $(\mu, v) \in \mathbb{R} \times X$ such that $v \neq 0$ and $(\mu, v/\|v\|_X^2) \in \mathcal{C}^\pm$, and the following properties hold:*

(a) *there exist sequences of pairs $(\mu_n, u_n) \in \mathcal{C}^+$ and $(\mu_n', u_n') \in \mathcal{C}^-$ such that $\mu_n \to \lambda_1$, $\mu_n' \to \lambda_1$, $\|u_n\|_X \to \infty$ and $\|u_n'\|_X \to \infty$, together with $u_n/\|u_n\|_X \to \varphi_1/\|\varphi_1\|_X$ and $u_n'/\|u_n'\|_X \to -\varphi_1/\|\varphi_1\|_X$ strongly in X;*

(b) *either both $\mathcal{C}^+$ and $\mathcal{C}^-$ are unbounded in the λ-direction, or else $\widetilde{\mathcal{C}}^+ \cap \widetilde{\mathcal{C}}^-$ contains a point other than $\{(\lambda_1, 0)\}$.*

REMARK 3.6. Let us point out that it follows from Theorem 3.5 combined with regularity results [**28**] that the convergence above $u_n/\|u_n\|_X \to \varphi_1/\|\varphi_1\|_X$, and $u_n'/\|u_n'\|_X \to -\varphi_1/\|\varphi_1\|_X$ as well, occurs strongly not only in $X = W_0^{1,p}(\Omega)$ but even in $C^{1,\beta'}(\overline{\Omega})$. Thus, $\|u_n\|_X \to \infty$ is equivalent to $\|u_n\|_{L^\infty(\Omega)} \to \infty$, and also to $\|u_n\|_{C^{1,\beta'}(\overline{\Omega})} \to \infty$, for $(\lambda_n, u_n) \in \mathcal{C}^+ \cup \mathcal{C}^-$.

REMARK 3.7. Our aim is to study the local behaviour of the bifurcation branches $\mathcal{C}^\pm$ near λ_1. For this reason we restrict our attention to $\lambda \in (-\infty, \Lambda)$ with some $\Lambda \in (\lambda_1, \lambda_2)$. Recall that λ_2 is the second eigenvalue of the negative Dirichlet p-Laplacian $-\Delta_p$ (cf. Theorem 3.1).

3.3. Case $\int_\Omega f\varphi_1 \, dx \neq 0$. We will now establish a priori bounds that allow us to detect whether λ belongs to the left or the right neighborhood of λ_1 provided $(\lambda, u) \in \mathcal{C}^\pm$ and the norm $\|u\|_{L^\infty(\Omega)}$ is large enough.

THEOREM 3.8. (cf. [**5**]) *Let $\Lambda \in (\lambda_1, \lambda_2)$. For every $f \in L^\infty(\Omega)$ with $\int_\Omega f\varphi_1 \, dx \neq 0$, there exists a constant $M > 0$ such that the following statements and implications hold.*

(i) *If $\int_\Omega f\varphi_1 \, dx < 0$ then every solution (λ, u) to (1.18) satisfies*
  (a) *$u < 0$ in Ω and $\lambda_1 \le \lambda \le \Lambda \implies \|u\|_{L^\infty(\Omega)} \le M$;*
  (b) *$u > 0$ in Ω and $\lambda \le \lambda_1 \implies \|u\|_{L^\infty(\Omega)} \le M$.*

(ii) *If $\int_\Omega f\varphi_1 \, dx > 0$ then every solution (λ, u) to (1.18) satisfies*
  (a) *$u < 0$ in Ω and $\lambda \le \lambda_1 \implies \|u\|_{L^\infty(\Omega)} \le M$;*
  (b) *$u > 0$ in Ω and $\lambda_1 \le \lambda \le \Lambda \implies \|u\|_{L^\infty(\Omega)} \le M$.*

3.4. Case $\int_\Omega f\varphi_1 \, dx = 0$. In this subsection we distinguish between the cases $1 < p < 2$ and $p > 2$.

THEOREM 3.9. *Let $\Lambda \in (\lambda_1, \lambda_2)$. For every $f^\top \in L^\infty(\Omega)$, $\int_\Omega f^\top \varphi_1 \, dx = 0$, $f^\top \not\equiv 0$, there exists a constant $M > 0$ such that every solution (λ, u) to (1.18) satisfies:*

(i) *if $1 < p < 2$, then $\lambda_1 \le \lambda \le \Lambda$ implies $\|u\|_{L^\infty(\Omega)} \le M$;*
(ii) *if $p > 2$, then $\lambda \le \lambda_1$ implies $\|u\|_{L^\infty(\Omega)} \le M$.*

A sketch of the bifurcation diagram which follows from a priori bounds in Theorem 3.9 is depicted in Figure 1.

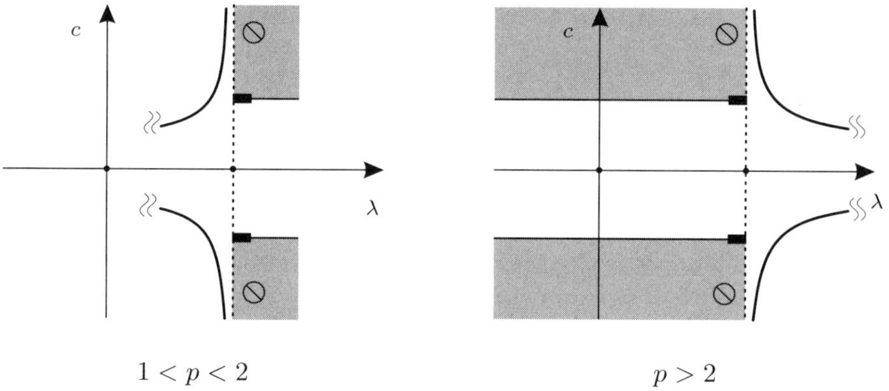

FIGURE 1. A priori bounds and bifurcations from infinity of solutions to (1.18) for $p > 1$, $p \neq 2$ and $a = 0$. There is no solution in the shaded regions (owing to a priori bounds).

3.5. Case $\int_\Omega f\varphi_1\,\mathrm{d}x \neq 0$ revisited. In this subsection we consider perturbations f of functions $f^\top \in L^\infty(\Omega)$, $\int_\Omega f^\top\varphi_1\,\mathrm{d}x = 0$, of the form $f = f^\top + a\varphi_1$ with $a \in \mathbb{R}$ small enough. Therefore, we need to distinguish between the cases $1 < p < 2$ and $p > 2$ again.

THEOREM 3.10. [**18**, Thm. 5.9] *Let $f^\top \in L^\infty(\Omega)$ be fixed, $f^\top \not\equiv 0$ and $\int_\Omega f^\top\varphi_1\,\mathrm{d}x = 0$. Given $a \in \mathbb{R}$, let $\mathcal{C}_a^\pm$ denote continua of solutions $(\lambda, u) \in \mathbb{R} \times X$* (in the sense of Theorem 3.5) *to*

$$\left\{ \begin{array}{rcll} -\Delta_p u - \lambda|u|^{p-2}u &=& f^\top + a\varphi_1 & \text{in } \Omega\,; \\ u &=& 0 & \text{on } \partial\Omega\,. \end{array} \right. \tag{3.8}$$

Let every solution u of (3.8) *be written in the form $u = t^{-1}(\varphi_1 + v^\top)$, $t \in \mathbb{R}$, $v^\top \in C^{1,\beta'}(\overline{\Omega})$. Then there exist $\varepsilon > 0$, $\underline{a} = \underline{a}(f^\top, p)$ and $\overline{a} = \overline{a}(f^\top, p)$, such that $\underline{a} < 0 < \overline{a}$ and for $a \in (\underline{a}, 0) \cup (0, \overline{a})$ we have $u > 0$ in Ω if $t \in (0, \varepsilon)$ and $u < 0$ if $t \in (-\varepsilon, 0)$. Furthermore,*

- (i) *if $1 < p < 2$ then*
  - (a) *for any $a \in (\underline{a}, 0)$ there exists $t_0 \in (0, \varepsilon)$ such that $u = t_0^{-1}(\varphi_1 + v^\top)$ and $(\lambda, u) \in \mathcal{C}_a^+$ imply $\lambda < \lambda_1$;*
  - (b) *for any $a \in (0, \overline{a})$ there exists $t_0 \in (-\varepsilon, 0)$ such that $u = t_0^{-1}(\varphi_1 + v^\top)$ and $(\lambda, u) \in \mathcal{C}_a^-$ imply $\lambda < \lambda_1$;*
- (ii) *if $p > 2$ then*
  - (a) *for any $a \in (\underline{a}, 0)$ there exists $t_0 \in (-\varepsilon, 0)$ such that $u = t_0^{-1}(\varphi_1 + v^\top)$ and $(\lambda, u) \in \mathcal{C}_a^-$ imply $\lambda > \lambda_1$;*
  - (b) *for any $a \in (0, \overline{a})$ there exists $t_0 \in (0, \varepsilon)$ such that $u = t_0^{-1}(\varphi_1 + v^\top)$ and $(\lambda, u) \in \mathcal{C}_a^+$ imply $\lambda > \lambda_1$.*

THEOREM 3.11. *Let $f^\top \in L^\infty(\Omega)$, $\mathcal{C}_a^\pm$, ε, $\underline{a}$, $\overline{a}$ and a be as in* Theorem 3.10 *above. Then*

- (i) *there exists $t_a \in (0, \varepsilon)$ and $u = t_a^{-1}(\varphi_1 + v^\top) > 0$ in Ω such that $(\lambda_1, u) \in \mathcal{C}_a^+$;*
- (ii) *there exists $t_a \in (-\varepsilon, 0)$ and $u = t_a^{-1}(\varphi_1 + v^\top) < 0$ in Ω such that $(\lambda_1, u) \in \mathcal{C}_a^-$.*

For a given $a \in (\underline{a}, 0) \cup (0, \overline{a})$, let us denote $\widehat{t_a} = \inf |t_a|$ where the infimum is taken over all t_a $(0 < |t_a| < \varepsilon)$ associated to a via Theorem 3.11.

THEOREM 3.12. *We have* $\lim_{a \to 0} \widehat{t_a} = 0$.

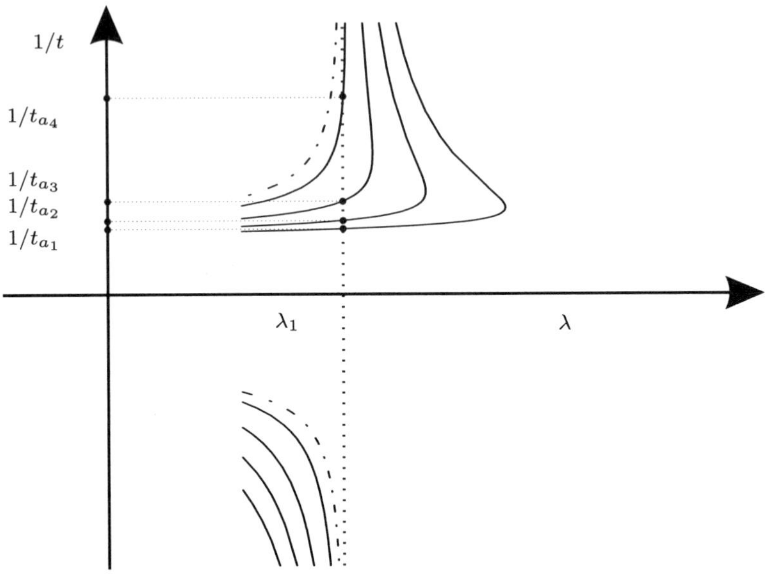

FIGURE 2. Illustration of Theorem 3.12 for $1 < p < 2$. The dash-dotted curves represent branches $\mathcal{C}^{\pm}$ for $a = 0$. One can see how the transcritical bifurcation for $a \neq 0$ transforms to a subcritical one for $a = 0$.

The statement of Theorem 3.12 is illustrated in Figure 2 (with $a_n \nearrow 0$). Furthermore, in Figure 3, we illustrate how the graphs depend on the value of a in case $1 < p < 2$.

4. Multiplicity results for $\lambda < \lambda_2$

In this section we will sumarize known results about the existence and multiplicity of solutions to (1.18). Many of these facts are obtained as applications and consequences of the asymptotic formulas derived in [**18**] and listed in the previous section.

In order to formulate our existence and multiplicity results, we rewrite problem (1.18) to the form (1.18') where $f^{\top} \in L^{\infty}(\Omega)$ is a given function, with $\int_{\Omega} f^{\top} \varphi_1 \, dx = 0$ and $f^{\top} \not\equiv 0$, and $\lambda, a \in \mathbb{R}$ are real parameters.

For the convenience of the reader, we organize these results in following three tables. Dependence of the existence, multiplicity and *a priori* bounds of the solutions on the spectral parameter λ can easily be deduced from these tables.

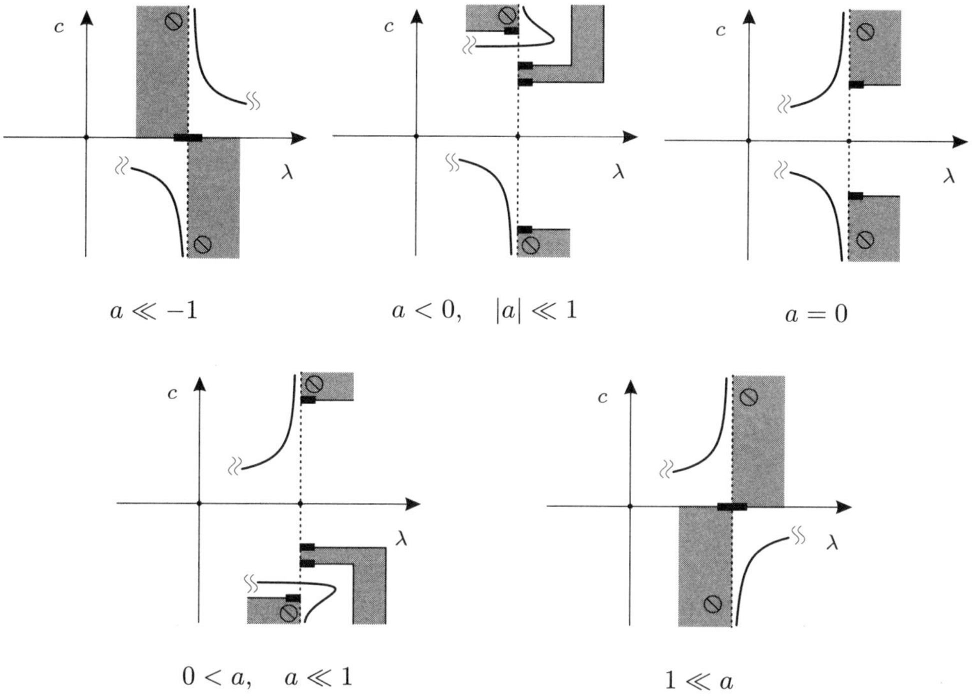

FIGURE 3. Dependence of a priori bounds and bifurcations from infinity of solutions to (1.18) on $a = \int_\Omega f\varphi_1 \, \mathrm{d}x$ for $1 < p < 2$. There is no solution in shaded regions.

$1 < p < 2,\ f \not\equiv 0$

$\lambda < 0$	$0 \le \lambda \le \lambda_1$	λ_1	$\lambda_1 < \lambda < \lambda_2 - \delta$
$\Vert u\Vert_{C^{1,\beta}(\overline{\Omega})} \le M(f^\top, a, p, \lambda, \Omega) < \infty$			
		$a = 0 \Rightarrow M(f^\top, a, p, \lambda, \Omega) < M < \infty \quad \forall\lambda$	
	$M(f^\top, a, p, \lambda, \Omega) \to \infty$ as $\lambda \to \lambda_{1-}$	$M(f^\top, a, p, \lambda_1, \Omega) < \infty \quad \forall a \in \mathbb{R}$	$a \ne 0 \Rightarrow M(f^\top, a, p, \lambda, \Omega) \to \infty$ as $\lambda \to \lambda_{1+}$
$S = 1$	$S \ge 1$		$S \ge 1$
	$f \prec 0$: $S, N = 1, P = 0$ *(LMP)* $a < \underline{A}$: *(LMP)* $\exists\eta > 0\,\forall\lambda \in (\lambda_1 - \eta, \lambda_1)$: $N \ge 1, S - N = 0$	$f \prec 0$ or $a < \underline{A}$: $S = 0$	$f \prec 0$ or $a < \underline{A}$: *(LAMP)* $\exists\eta > 0\,\forall\lambda \in (\lambda_1, \lambda_1 + \eta)$: $P \ge 1, S - P = 0$
	$\underline{A} < a < \underline{a}$: $N \ge 1$ PAB	$\underline{A} < a < \underline{a}$: $S \ge 2$ *(UpLow)*	$\underline{A} < a < \underline{a}$: $P \ge 1$, NAB
	$\forall\varepsilon' \in (0, -\underline{a})\exists\eta = \eta(f^\top, \underline{a}, \varepsilon')\ \forall\lambda$: $\lambda_1 - \eta < \lambda < \lambda_1$ $\underline{a} < a < -\varepsilon'$: $S \ge 3, P, N \ge 1$	$\underline{a} < a < 0$: $S \ge 2, P \ge 1$ $M(f^\top, a, p, \Omega) \to \infty$ as $a \to 0_-$	$\forall\varepsilon' \in (0, -\underline{a})\exists\eta = \eta(f^\top, \underline{a}, \varepsilon')\ \forall\lambda$: $\lambda_1 < \lambda < \lambda_1 + \eta$ $\underline{a} < a < -\varepsilon'$: $S \ge 2, P \ge 2$
	$\lambda_1 - \eta < \lambda < \lambda_1$ $a = 0$: $S \ge 3, P, N \ge 1$	$a = 0$: $S \ge 1$ $\Vert u\Vert_{C^{1,\beta}(\overline{\Omega})} \le M(f^\top, 0, p, \Omega) < \infty$	$\lambda_1 < \lambda < \lambda_1 + \eta$ $a = 0$: $S \ge 1$
	$\forall\varepsilon' \in (0, \overline{a})\exists\eta = \eta(f^\top, \overline{a}, \varepsilon')\ \forall\lambda$: $\lambda_1 - \eta < \lambda < \lambda_1$ $\varepsilon' < a < \overline{a}$: $S \ge 3, P, N \ge 1$	$0 < a < \overline{a}$: $S \ge 2, N \ge 1$ $M(f^\top, a, p, \Omega) \to \infty$ as $a \to 0_+$	$\forall\varepsilon' \in (0, \overline{a})\exists\eta = \eta(f^\top, \overline{a}, \varepsilon')\ \forall\lambda$: $\lambda_1 < \lambda < \lambda_1 + \eta$ $\varepsilon' < a < \overline{a}$: $S \ge 2, N \ge 2$
	$\overline{a} < a < \overline{A}$: $P \ge 1$, NAB	$\overline{a} < a < \overline{A}$: $S \ge 2$ *(UpLow)*	$\overline{a} < a < \overline{A}$: $N \ge 1$, PAB
	$\overline{A} < a$: *(LMP)* $\exists\eta > 0\,\forall\lambda \in (\lambda_1 - \eta, \lambda_1)$: $P \ge 1, S - P = 0$ $f \succ 0$: $S, P = 1, N = 0$ *(LMP)*	$\overline{A} < a$ or $f \succ 0$: $S = 0$	$\overline{A} < a$ or $f \prec 0$: *(LAMP)* $\exists\eta > 0\,\forall\lambda \in (\lambda_1, \lambda_1 + \eta)$: $N \ge 1, S - N = 0$

Legend: S (N, P) number of (negative, positive) solution; $S \ge n$ at least n solutions; $N \ge 1$ at least one negative solution; $S - N = 0$ all solutions are negative; NAB (PAB) negative (positive) solutions are *a priori* bounded; *(LMP)* Local Maximum Principle, *(LAMP)* Local Anti-Maximum Principle, *(UpLow)* by upper and lower solutions argument.

$p = 2$

$\lambda \le 0$	$0 < \lambda \le \lambda_1$	λ_1	$\lambda_1 < \lambda < \lambda_2 - \delta$
	$\|u\|_{C^{1,\beta(\overline{\Omega})}} \le M(f, p, \lambda, \Omega) < \infty$		$\|u\|_{C^{1,\beta(\overline{\Omega})}} \le M(f, p, \lambda, \Omega) < \infty$
	$M(f, p, \lambda, \Omega) \to \infty$ as $\lambda \to \lambda_{1-}$	$a = 0$ no bound exists $a \neq 0$ no solution exists	$M(f^\top, a, p, \lambda, \Omega) \to \infty$ as $\lambda \to \lambda_{1+}$
$\mathsf{S} = 1$	$\mathsf{S} = 1$		$\mathsf{S} = 1$
	$a < 0$ $\mathsf{N} = 1$ for $\lambda \to \lambda_{1-}$	$a < 0$ $\mathsf{S} = 0$	$a < 0$ $\mathsf{P} = 1$ for $\lambda \to \lambda_{1+}$
	$a = 0$ $\mathsf{S} = 1$, $\mathsf{N}, \mathsf{P} = 0$	$a = 0$ $\mathsf{N}, \mathsf{P} = \infty$, $\mathsf{S} - \mathsf{N} - \mathsf{P} = \infty$ $\exists u_0 \forall c \in \mathbb{R} : u = u_0 + c\varphi_1$ is a solution	$a = 0$ $\mathsf{S} = 1$, $\mathsf{P}, \mathsf{N} = 0$
	$0 < a$ $\mathsf{P} = 1$ for $\lambda \to \lambda_{1-}$	$0 < a$ $\mathsf{S} = 0$	$0 < a$ $\mathsf{P} = 1$ for $\lambda \to \lambda_{1-}$

$p > 2,\ f \not\equiv 0$

$\lambda < 0$	$0 \le \lambda \le \lambda_1$	λ_1	$\lambda_1 < \lambda < \lambda_2 - \delta$
	$\|u\|_{C^{1,\beta(\overline{\Omega})}} \le M(f^\top, a, p, \lambda, \Omega) < \infty$		
	$a = 0 \Rightarrow M(f^\top, a, p, \lambda, \Omega) < M < \infty \quad \forall \lambda$		$M(f^\top, a, p, \lambda, \Omega) \to \infty$ as $\lambda \to \lambda_{1+}$
	$a \neq 0 \Rightarrow M(f^\top, a, p, \lambda, \Omega) \to \infty$ as $\lambda \to \lambda_{1-}$	$M(f^\top, a, p, \lambda_1, \Omega) < \infty \quad \forall a \in \mathbb{R}$	
$\mathsf{S} = 1$	$\mathsf{S} \ge 1$		$\mathsf{S} \ge 1$
	$f \prec 0$ $\mathsf{S}, \mathsf{N} = 1, \mathsf{P} = 0$ *(LMP)* $a < \underline{A}$ *(LMP)* $\exists \eta > 0 \forall \lambda \in (\lambda_1 - \eta, \lambda_1)$: $\mathsf{N} \ge 1, \mathsf{S} - \mathsf{N} = 0$	$f \prec 0$ or $a < \underline{A}$ $\mathsf{S} = 0$	$f \prec 0$ or $a < \underline{A}$ *(LAMP)* $\exists \eta > 0 \forall \lambda \in (\lambda_1, \lambda_1 + \eta)$: $\mathsf{P} \ge 1, \mathsf{S} - \mathsf{P} = 0$
	$\underline{A} < a < \underline{a}$ $\mathsf{N} \ge 1$ PAB	$\underline{A} < a < \underline{a}$ $\mathsf{S} \ge 2$ *(UpLow)*	$\underline{A} < a < \underline{a}$ $\mathsf{P} \ge 1$, NAB
	$\forall \varepsilon' \in (0, -\underline{a}) \exists \eta = \eta(f^\top, \underline{a}, \varepsilon') \quad \forall \lambda :$ $\lambda_1 - \eta < \lambda < \lambda_1$ $\underline{a} < a < -\varepsilon'$ $\mathsf{S} \ge 2, \mathsf{N} \ge 2$	$\underline{a} < a < 0$ $\mathsf{S} \ge 2, \mathsf{P} \ge 1$ $M(f^\top, a, p, \Omega) \to \infty$ as $a \to 0_-$	$\forall \varepsilon' \in (0, -\underline{a}) \exists \eta = \eta(f^\top, \underline{a}, \varepsilon') \quad \forall \lambda :$ $\lambda_1 < \lambda < \lambda_1 + \eta$ $\mathsf{S} \ge 3, \mathsf{P}, \mathsf{N} \ge 1$
	$\lambda_1 - \eta < \lambda < \lambda_1$ $a = 0$ $\mathsf{S} \ge 1$	$a = 0$ $\mathsf{S} \ge 1$ $\|u\|_{C^{1,\beta(\overline{\Omega})}} \le M(f^\top, 0, p, \Omega) < \infty$	$\lambda_1 < \lambda < \lambda_1 + \eta$ $a = 0$ $\mathsf{S} \ge 3, \mathsf{P}, \mathsf{N} \ge 1$
	$\forall \varepsilon' \in (0, \overline{a}) \exists \eta = \eta(f^\top, \overline{a}, \varepsilon') \quad \forall \lambda :$ $\lambda_1 - \eta < \lambda < \lambda_1$ $\varepsilon' < a < \overline{a}$ $\mathsf{S} \ge 2, \mathsf{P} \ge 2$	$0 < a < \overline{a}$ $\mathsf{S} \ge 2, \mathsf{N} \ge 1$ $M(f^\top, a, p, \Omega) \to \infty$ as $a \to 0_+$	$\forall \varepsilon' \in (0, \overline{a}) \exists \eta = \eta(f^\top, \overline{a}, \varepsilon') \quad \forall \lambda :$ $\lambda_1 < \lambda < \lambda_1 + \eta$ $\varepsilon' < a < \overline{a}$ $\mathsf{S} \ge 3, \mathsf{P}, \mathsf{N} \ge 1$
	$\overline{a} < a < \overline{A}$ $\mathsf{P} \ge 1$, NAB	$\overline{a} < a < \overline{A}$ $\mathsf{S} \ge 2$ *(UpLow)*	$\overline{a} < a < \overline{A}$ $\mathsf{N} \ge 1$, PAB
	$\overline{A} < a$ *(LMP)* $\exists \eta > 0 \forall \lambda \in (\lambda_1 - \eta, \lambda_1)$: $\mathsf{P} \ge 1, \mathsf{S} - \mathsf{P} = 0$ $f \succ 0$ $\mathsf{S}, \mathsf{P} = 1, \mathsf{N} = 0$ *(LMP)*	$\overline{A} < a$ or $f \succ 0$ $\mathsf{S} = 0$	$\overline{A} < a$ or $f \prec 0$ *(LAMP)* $\exists \eta > 0 \forall \lambda \in (\lambda_1, \lambda_1 + \eta)$: $\mathsf{N} \ge 1, \mathsf{S} - \mathsf{N} = 0$

Legend: S (N, P) number of (negative, positive) solution; $\mathsf{S} \ge n$ at least n solutions; $\mathsf{N} \ge 1$ at least one negative solution; $\mathsf{S} - \mathsf{N} = 0$ all solutions are negative; NAB (PAB) negative (positive) solutions are *a priori* bounded;
(LMP) Local Maximum Principle, *(LAMP)* Local Anti-Maximum Principle, *(UpLow)* by upper and lower solutions argument.

5. Asymptotic bifurcation for one-dimesional p-Laplacian

Let us consider the following boundary value problem

$$(5.1)\qquad \begin{aligned} -\left(|u'|^{p-2}u'\right)' - \lambda|u|^{p-2}u = f \qquad & \text{in } (0,\pi_p)\,; \\ u(0) = u(\pi_p) = 0\,, \qquad & \end{aligned}$$

where $p > 1$, $\pi_p = 2\int_0^1 \frac{ds}{(1-s^p)^{1/p}} = \frac{2\pi}{p\sin\frac{\pi}{p}}$ ($\pi_2 = \pi = 3.1415926\ldots$), λ is a real number (bifurcation parameter) and $f \in L^\infty(0,\pi_p)$. In particular, we are interested in values $\lambda < \lambda_2$ where λ_2 is the second eigenvalue of the problem

$$(5.2)\qquad \begin{aligned} -\left(|u'|^{p-2}u'\right)' - \lambda|u|^{p-2}u = 0 \qquad & \text{in } (0,\pi_p)\,; \\ u(0) = u(\pi_p) = 0\,. \qquad & \end{aligned}$$

Note that the set of all eigenvalues of (5.2) is described by

$$(5.3)\qquad \lambda_k = (p-1)k^p \qquad k \in \mathbb{N},$$

each eigenvalue being simple with corresponding eigenfunction $\varphi_k(x) = \sin_p(kx)$. In GARCÍA and SABINA DE LIS [**25**] bifurcation from zero at radial eigenfunctions of

$$\begin{aligned} -\Delta_p u - \lambda|u|^{p-2}u &= g(\lambda,u,x) \quad \text{in } B_r(0); \\ u &= 0 \quad \text{on } \partial B_r(0) \end{aligned}$$

was studied, where $B_r(0)$ denotes a ball with radius r and centred at the origin. By using of the Implicit Function Theorem, a quasilinear variant of Crandall-Rabinowitz Local Bifurcation Theorem (see CRANDALL and RABINOWITZ [**10**]) was proved in [**25**]. In particular, they proved that a curve of radially symmetric solutions emanates at each eigenvalue with radially symmetric eigenfunction.

If we limit ourselves to one-dimensional case, any ball is an interval and *vice versa*. Therefore all solutions to (5.1) are radially symmetric in one-dimensional case. By using the standard trick from Subsection 2.3, we can transform bifurcation from (λ_k,∞) to bifurcation from $(\lambda_k,0)$. In particular, we obtain that the sets $\mathcal{C}^\pm$ become *curves* for λ close to λ_1 and $\|u\|_{W_0^{1,p}}$ large enough. More precisely combining our asymtotic estimate (3.3) and the results from [**25**] we get the following assertion:

There exists $\varepsilon > 0$ and $\delta > 0$ small enough and continuous mappings μ : $(-\varepsilon,\varepsilon) \to \mathbb{R}$, $\mathcal{V}^\top : (-\varepsilon,\varepsilon) \to C_0^1(0,\pi_p) \cap W^{1,p}(0,\pi_p)^\top$ such that

$$\mathcal{C}^+ \cap \{(\lambda,u) \in \mathbb{R}\times W_0^{1,p}(0,\pi_p) : \lambda_1 - \delta < \lambda < \lambda_1 + \delta, \quad \|u\|_{W_0^{1,p}(0,\pi_p)} > 2/\varepsilon\} =$$

$$\left\{(\lambda(t),u(t)) : \lambda = \lambda_1 + \mu(t)\,, \quad u(t) = t^{-1}(\varphi_1 + \mathcal{V}^\top(t))\,, \quad t \in (0,\varepsilon/2)\right\}\,,$$

$$\mathcal{C}^- \cap \{(\lambda,u) \in \mathbb{R}\times W_0^{1,p}(0,\pi_p) : \lambda_1 - \delta < \lambda < \lambda_1 + \delta, \quad \|u\|_{W_0^{1,p}(0,\pi_p)} > 2/\varepsilon\} =$$

$$\left\{(\lambda(t),u(t)) : \lambda = \lambda_1 + \mu(t)\,, \quad u(t) = t^{-1}(\varphi_1 + \mathcal{V}^\top(t))\,, \quad t \in (-\varepsilon/2,0)\right\}\,,$$

where $\lim_{t\to 0}\mathcal{V}^\top(t) = 0$ *in* $C^{1,\beta}([0,\pi_p])$ *for some* $\beta \in (0,1)$ *by regularity,* $\lim_{t\to 0}|t|^{-p}t\,\mathcal{V}^\top(t) = V^\top$ *strongly in* $\mathcal{D}_{\varphi_1}$ *for* $p > 2$ *and strongly in* $W_0^{1,2}(0,\pi_p)$ *if*

$1 < p < 2$, and

$$\begin{aligned}\mu(t) = &- |t|^{p-2}t \int_0^{\pi_p} f\,\varphi_1\,\mathrm{d}x + (p-2)|t|^{2(p-1)}\,\mathcal{Q}_0(V^\top, V^\top)\\ &+ (p-1)|t|^{2(p-1)}\left(\int_0^{\pi_p} f\,\varphi_1\,\mathrm{d}x\right)\left(\int_0^{\pi_p} \varphi_1^{p-1}\,V^\top\,\mathrm{d}x\right) + o\left(|t|^{2(p-1)}\right).\end{aligned} \tag{5.4}$$

(for the definition of $\mathcal{Q}_0$ see Theorem 3.1). The limit function $V^\top$ satisfies the boundary value problem (3.4) the divergent form of which for one-dimensional case reads as follows

$$\begin{aligned}\frac{\mathrm{d}}{\mathrm{d}x}\left(\left(\frac{\mathrm{d}\varphi_1}{\mathrm{d}x}\right)^{p-2}\frac{\mathrm{d}V^\top}{\mathrm{d}x}\right) - \lambda_1\varphi_1^{p-2}V^\top &= f - \int_0^{\pi_p}(f\varphi_1)\varphi_1^{p-1} \quad \text{in } (0,\pi_p)\,;\\ V^\top(0) = 0 \quad &; \quad V^\top(\pi_p) = 0;\\ \int_0^{\pi_p} V^\top\varphi_1 &= 0\,.\end{aligned}$$

We present some numerical results concerning the last boundary value problem in Section 6.

6. Numerical treatment of the one-dimesional p-Laplacian

In this section we will introduce three different numerical experiments applied to one-dimensional p–Laplace operator, where a "top view" geometrical projection is used. We focus on typical properties of the p–trigonometric functions, its scalar products and suggesting similarities with statements of the linear Fredholm alternative.

6.1. Scalar products of p–trigonometric functions.

Let $Q = \langle 1, \infty) \times \langle 1, \infty) \subset \mathbb{R}^2$. For fixed $k, l \in \mathbb{N}$, define scalar product $I_{k,l} : Q \to \mathbb{R}$ as a function of p, q:

$$I_{k,l}(p,q) = \int_0^1 \sin_p(k\pi_p t)\cdot\sin_q(l\pi_q t)\,dt, \qquad k,l \in \mathbb{N},\ (p,q) \in Q. \tag{6.1}$$

From (6.1) and from basic properties of $\sin_p(\cdot)$ it follows that

$$\begin{aligned} I_{k,l}(p,q) &= I_{l,k}(q,p), && k,l \in \mathbb{N}, && (p,q) \in Q,\\ I_{k,l}(p,q) &= I_{nk,nl}(p,q), && k,l,n \in \mathbb{N}, && (p,q) \in Q,\\ I_{k,k}(p,q) &> 0, && k \in \mathbb{N}, && (p,q) \in Q,\\ I_{k,l}(p,q) &= 0, && k+l \text{ is odd}, && (p,q) \in Q. \end{aligned} \tag{6.2}$$

Any function $I_{k,l}$ can be depicted in 3D space or in 2D "top view". The "top view" applied to sign $(I_{k,l})$ shows us the zero points of sign $(I_{k,l})$. Figure 4 illustrates the case $k = 3$ and $l = 5$.

The values of $I_{k,l}(p,q)$ for k, $l \in \mathbb{N}$ and $(p,q) \in Q$ can be represented in the space $\mathbb{N}^2 \times Q$ in term of a matrix $\mathbf{A} = [\text{sign}\,(I_{k,l})]$. In this example we represent this space by 10^8 scalar products of the type (6.1).

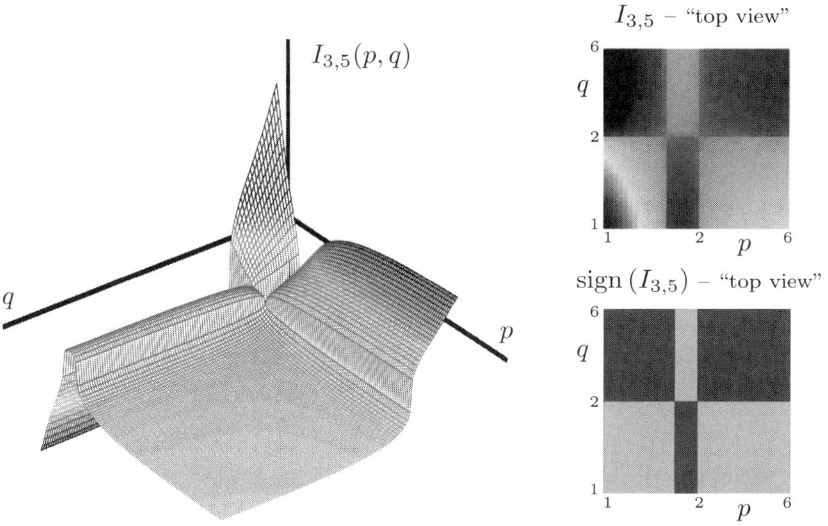
$I_{3,5}(p,q)$
q
p
$I_{3,5}$ – "top view"
q
p
1
2
6
1
2
6
sign $(I_{3,5})$ – "top view"
q
p
1
2
6
1
2
6

FIGURE 4

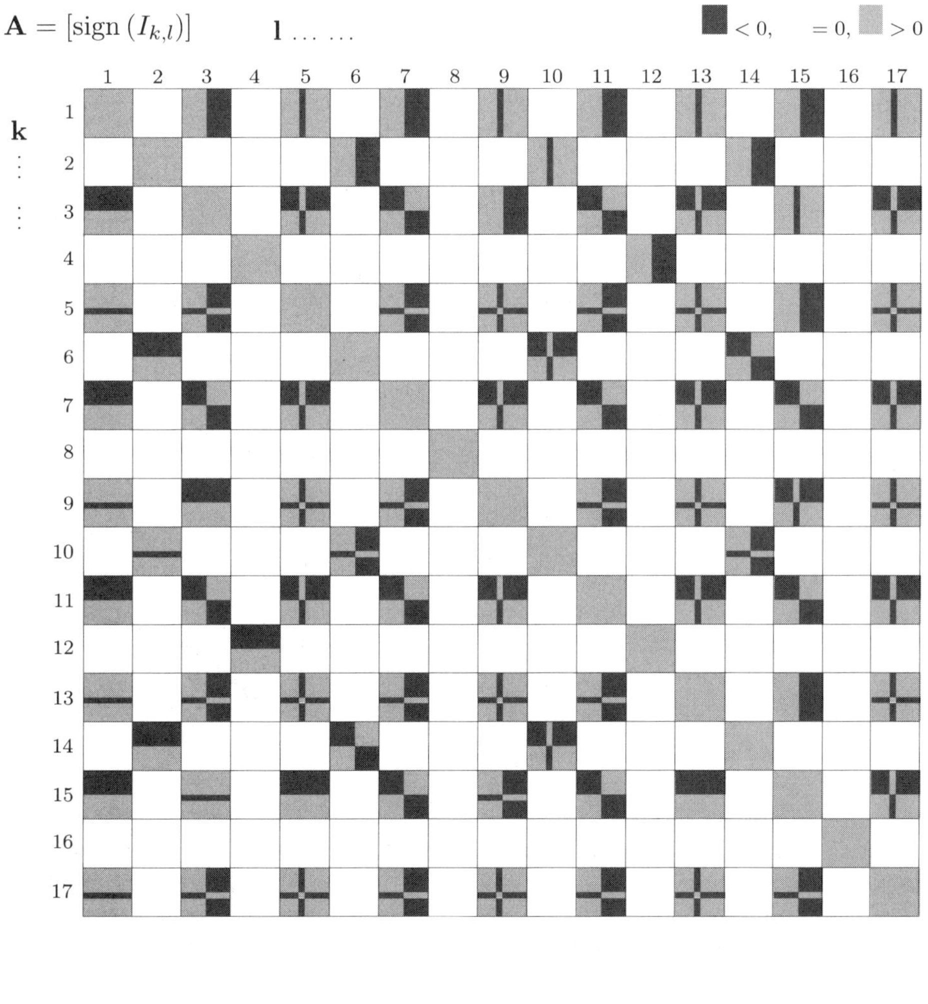
$\mathbf{A} = [\text{sign}\,(I_{k,l})]$
l … …
k
< 0, = 0, > 0
1 2 3 4 5 6 7 8 9 10 11 12 13 14 15 16 17
1 2 3 4 5 6 7 8 9 10 11 12 13 14 15 16 17

We can observe that there is only seven different types of sign $(I_{k,l})$ "top views". We specify them as follows:

T_0	T_1	T_2	T_3	T_4	T_5	T_6

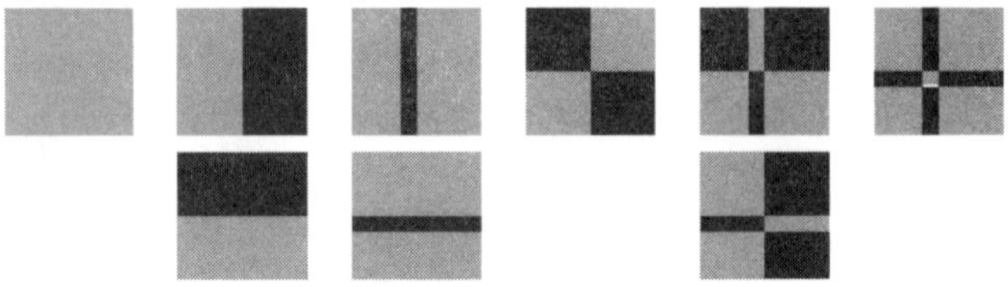

Some properties of matrix $\mathbf{A}$ follow directly from the basic properties of scalar product and from symmetry of $\sin_p(\cdot)$ (see (6.2)):

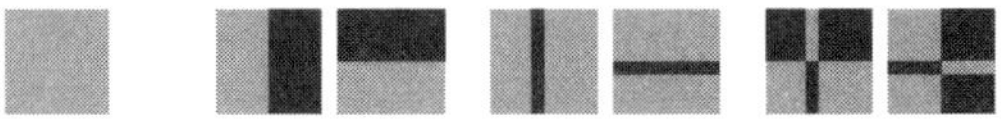

Besides the basic properties listed in (6.2) the scalar products $I_{k,l}(p,q)$ have more involved properties which we present in the following two assertions:

LEMMA 6.1. *Let $k,\ l \in \mathbb{N}$ and $l \neq (2s-1)k$ for all $s \in \mathbb{N}$. Then*

$$\int_0^1 \sin_p(k\pi_p t) \cdot \sin(l\pi t)\, dt = 0, \qquad \forall p \geq 1.$$

LEMMA 6.2. *Let $n \in \{5, 9, 11, \dots\}$. Then there exist $p(n) \in (1,2)$ such that*

$$\int_0^1 \sin_p(\pi_p t) \cdot \sin(n\pi t)\, dt = 0.$$

Nodal lines in T_2 and T_4 correspond to the assertion of Lemma 6.1. And finally the nodal lines observed in types T_3, T_5 and T_6 correspond both to assertions of Lemma 6.1 and Lemma 6.2.

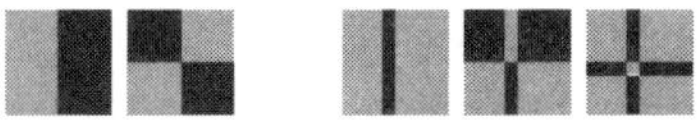

The Figure 5 illustrates the restrictions of $I_{k,l}(p,q)$ to the level $q = 2$ for some special choices of k and l.

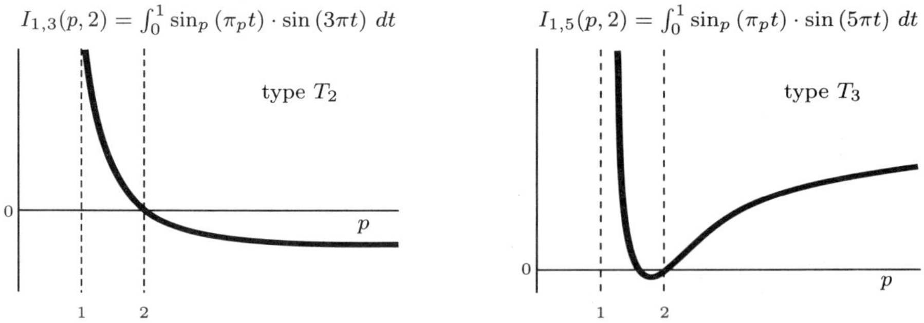

FIGURE 5

Prof. P.A. BINDING from University of Calgary has formulated conjecture about the completeness of sequence $\{\sin_p(nt)\}_{n=1}^{\infty}$ in $L^p(0,\pi_p)$. Unfortunately, we are still far from its proof. On the other hand, we hope that some results of our numerical experiments, can be useful and can help to find the proof in future.

6.2. Linearization.
Let us consider the linearized boundary value problem:

$$\begin{aligned} &\left(|\varphi_n'(t)|^{p-2}V'\right)' + \bar{\lambda}|\varphi_n(t)|^{p-2}V = 0, \quad t \in (0,\pi_p)\,; \\ &V(0) = 0, \quad V(\pi_p) = 0. \end{aligned} \tag{6.3}$$

for fixed $\varphi_n(t) = \sin_p nt$, $n \in \mathbb{N}$.

We would like to compare the eigenvalues λ_k of nonlinear p-Laplace equation with eigenvalues $\bar{\lambda}_{k,n}$ of linearized equation (6.3).

We apply a modified shooting method to solve boundary value problem (6.3). Let us define a function $F_n : (1,+\infty) \times (0,+\infty) \to \mathbb{R}$ as follows:

$$F_n(p,\bar{\lambda}) = \int_0^{\pi_p} \left(|\varphi_n'(t)|^{2-p} \cdot \left(1 - \bar{\lambda} \int_0^t |\varphi_n(\tau)|^{p-2}\, V_{0,1}(\tau)\, d\tau \right)\right) dt$$

where $V_{0,1} \in L^1(0,\pi_p)$, $V_{0,1}(0) = 0$ and $V_{0,1}'(0) = 1$ is a solution of initial value problem corresponding to the equation in (6.3). Figure 6 illustrates the solution $V_{0,1}$ in case $n = 2$ and $p = 3$.

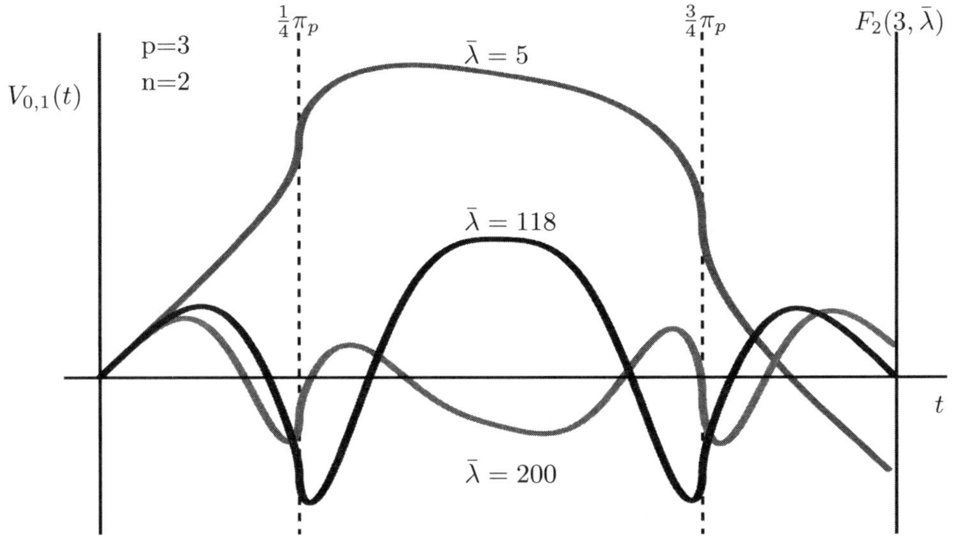

FIGURE 6

If $F_n(p,\bar{\lambda}) = 0$ then $\bar{\lambda} = \bar{\lambda}(p)$ is an eigevalue of linearized p-Laplace operator. That is why we use the “top view” projection applied to sign$\,(F_n)$. The nodal lines show us the eigenvalues $\bar{\lambda}_{n,k}$ and its dependence on parametr p. Figure 7 illustrates the case $n = 2$.

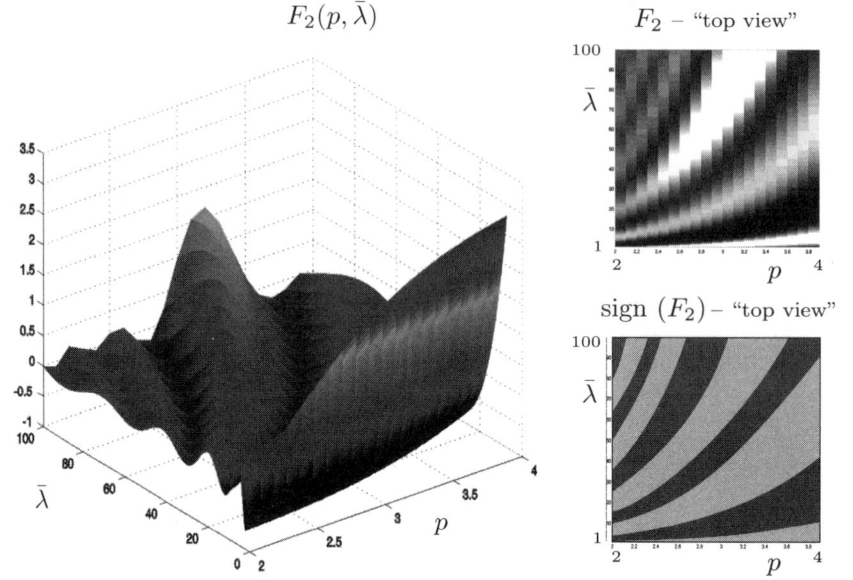

FIGURE 7

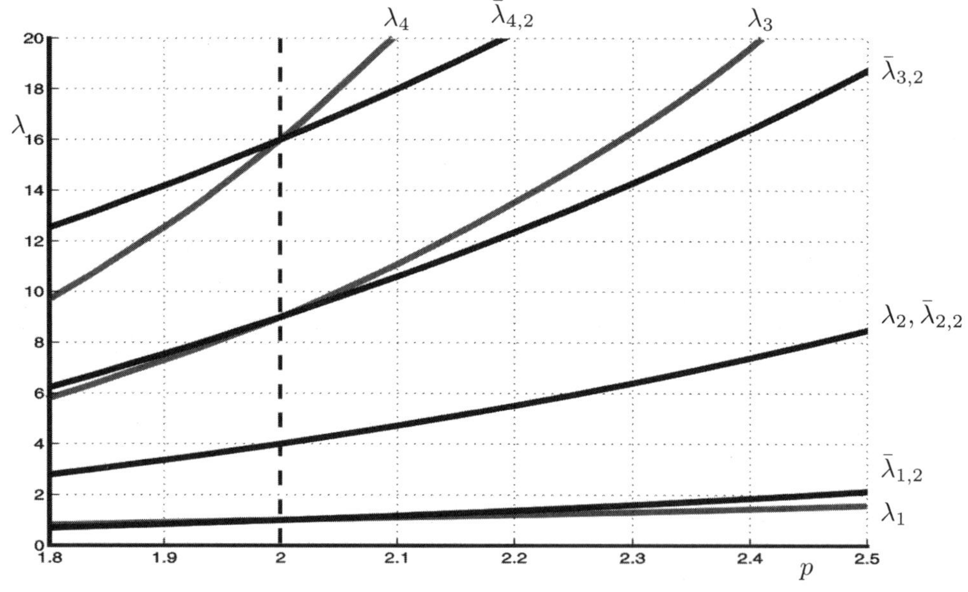

FIGURE 8

With respect to numerical results we can formulate the following conjecture: *Let* $\lambda_k = (p-1)k^p$ *and let* $\bar{\lambda}_{k,n} = \bar{\lambda}_{k,n}(p)$, $k, n \in \mathbb{N}$ *be eigenvalues of linearized one-dimensional* p*-Laplace equation* (6.3). *Then for* $p = 2$ *we have* $\bar{\lambda}_{k,n} = \lambda_k$, $\forall k \in \mathbb{N}$ *and otherwise:*

$$\left.\begin{array}{ll} \bar{\lambda}_{k,n} < \lambda_k, & k < n, \\ \bar{\lambda}_{k,n} = \lambda_k, & k = n, \\ \bar{\lambda}_{k,n} > \lambda_k, & k > n, \end{array}\right\} \text{ for } 1 < p < 2, \qquad \left.\begin{array}{ll} \bar{\lambda}_{k,n} > \lambda_k, & k < n, \\ \bar{\lambda}_{k,n} = \lambda_k, & k = n, \\ \bar{\lambda}_{k,n} < \lambda_k, & k > n. \end{array}\right\} \text{ for } p > 2.$$

6.3. Bifurcation diagrams.
Let us consider one-dimensional p-Laplace equation:

$$\begin{aligned} &-\left(|u'|^{p-2}u'\right)' - \lambda|u|^{p-2}u = f, \quad t \in (0, \pi_p); \\ &u(0) = 0, \quad u(\pi_p) = 0, \end{aligned} \tag{6.4}$$

where $p > 1$, $\lambda > 0$ and $f \in L^\infty(0, \pi_p)$.

As in the previous section we apply the shooting method to solve boundary value problem (6.4) and define a projection $M_f : \mathbb{R}^3 \to \mathbb{R}$ as follows:

$$M_f(\alpha, \lambda, p) = u_{0,\alpha}(\pi_p),$$

where $u_{0,\alpha} \in L^\infty(0, \pi_p)$, $u_{0,\alpha}(0) = 0$ and $u'_{0,\alpha}(0) = \alpha$ is a solution of initial value problem corresponding to equation in (6.4).

The *bifurcation diagram* is a set of zero points of projection M_f. We call this a *solution diagram* if it contains the information which is sufficient for backward construction of the solution. In Figure 9 we explain how to construct the *solution-diagram.*

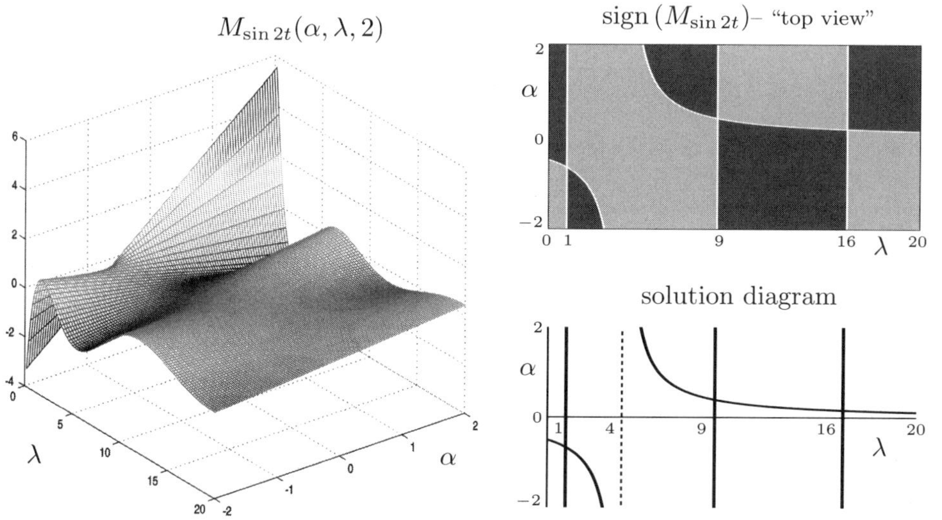

FIGURE 9

The *solution diagram* in Figure 9 is a geometrical representation of the linear Fredholm alternative for Dirichlet boundary value problem (6.4) with corresponding parameter $p = 2$.

In the case $p \neq 2$, the "top view" applied to sign (M_f) shows us the geometrical representation of non-linear Fredholm alternative. Because nonlinear problem (6.4) does not have unique solution we obtain the *bifurcation diagram* only (*i.e.* not the *solution diagram*).

Because $\lambda_n = \lambda_n(p)$, we may construct so called *r-spectral bifurcation diagrams.* The value $r \in \mathbb{R}$ parametrizes λ in equation (6.4) in such a way that $\lambda(r) = (p-1)r^p$. If r is a natural number (i.e. $r = n$) then $\lambda(r)$ is equal to eigenvalue λ_n. The scale on the horizontal axis does not depend on the value of p.

The linear and nonlinear Fredholm alternatives have many common features at least from this geometrical point of view. Figure 10 illustrates these similarities.

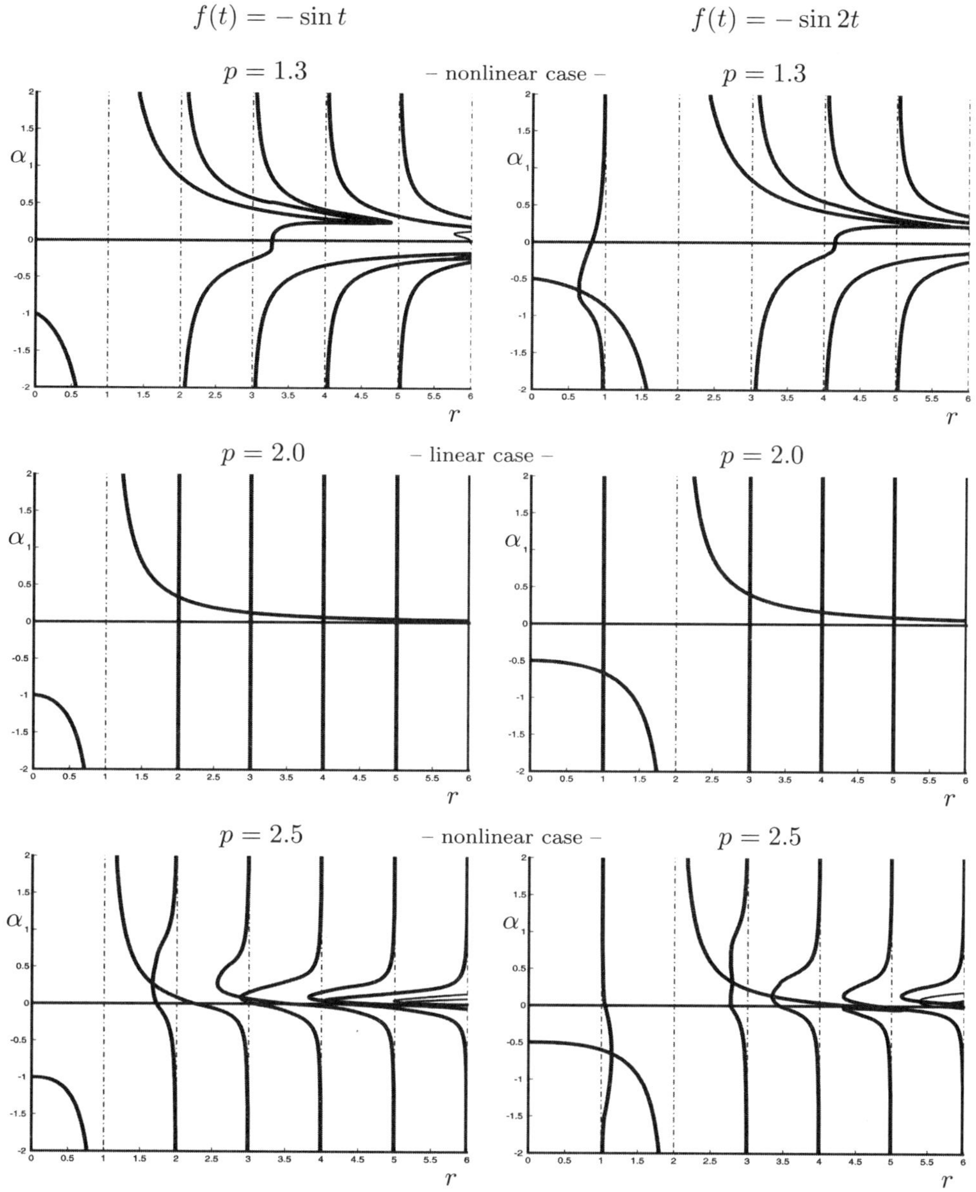

$f(t) = -\sin t$
$f(t) = -\sin 2t$
$p = 1.3$
– nonlinear case –
$p = 1.3$
$p = 2.0$
– linear case –
$p = 2.0$
$p = 2.5$
– nonlinear case –
$p = 2.5$
α_1
r

FIGURE 10

References

[1] A. Anane, *Simplicité et isolation de la première valeur propre du p-laplacien avec poids*, Comptes Rendus Acad. Sc. Paris Série I, **305** (1987), pp. 725–728.

[2] A. Anane, J.P. Gossez, *Strongly nonlinear elliptic problems near resonance: A variational approach,* Comm. Partial Differential Equations, **15** (1990), pp. 1141–1159.

[3] A. Anane, N. Tsouli, *On the second eigenvalue of the p-Laplacian*, In: "Nonlinear partial differential equations", A. Benkirane - J-P. Gossez (Eds.), Pitman Research Notes in Mathematics Series **343**, Addison Wesley Longman, Essex, U.K., 1996, pp. 19.

[4] D. Arcoya, L. Orsina, *Landesmann–Lazer conditions and quasilinear elliptic equations,* Nonlin. Anal., Theory, Methods & Appl. **28** (1997), 1623–1632.

[5] D. Arcoya, J.-L. Gámez, *Bifurcation Theory and Related Problems: Anti-Maximum Principle and Resonance*, Comm. Partial Diff. Equations, **26**(9–10) (2001), pp. 1879–1911.

[6] L. Boccardo, P. Drábek, D. Giachetti and M. Kučera, *Generalization of Fredholm alternative for nonlinear differential operators,* Nonlin. Anal., Theory, Meth. & Appl. 10 (10), (1986), pp. 1083–1103.

[7] L. Boccardo, P. Drábek and M. Kučera, *Landesman-Lazer conditions for strongly nonlinear boundary value problems,* Comment. Math. Univ. Carolinae **30** (1989), pp. 411–427.

[8] P. Binding, P. Drábek and Y.X. Huang, *On the Fredholm alternative for the p-Laplacian,* Proc. Amer. Math. Soc. **125** (1997), pp. 3555–3559.

[9] F.E. Browder, W.V. Petryshin, *Approximation methods and the generalized topological degree for nonlinear mappings in Banach Spaces*, J. Func. Analysis, **3** (1969), pp. 217–245.

[10] M.G. Crandal, P.H. Rabinowitz, *Bifurcation from simple eigenvalues*, J. Funct. Anal. **8** (1971), pp. 321–340.

[11] J. Čepička, *Numerical Experiments for Nonlinear Problems*, Ph.D. Thesis, University of West Bohemia, Plzeň, 2001, pp. 97.

[12] Dancer, E., *On the Structure of Non-Linear Eigenvalue Problems*, Indiana Univ. Math. J. **23** (1974), pp. 1069–1076.

[13] Drábek, P., *Solvability and Bifurcations of Nonlinear Equations*, Pitman Research Notes in Mathematics Series, Vol. 264, Longman Scientific & Technical, Essex, 1992.

[14] P. Drábek, *On Fredholm alternative for certain quasilinear boundary value problems,* Math. Bohemica **127** (2) (2002), pp. 197–2002.

[15] P. Drábek, *Geometry of the energy functional and the Fredholm alternative for the p-Laplacian in more dimensions*, Electronic J. Diff. Equations, Conference 08 (2002), pp. 103–120.

[16] M. Del Pino, P. Drábek and R. F. Manásevich, *The Fredholm alternative at the first eigenvalue for the one-dimensional p-Laplacian*, J. Differential Equations, **151** (1999), pp. 386–419.

[17] P. Drábek, P. Girg and R. F. Manásevich, *Generic Fredholm alternative for the one dimensional p-Laplacian*, Nonlin. Diff. Equations and Applications **8** (2001), pp. 285–298.

[18] P. Drábek, P. Girg, P. Takáč and M. Ulm, *The Fredholm alternative for the p-Laplacian: bifurcation from infinity, existence and multiplicity,* Indiana Univ. Math. J., to appear.

[19] P. Drábek, G. Holubová, *Fredholm alternative for the p-Laplacian in higher dimensions*, J. Math. Anal. Appl. **263** (2001), pp. 182–194.

[20] P. Drábek, S. Robinson, *Resonance problems for the p-Laplacian,* J. Funct. Anal. **169** (1999), pp. 189–200.

[21] P. Drábek, P. Takáč, *A counterexample to the Fredholm alternative for the p-Laplacian*, Proc. Amer. Math. Soc. **127** (1999), pp. 1079–1087.

[22] S. Fučík, J. Nečas, J. Souček and V. Souček, *Spectral Analysis of Nonlinear Operators*, Lecture Notes in Mathematics **346**, Springer-Verlag, New York-Berlin-Heidelberg, 1973.

[23] J. Fleckinger, P. Takáč, *An improved Poincaré inequality and the p-Laplacian at resonance for $p > 2$*, Advances Diff. Equations **7** (2002), pp. 951–971.

[24] J. L. Gámez, P. Girg, *Fredholm Alternative for the p-Laplacian at the first eigenvalue and bifurcations from the infinity*, preprint.

[25] J. García, J. Sabina de Lis, *A local bifurcation theorem for degenerate elliptic equations with radial symmetry*, J. Differential Equations **179** (2002), pp. 27–43.

[26] E. M. Landesman, A. C.Lazer, *Nonlinear perturbations of linear elliptic boundary value problems at resonance,* J. Math. Mech. **19** (1970), pp. 609–623.

[27] J. Lerray, J. Schauder, *Topologie et équations fonctionelles*, Ann. Sci. Ecole Norm. Sup. **51** (1934), pp. 45–78.

[28] G. Lieberman, *Boundary regularity for solutions of degenerate elliptic equations*, Nonlinear Anal., **12** (1988), pp. 1203–1219.

[29] R. F. Manásevich, P. Takáč, *On the Fredholm alternative for the p-Laplacian in one dimension*, Proc. London Math. Soc. **84** (2002), pp. 324–342.

[30] J. Nečas, *Sur e'alternative de Fredholm pour les operateurs non-lineaires avec applications aux problémes aux limites,* Anali Sem. Norm. Sup. Pisa **23** (1969), pp. 331–345.

[31] S. I. Pohozaev, *On the solvability of nonlinear equations involving odd operators,* Funkc. Anal. i Priloz. **1** (1967), pp.66–72 (in Russian).

[32] P. Rabinowitz, *Some Global Results for Nonlinear Eigenvalue Problems*, J. Functional Analysis **7** (1971), pp. 487–513.

[33] I.V. Skrypnik, *Nonlinear Elliptic Boundary Value Problems* (Russian), Naukovaja Dumka, Kyiev 1973, English Translation: Teubner, Leipzig 1986.

[34] P. Takáč, *On the Fredholm alternative for the p-Laplacian at the first eigenvalue*, Indiana Univ. Math. J. **51** (2002), pp. 187–237.

[35] P. Takáč, *On the number and structure of solutions for a Fredholm alternative with the p-Laplacian*, J. Differ. Equations **185** (2002), pp. 306–347.

[36] P. Takáč, *A variational approach to the Fredholm alternative for the p-Laplacian near the first eigenvalue,* preprint.

Centre of Applied Mathematics, University of West Bohemia, Univerzitní 22, CZ-306 14 Plzeň, Czech Republic
E-mail address: cepicka@kma.zcu.cz

Centre of Applied Mathematics, University of West Bohemia, Univerzitní 22, CZ-306 14 Plzeň, Czech Republic
E-mail address: pdrabek@kma.zcu.cz

Centre of Applied Mathematics, University of West Bohemia, Univerzitní 22, CZ-306 14 Plzeň, Czech Republic
E-mail address: pgirg@kma.zcu.cz

Contemporary Mathematics
Volume **357**, 2004

Eigenvalue Problems, Resonance Problems and Open Problems

Pavel Drábek and Stephen B. Robinson

ABSTRACT. We discuss several open problems related to the eigenvalues and eigenfunctions of the p-Laplacian. We emphasize the fundamental role played by the variational characterization of eigenvalues and eigenfunctions.

1. Introduction

The purpose of this paper is to provide a brief introduction to several open problems and to describe how they arise within the context of the authors' recent work. Hopefully this will encourage their further study and eventual solution.

The central theme is to understand the nonlinear eigenvalue problem

$$(1)\qquad \begin{aligned} -\Delta_p u - \lambda |u|^{p-2}u = 0 \text{ in } \Omega, \\ u|_{\partial\Omega} = 0, \end{aligned}$$

and related problems such as

$$(2)\qquad \begin{aligned} -\Delta_p u - \lambda |u|^{p-2}u + f(x,u) = 0 \text{ in } \Omega, \\ u|_{\partial\Omega} = 0, \end{aligned}$$

where $p > 1$, $\Delta_p u := \nabla\cdot(|\nabla u|^{p-2}\nabla u)$, $\Omega \subset \mathbb{R}^N$ is a smooth bounded domain, and f is a bounded function satisfying standard Caratheodory conditions.

Solutions of problem (1) correspond to critical points of the *Rayleigh Quotient* functional

$$I(u) := \frac{\int_\Omega |\nabla u|^p}{\int_\Omega |u|^p} \quad \text{for } u \in W_0^{1,p}(\Omega) \setminus \{0\}.$$

1991 *Mathematics Subject Classification.* Primary 35J20,35J65; Secondary 35P30.

Research sponsored by the Grant Agency of the Czech Republic, Project no. 201/03/0671, partly by the Ministery of Education of the Czech Republic, Project no. MSM 235200001.

It is clear that $I(tu) = I(u)$ for all $t \in \mathbb{R}$, so the search for critical points can be restricted to the L^p unit sphere $\mathcal{S} = \{u \in W_0^{1,p}(\Omega) \setminus \{0\} : \int_\Omega |u|^p = 1\}$. In section 2 we will address what is known about the critical values of this functional, i.e. the eigenvalues of $-\Delta_p$, and in section 3 we will discuss the properties of the critical points, i.e. the eigenfunctions of $-\Delta_p$.

Solutions of problem (2) correspond to critical points of the functional

$$J(u) = \frac{1}{p}\int_\Omega |\nabla u|^p - \frac{1}{p}\int_\Omega |u|^p - \int_\Omega F(x,u) \quad \text{for } u \in W_0^{1,p}(\Omega),$$

where $F(x,u) := \int_0^u f(x,t)dt$. In section 4 we discuss existence theorems for this problem and how they relate to our incomplete understanding of problem (1). Of particular interest to us are the Landesman-Lazer type results that have motivated so many papers since [**18**].

Since this is an expository work, we generally just sketch what we see as the important parts of a proof and leave the technical details for the references. Most of the theoretical background for the variational methods we apply can be found in [**23**] and [**15**].

2. A Variational Characterization of the Eigenvalues

How one understands the variational characterization of eigenvalues plays a fundamental role in understanding and solving related problems. This has clearly been true in the linear case where the eigenvalues of $-\Delta_2$ are obtained by minimizing the Raleigh quotient over $W_0^{1,2}(\Omega)$ to get λ_1, restricting to an orthogonal subspace and minimizing again to get λ_2, and so on. This provides a beautiful natural structure for solving problems such as (2) by variational methods. We can decompose $W_0^{1,2}(\Omega)$ into orthogonal subspaces, one whose basis is composed of eigenfunctions with energy less than λ and the other whose basis is composed of eigenfunctions with energy greater than λ, and then apply saddle point arguments. (Saddle point arguments will be discussed in greater detail in section 4.)

Unfortunately, such properties as orthogonality are not available when $p \neq 2$. However, through the work of Ljusternik, Schnirelmann, Krasnoselskii, and others we can exploit the symmetry of I and $\mathcal{S}$ to describe the so-called *variational* eigenvalues of $-\Delta_p$. We rely on a somewhat more intuitive, to us, variant of the usual approach. For $k \in \mathbb{N}$ let $\mathcal{S}^{k-1}$ represent the unit sphere in $\mathbb{R}^k$ and let

$$\mathcal{F}_k := \{\mathcal{A} \subset \mathcal{S} : \mathcal{A} \text{ is the image of a continuous odd function } h : \mathcal{S}^{k-1} \to \mathcal{S}\}.$$

Define

$$\lambda_k := \inf_{\mathcal{A} \in \mathcal{F}_k} \sup_{u \in \mathcal{A}} I(u).$$

Notice that for $k = 1$ each symmetric embedding of $\mathcal{S}^0 = \{-1, 1\}$ is just a pair of antipodal points $\{u, -u\}$ in $\mathcal{S}$, and $I(-u) = I(u)$, so we are simply minimizing the Rayleigh quotient over $\mathcal{S}$ as in the linear case. λ_1 has many of the desirable properties that we have become accustomed to in the linear case. It is simple, positive, and has an eigenfunction, ϕ_1, which is positive in Ω (See [**20**].). For $k = 2$ we embed circles in $\mathcal{S}$. Perhaps the best mental image of such an embedding is to think of creating a curve on $\mathcal{S}$ connecting $\{\pm\phi_1\}$, and then close the curve by including its reflection across the origin. Our linear intuition imagines that a well chosen curve will achieve its maximum energy as it crosses a subspace *orthogonal* to ϕ_1. Of course, this isn't the case for $p \neq 2$, but the analogy is not bad. It is known that λ_2 is the minimum of all eigenvalues, variational or otherwise, that are strictly greater than λ_1, (see [**3**]), and it is known that any eigenfunction associated with λ_2 has exactly two nodal domains (See [**9**] or [**14**].). (More about this in section 3.) In the ODE case, where $N = 1$, it is known that $\{\lambda_k\}_{k=1}^{\infty}$ describes all of the eigenvalues, each eigenvalue is simple, and the eigenfunction associated with λ_n has exactly n nodal domains. (See [**11**].) However, in the PDE case many questions remain unanswered regarding the properties of the spectrum beyond λ_2. This brings us to a well known open problem, which is perhaps the most profound problem mentioned in this paper.

OPEN PROBLEM 2.1. *Are there any* nonvariational *eigenvalues? If there are, then under what conditions do they arise, and how are they distributed?*

Given the discussion above it is not surprising that most of the literature on problems such as (2) has focused on the case where $\lambda < \lambda_1$, $\lambda = \lambda_1$, or perhaps $\lambda_1 < \lambda < \lambda_2$. In section 4 we will state a theorem that allows a broader choice of λ and suggests a reasonable followup of our first open problem. But first we will investigate the properties of eigenfunctions in a little more detail. In both of the sections that follow it is important to notice the fundamental role that the given characterization of eigenvalues plays in understanding and exploiting the inherent geometry of the problems.

3. Describing the Eigenfunctions

A classic theorem of Courant for the case $p = 2$ states that any eigenfunction associated with λ_n has at most n nodal domains, where the nodal domains of an eigenfunction ϕ_n are defined as the connected components of $\{x : \phi_n(x) \neq 0\}$. (See [**7**].) Simple examples show that this estimate is optimal. For example, a long skinny rectangle will have a high energy eigenfunction with only two nodal domains, and lower energy eigenfunctions with many more nodal domains.

Relying on the variational characterization in section 2 we have proved the following generalizations of Courant's theorem. (See [**14**].)

THEOREM 1. *Suppose $\lambda < \lambda_{n+1}$ is an eigenvalue with associated eigenfunction ϕ_λ. Then ϕ_λ has at most n nodal domains.*

THEOREM 2. *Let ϕ_n be an eigenfunction associated with λ_n. Then ϕ_n has at most $2n - 2$ nodal domains.*

Notice that Theorem 1 allows for the possibility that λ is either variational or nonvariational. If all variational eigenvalues were simple, then there would be no necessity for Theorem 2, but in a case of multiplicity such as $\lambda_3 = \lambda_4 = \lambda_5$, then Theorem 2 guarantees that the associated eigenfunctions have no more than 4 nodal domains. For full proof we refer to [**14**], but we will sketch enough of the argument here to demonstrate the importance of the characterization given in section 2, and to motivate the open problems mentioned below.

Consider the case where ϕ_2 is an eigenfunction associated with λ_2. Theorem 2 claims that ϕ_2 should have only two nodal domains. Our argument is by contradiction, so suppose that ϕ_2 has three nodal domains Ω_1, Ω_2, and Ω_3, and suppose that $\phi_2 = \phi_{21} + \phi_{22} + \phi_{23}$, where ϕ_{2i} is the component of ϕ_2 over Ω_i, i.e. $\phi_{2i} \equiv \phi_2$ on Ω_i and $\phi_{2i} \equiv 0$ on the complement of Ω_i. Using these components we can construct a symmetric embedding of $\mathcal{S}^1$ into $\mathcal{S}$. Let $\mathcal{A} := \{\alpha\phi_{21} + \beta\phi_{22} : |\alpha|^p \int_\Omega |\phi_{21}|^p + |\beta|^p \int_\Omega |\phi_{22}|^p = 1\}$. Notice that the component over one nodal domain, Ω_3, has been suppressed. It is straightforward to check that $\mathcal{A}$ is a symmetric embedding of $\mathcal{S}^1$ into $\mathcal{S}$ and that $I|_{\mathcal{A}} \equiv \lambda_2$. A crucial point that must now be argued is that $\mathcal{A}$ contains no critical points of I. Assuming for the moment that this has been established, then we quickly reach a contradiction as follows: Using the negative gradient of I on $\mathcal{S}$, let the set $\mathcal{A}$ *flow downhill* to become the set $\mathcal{A}'$ which is a symmetric embedding of $\mathcal{S}^1$ into $\mathcal{S}$ such that $I|_{\mathcal{A}'} \leq \lambda_2 - \epsilon$ for some $\epsilon > 0$. This contradicts the characterization of λ_2.

We know of two ways to argue that $\mathcal{A}$ contains no critical points. First, suppose that $-\Delta_p$ satisfies the unique continuation property, i.e. its nontrivial eigenfunctions cannot be zero on a set of positive measure, then it is clear that the elements of $\mathcal{A}$ cannot be eigenfunctions, because they are all nontrivial and are all identically zero on Ω_3. Second, suppose that the nodal domains can be connected in a chain so that Ω_1 and Ω_2 are separated by a smooth boundary where $\nabla\phi_2 \neq 0$, and where Ω_2 and Ω_3 are separated similarly. In this case any element of $\mathcal{A}$ will be nondifferentiable on one of the separating boundaries, but all eigenfunctions of $-\Delta_p$ are in $C^{1,\mu}(\overline{\Omega})$ for some $\mu > 0$. See [**24**] or [**10**], for example. Thus, $\mathcal{A}$ contains no critical points.

The unique continuation property for $-\Delta_p$ has recently been studied in [**17**] and [**2**], where the second paper goes on to obtain a sharp generalization of Courant's theorem. However, there are still cases where the question has not been resolved. For example, in [**2**] it is assumed that $p > 2$ and $N > 2$.

The nodal domain chain property mentioned above can be verified for the case $n = 2$ via an application of the maximum principle, and so we get an optimal generalization of Courant's Theorem for this case.(See [**21**] and [**14**].) For higher eigenvalues we can show that each nodal domain is paired up in a chain with at least one other nodal domain, and a careful application of this fact is used in the proofs of Theorems 1 and 2. However, it is not clear that the nodal domains can all be connected in a single chain. Notice that the chain does not have to be linear in order for the proof to work. It would suffice, for example, if each nodal domain was connected to some central nodal domain, or if they were all connected to create a *nodal domain graph.* It seems that this property could only be prevented if the nodal curves or surfaces separating domains are allowed to exhibit rather bizarre behavior, which would seem to contradict the idea of energy minimization assumed in the definition.

In order to obtain a sharp generalization of the Courant Theorem for higher eigenvalues we have been led to the following unresolved questions.

OPEN PROBLEM 3.1. *Do the eigenfunctions of* $-\Delta_p$ *satisfy the unique continuation property?*

OPEN PROBLEM 3.2. *Given an eigenfunction of* $-\Delta_p$, *can the nodal domains be linked in a connected graph? Further, do the curves or surfaces separating nodal domains satisfy some smoothness or minimal energy criterion that would help resolve this and related questions?*

Nodal domain theorems have a long and interesting history. For $-\Delta_p$ the first similar contribution that we are aware of is in [**3**] where Proposition 1 is essentially the same as Theorem 1 above, and a variant of Theorem 2 is proved with upper bound $2n-1$ rather than $2n-2$. More recently, in [**8**] and [**9**] and related works, the fact that the second eigenfunction has exactly two nodal domains is proved and generalized to the analogous Fučik spectrum problem.

Studying the nodal curve properties for $-\Delta_2$ presents some fascinating and very difficult problems. For example, see the voluminous literature regarding the nodal curve conjecture of Payne and the Faber-Krahn inequalities. Some results that come closer in spirit to the open problem mentioned above can be found in [**16**] and [**22**] and related works. Improvement of these results and generalization to the case $p \neq 2$ are left as exercises for the reader.

4. Solvability of Related Resonance Problems

Problem (2) is called a resonance or a nonresonance problem, depending on whether λ is or is not an eigenvalue, respectively. For the case $p = 2$ these problems have been studied rather thoroughly over the last thirty years. A variety of nonlinear methods have been motivated or refined while studying such problems, and saddle point theorems are a particularly important example of this.

Before stating Theorem 3, and sketching its proof, it is instructive to outline a standard saddle point existence proof for problem (2) in the case $p = 2$, so that we can be reminded of how heavily the argument relies on a detailed knowledge of the spectral properties of $-\Delta_2$. Consider the case $\lambda_n < \lambda < \lambda_{n+1}$. Our first step is to establish a saddle-type geometry for the functional J. To do this we split $W_0^{1,2}(\Omega)$ into the orthogonal subspaces $U = \text{span}\{\phi_1, ..., \phi_n\}$ and $V = \text{span}\{\phi_{n+1}, \phi_{n+2}, ...\}$, show that J is downward coercive on U using the fact that $\int_\Omega |\nabla u|^2 \le \lambda_n \int_\Omega |u|^2$ on U, and then show that J is upward coercive on V using the fact that $\int_\Omega |\nabla v|^2 \ge \lambda_{n+1} \int_\Omega |v|^2$ on V. The saddle geometry allows us to locate approximate solutions by connecting low points in U by paths that must cross the intermediate *mountain range* V. More formally, we choose a ball in U, centered at the origin, whose radius is so large that the values of J on the boundary are well below the infimum of J on V. Then we consider deformations of this ball which leave the boundary fixed. Every such deformation must cross V at some point. To get approximate solutions we look for the deformations with the lowest crossing points.

Our second step is to show that sequences of approximate solutions cannot slip away to infinity. (More formally, we must prove a compactness condition such as the Palais-Smale condition.) Suppose that $J'(u_n) \to 0$. (Note: In general it is also important to assume that $J(u_n)$ is bounded.) We argue by contradiction, so suppose that $||u_n|| \to \infty$. A few standard estimates and a careful examination of the limit $\frac{J'(u_n)}{||u_n||} \to 0$ shows that $\{\frac{u_n}{||u_n||}\}$, or an appropriate subsequence, converges to a nontrivial eigenfunction ϕ_λ associated with λ. But when $p = 2$ there are no eigenvalues other than the variational eigenvalues, so λ is not an eigenvalue and hence $\phi_\lambda \equiv 0$. We have a contradiction, and the second element of the saddle point proof is finished. Notice that the particular form of f played no role in this argument. If we had assumed the resonance case, $\lambda = \lambda_n$, then more information would be needed about f before either the saddle geometry or the compactness could be proved.

Given the argument above and given what little is known about the spectral properties of $-\Delta_p$ when $p \ne 2$, it is not surprising that most of the existing literature focuses on the various cases where $\lambda \le \lambda_2$. (See [**5**] and [**4**], for example.) Here is a theorem that provides a Landesman-Lazer type result for arbitrary λ. (See [**13**].)

THEOREM 3. *Suppose that $f(x,t)$ is a Caratheodory function and that there is an $\overline{f} \in L^q(\Omega)$, where $q = \dfrac{p}{p-1}$, such that $|f(x,t)| \le \overline{f}(x)\ \forall (x,t) \in \Omega \times \mathbb{R}$. Further, assume that either*

$$(LL)_\lambda^+ : \int_{v>0} f^+ v + \int_{v<0} f^- v > 0\ \forall v \in \ker(-\Delta_p - \lambda) \setminus \{0\},$$

or

$$(LL)_\lambda^- : \int_{v>0} f^+ v + \int_{v<0} f^- v < 0\ \forall v \in \ker(-\Delta_p - \lambda) \setminus \{0\},$$

where $f^\pm(x) := \lim_{t \to \pm\infty} f(x,t)$, a limit that is assumed to exist for a.e. $x \in \Omega$. Then (2) has a weak solution.

Notice that the Landesman-Lazer conditions are vacuously satisfied if λ is not an eigenvalue. Also, it is relatively simple to create examples that satisfy the given conditions, such as $f(x,t) = \arctan(t) + h(x)$ where $||h||_{L^\infty} < \frac{\pi}{2}$.

It is interesting to note that, even if there are many nonvariational eigenvalues with unknown distribution, the variational eigenvalues provide sufficient structure to adapt the proof discussed above. Once again supposing that $\lambda_n < \lambda < \lambda_{n+1}$ we can prove the saddle point geometry by decomposing the space $W_0^{1,p}(\Omega)$ in an appropriate way. Rather than decomposing into orthogonal subspaces whose bases are eigenfunctions, we use our understanding of the characterization in section 2 to decompose into appropriate sets that *link*. The subspace V is replaced by the set $V' := \{u : I(u) \geq \lambda_{n+1}\}$. The boundary of the ball in the subspace U is replaced by a set $\mathcal{A}$, which is a symmetric embedding of $\mathcal{S}^{n-1}$ such that $J|_{\mathcal{A}} \leq \lambda_n + \epsilon < \lambda$. The ball itself is replaced by the set $\mathcal{B} := \{ta : 0 \leq t \leq 1, a \in \mathcal{A}\}$. The sets $\mathcal{B}$ and V' share the same essential relationship as did the ball in U and the subspace V, i.e. every symmetric deformation of $\mathcal{B}$ that leaves $\mathcal{A}$ fixed must intersect V'. This follows from the fact that a deformation of $\mathcal{B}$, joined with its symmetric reflection, represents an embedding of $\mathcal{S}^n$, so the characterization of λ_{n+1} implies that $J(u) \geq \lambda_{n+1}$ for some point u in the deformation.

Proving the compactness condition is more delicate. Assume that $J(u_n)$ is bounded and that $J'(u_n) \to 0$. If $||u_n|| \to \infty$ we arrive at a contradiction of the (LL) condition by examining the limit of $\frac{1}{||u_n||}(pJ(u_n) - J'(u_n))$. (See **[13]** for details.)

The case where $\lambda = \lambda_n$ for some n is treated by an approximation argument using the result for nonvariational λ. The given Landesman-Lazer conditions guarantee that both the geometry and the compactness conditions are preserved in the limit.

Theorem 3 and its proof suggest that the variational eigenvalues provide much of the structure necessary to investigate perturbations of the eigenvalue problem (1). In some sense the nonvariational eigenvalues are less important. Admittedly, this is a vague statement, but perhaps the person who can pose the questions more rigorously will also be the person who answers the questions. In any case, we close with the following problem, which we feel is worth more study.

OPEN PROBLEM 4.1. *Are the variational eigenvalues complete in some appropriate sense?*

References

[1] A. Anane, *Simplicité et isolation de la premiére valeur propre du p-laplacien avec poids*, C.R.Ac. Paris 305 (1987), 725-728.

[2] A. Anane, I. Hadi, M. Moussa, N. Tsouli, *The unique continuation property of the p-Laplacian Operator and the Courant Nodal Domain Theorem*, preprint

[3] A. Anane, N. Tsouli, *On the second eigenvalue of the p-Laplacian*, in: Nonlinear Partial Differential Equations (From a conference in Fes, Morocco, 1994), A Benkirane and J.-P. Gossez Ed., Pitman Research Notes in Math. 343, Longman 1996, pp.1-9.

[4] A. Anane, N. Tsouli, *On a nonresonance condition between the first and the second eigenvalues for the p-Laplacian* , Int. J. Math. Math. Sci 26 (2001), no.10, 625-634.

[5] D. Arcoya and L. Orsina, *Landesman-Lazer conditions and quasilinear elliptic equations*, Nonlinear Analysis T.M.A. 28 (1997), 1623-1632.

[6] J. P. Garcia Azorero, I. Peral Alonso, *Existence and nonuniqueness for the p-Laplacian: nonlinear eigenvalues*, Comm. Partial Differential Equations 12(12) (1987), 1389-1430.

[7] R. Courant and D. Hilbert, *Method of Mathematical Physics*, Volume II, Interscience, New York, 1962.

[8] M. Cuesta, D.G. de Figueiredo, J.-P. Gossez, *The beginning of the Fučik Spectrum for the p-Laplacian*, J. Differential Equations 159(1) (1999), 212-238.

[9] M. Cuesta, D. G. de Figueiredo, J.-P. Gossez, *A nodal domain property for the p-Laplacian*, C.R. Acad. Sci. I - Math. 330(8)(2000), 669-673.

[10] E. Di Benedetto, $C^{1+\alpha}$ local regularity of weak solutions of degenerate elliptic equations, Nonlinear Anal. 7(8) (1983), 827-850.

[11] P. Drábek, *Solvability and Bifurcations of Nonlinear Equations*, Pitman Research Notes in Mathematics Series 264, Longman, Harlow 1992.

[12] P. Drábek, A Kufner, F. Nicolosi, *Quasilinear Elliptic Equations with Degenerations and Singularities*, De Gruyter Series in Nonlinear Analysis and Applications 5, Walter de Gruyter, Berlin - New York 1997.

[13] P. Drábek, S. Robinson,*Resonance problems for the p-Laplacian*, J. Functional Analysis 169 (1999), 189-200.

[14] P. Drábek, S. Robinson, *On the generalization of the Courant nodal domain theorem*, J. Differential Equations 181 (2002), no.1, 58-71.

[15] N. Ghoussoub, *Duality and Perturbation Methods in Critical Point Theory* , Cambridge University Press, 1993.

[16] Greiser, Daniel and Jerison, David, *Asymptotics of the first nodal line of a convex domain*, Invent. Math. 125 (1996), no.2, 197-219.

[17] Hadi, Islam Edine and Tsouli, N., *Strong Unique continuation of eigenfunctions for p-Laplacian operator*, Int. J. Math. Math. Sci, 25 (2001), no.3, 213-216

[18] E. M. Landesman and A. C. Lazer, *Nonlinear perturbations of linear elliptic boundary value problems at resonance*, J. Math. Mech. 19, 609-623 (1970).

[19] G. Lieberman, *Boundary regularity for solutions of degenerate elliptic equations*, Nonlinear Anal. 12(11) (1988), 1203-1219.

[20] P. Lindqvist, *On the equation* $\nabla(|\nabla u|^{p-2}\nabla u) + \lambda|u|^{p-2}u = 0$, Proc. Amer. Math. Society 109 (1990), 157-164.

[21] M. Otani, T. Teshima, *On the first eigenvalue of some quasilinear elliptic equations*, Proc. Japan Acad. 64, ser. A (1988), 8-10.

[22] Savo, Allessandro, *Lower bounds for the nodal length of eigenfunctions of the Laplacian*, Ann. Global Anal. Geom. 19 (2001), no.2, 133-151

[23] M. Struwe, *Variational Methods; Applications to Nonlinear Partial Differential Equations and Hamiltonian Systems,* Springer-Verlag, New York, 1990.

[24] P. Tolksdorf, *Regularity for a more general class of quasilinear elliptic equations*, J. Differential Equations 51 (1984), 126-150.

[25] N. Tsouli, *Etude de l'ensemble nodale des fonctions propres et de la nonrésonance pourl'operateur p-Laplacien.* Th. Doct. Unversit. Mohammed Ier. Oujda

Department of Mathematics, University of West Bohemia, P.O. Box 314, 306 14 Pilsen

E-mail address: pdrabek@kma.zcu.cz

Department of Mathematics, Wake Forest University, Winston-Salem, NC 27109

E-mail address: sbr@wfu.edu

Contemporary Mathematics
Volume **357**, 2004

Variational, Dynamic and Geometric Aspects of some Nonlinear Problems

Pablo Padilla

God and Nature do nothing without reason
Greek proverb

1. Introduction

Many nonlinear problems have variational structure; this work is an attempt to present some related open questions in a unified way in order to understand some important qualitative features of their solutions. In particular, examples arising in biology, economics, engineering and physics have shown that there is an underlying geometric or dynamical problem associated with the infinite dimensional variational formulation that governs the behavior of the whole system. Usually, this underlying problem becomes more easily apparent when a natural small parameter is introduced.

In fact, much effort has been devoted to the study of how the presence of this parameter induces the formation of localized structures or concentrations in these situations: spikes, layers, vortices or more complicated objects. Once these structures have been identified, the study of the original variational problem is transformed into a different one, namely the study of their behavior and interactions, equilibrium configurations and/or dynamics in general. It could be the interaction of vortices in superconductors, the formation of patterns (spots or stripes) in plants or other organisms, the description of the cooling down of an alloy, or the properties of a ferromagnet or of a triple point in a capillary tube. From the mathematical point of view, all these situations can be put in the framework of a variational problem in which a small parameter induces some sort of concentration. The original variational structure is inherited by the reduced problem at the level of localized structures, and it is here where the connection with a geometric or dynamic problem is made apparent. Once the structures are formed and interacting, their behavior can be described by a finite dimensional dynamical system related to a geometric or optimization problem. The previous discussion suggests a general approach to these problems. Once the problem has been formulated as a variational one in which a small parameter has been identified, it is necessary to study the existence of equilibrium configurations, in particular, the simplest ones (for instance, minima). This can be done using variational methods or sometimes by directly working with the corresponding Euler-Lagrange equations.

It is important to analyze what happens to these simple solutions when the parameter tends to zero, since by doing this, the simplest localized structures are obtained. For example, in the models for pattern formation, these solutions would correspond to the simplest ones exhibiting a nontrivial (non spatially homogeneous) pattern, e.g., a single spot or spike. This step usually requires both asymptotic and analytic techniques. The former methodology is used to study the limiting case when the parameter tends to zero in order to establish the shape of the localized structure. This usually reduces the problem to solving an equation in the whole domain and with additional symmetries, for example, radial symmetry. Here, modulation theory, WKB approximations, Laplace's method and other tools are useful. Analytic techniques are used to rigorously establish that the same concentration occurs for small values of the parameter and in the general original situation. From the technical point of view, this step is usually rather involved and requires various techniques depending on the problem: concentration compactness or compensated compactness methods, or geometric measure theoretic methods such as Γ-convergence, varifold techniques, etc. A mathematically precise notion of what we mean by saying that one variational problem converges to another is required at this point.

The simple solutions described above can then be used as building blocks for obtaining more complicated patterns, i.e., configurations with several spots in our pattern formation example. The main idea is that even when the problem is nonlinear and therefore no superposition is strictly valid, for small values of the parameter an approximate superposition principle holds when the localized structures of the solutions do not strongly interact with each other. By using variational methods or implicit function arguments, the existence of qualitatively more complicated solutions can be proved. There is an interesting connection between these ideas about localized structures, the calculus of variations, and other subjects in nonlinear analysis such as integrable systems, soliton interaction, and so on. The reader interested in these topics will find the book *Solitons in Field Theory and Nonlinear Analysis* by Y. Yang an excellent reference both as an introduction as well as an advanced exposition of current research [**60**].

So far, we have only mentioned the equilibrium situation. In fact, in many cases the interest lies in understanding the dynamics (including the stability) of the problem. Again, the fact that the parameter is small suggests that we consider the equilibrium configurations already obtained or perturbations of them as initial conditions. In the problems we will discuss, the dynamics of the problem will be given by either a parabolic or hyperbolic partial differential equation. If the general methodology is to be applied to a particular problem, a number of ideas from different areas must be used, including analytical, asymptotic, geometric, numerical and topological techniques.

In order to be more concrete, we consider two specific examples arising in study of biological pattern formation and morphogenesis. We will provide more detail later in this article. Following Turing [**58**], this process can be modeled by a system of reaction-diffusion equations of the form

$$\begin{aligned} u_t &= D_1 \Delta u + g(u,v) \\ v_t &= D_2 \Delta v + h(u,v), \end{aligned} \tag{1}$$

where u and v represent the density of the so called morphogens, D_1 and D_2 the corresponding diffusion coefficients, and f and g the reactive terms determining the specific way in which the morphogens interact.

In many examples, it is common that D_1 is much smaller than D_2. For example, in the case where u and v are two chemicals, the first may corresponding to an activator and the second to an inhibitor in an autocatalytic reaction. See [**28**] for a discussion of activator-inhibitor models and [**43**] for a comprehensive introduction to this subject and to mathematical biology in general. In this situation, it is normally a good approximation to consider v as practically constant, so that the system reduces to

$$u_t = \varepsilon^2 \Delta u + f(u), \tag{2}$$

where ε is a small parameter (for instance, the ratio $\frac{D_1}{D_2}$) and f is a nonlinear term. See (for example) [**49**] and references therein for a more complete discussion on the validity of such an approximation; the article [**17**] describes a situation in which the behavior of the system and of the reduced equation are completely different.

Determining the equilibrium configurations (time independent states) is equivalent to solving the equation

$$\varepsilon^2 \Delta u + f(u) = 0.$$

This is the Euler-Lagrange equation of the functional

$$E(u) = \int_\Omega \frac{\varepsilon^2}{2} |\nabla u|^2 + F(u)\, dx,$$

where $F(u) = \int^u f(s)\, ds$. The domain Ω is where the pattern formation process is taking place. Typically this domain is considered to be a subset of the plane, but in reality it is interesting and often necessary to consider more complex geometries. For the time being, it suffices to say that many organisms exhibit spherical or nearly spherical symmetry and that it is very natural to consider Ω to be a surface like a sphere or an ovaloid. Moreover, the absence of a boundary makes some technicalities more transparent, which is convenient from the analytical point of view. We will focus on two specific and prototypical nonlinearities in Sections 2 and 3 where the localized structures that appear are spikes and layers, respectively. It is worth mentioning that spots, stripes and other patterns such as spirals or laberinthic type formations are prototypical solution of system 1. The challenging an fundamental problem of predicting the type of structures that will be formed for general nonlinearities remains open (see Paragraph g in Section 6).

In the first case, it is possible to construct higher energy solutions that exhibit multiple concentrations (spots or spikes) and to describe their behavior in a geometrical way. That is, if one considers solutions with k concentrations, then the ones that have least energy are such that the spikes tend to be located as far as possible from each other in the limit as ε tends to zero, thus solving a sphere packing problem on Ω. For the second problem, when layers are formed even the description of the least energy nontrivial solutions leads to an interesting geometric problem. Indeed, it can be shown that as ε tends to zero the concentration layer is located on a closed geodesic of minimal length. For these two problems, the interplay between analytic and geometric ideas is in a sense very direct, since the identification of an underlying geometric problem (sphere packing or the existence of closed geodesics) is more or less direct. Precise statements and references are given in the corresponding sections.

In Section 4 we consider a problem arising in capillarity theory and or in phase transitions models in general, namely the study of triple points. From the mathematical point of view, this leads to the study of a system, rather than a single equation. Fortunately, the variational structure of the problem is the same as in the previous two. In order to establish the existence of higher energy solutions we use ideas from dynamical systems applied to the negative gradient flow and combine them with minimax methods in the calculus of variations. The qualitative features of the solutions as the parameter tends to zero have not been established analytically; instead we refer to the recent numerical work by J. Neuberger, D. Rice and J. Swift [**48**]. In Section 5 we discuss a question arising in particle physics as the simplest model of gauge theories, the so called baby Skyrme model, which reduces to the sine-Gordon equation in two spatial dimensions. There, we formulate the problem of the existence of rotating traveling waves by transforming the original equation, which is hyperbolic, to an elliptic one by assuming a special form of the solution. When the parameter is small, these solutions concentrate on transition layers and could be identified with quasiparticles having constant angular momentum. In this case the use of asymptotic methods provides the necessary information in order to establish some qualitative features of the solutions. This allows the construction of specific trial functions in the variational formulation, i.e., the construction of a path in the mountain pass lemma, so that we can rigorously establish existence. In Section 6 we briefly present additional problems, some exhibiting similarities and others which are less related to the approach used in the study of the previously discussed problems.

While attempting to give a coherent and understandable presentation, we also intend to provide a readable introduction to several of these topics. Therefore, the style is informal. Instead of giving technical details and being rigorous and exhaustive, we provide references (by no means a complete list) so that the interested reader can continue. For the sake of some rigor, we include in Section 3 a detailed statement of a key result.

Our emphasis is precisely in providing a global perspective: motivations, methods, results and open problems. In this way, we hope this article will be useful to graduate students and researchers new to this subject area. The material presented here is an expanded version of the lecture given in Flagstaff, Arizona, in the *Variational Methods* congress held during June of 2002. It is not meant to be a research paper, but a personal perspective of possible interesting research directions. It is my pleasure to thank J. M. Neuberger and J. W. Swift for having invited me to participate in the conference, and also for the interest they (together with their student D. Rice) have shown in the triple well potential system. I have benefitted in many ways from their understanding and numerical implementations of the problem.

It is our hope that this work will provide a unified point of view of several variational problems and their connections with geometry and dynamical systems. Ideas and methods in one of these areas might clarify problems in another. After all, according to Poincaré, doing mathematics consists in giving the same name to different things.

Pablo Padilla
Seoul, July 2002.

2. A pattern formation problem on the sphere

We begin our discussion of singularly perturbed problems by choosing a specific nonlinearity for (2), namely:

$$f(u) = -\lambda u + u^p,$$

where λ is a positive number and $p > 1$, winch roughly means that the reaction is autocatalytic (later on we will impose further restrictions on p). Then looking for equilibrium solutions of this equation is equivalent to solving

$$-\varepsilon^2 \Delta u + \lambda u = u^p.$$

Moreover, in the applications, u usually represents a concentration, so it is natural to look for positive solutions. Usually, this problem is considered in a plane domain in R^n, $n = 2$ being the most common case. However, as already mentioned in the introduction, we will consider this problem on a sphere, S^m. Besides the fact that from the modeling point of view it is natural and probably more realistic to study the equation on a surface, it is also interesting to observe the effects of the absence of boundary conditions. In this case, since the sphere is a compact manifold, there is no boundary and therefore no boundary conditions are necessary. Similar situations on the sphere have been considered in the context of pattern formation and cancer modeling (see [**59**] and [**16**]). This particular non linearity is important, since it can be obtained as a reduction of the activator-inhibitor Gierer-Meinhardt system and is in that respect prototypical (see [**49**]). From the variational point of view, the solutions of Equation 2 correspond to critical points of the functional

$$E(u) = \int_{S^m} \frac{\varepsilon^2}{2}|\nabla u|^2 + \frac{\lambda}{2}u^2 - \frac{1}{p+1}u^{p+1}.$$

A few technical remarks are in order. First, for convenience, we will consider the case $m > 2$ in order to use some well known results that describe how approximate solutions behave. See [**39**],[**4**], [**5**], and [**54**]). These results have been extended to the case $m = 2$ (see [**1**]), but are more involved and not included here. In order to easily apply these results, we will modify the equation and consider instead:

$$-\Delta u + \lambda u = u^{p^*-\mu}, \tag{3}$$

with the corresponding change of functional

$$E(u) = \int_{S^m} \frac{1}{2}|\nabla u|^2 + \frac{\lambda}{2}u^2 - \frac{1}{p^*+1-\mu}u^{p^*+1-\mu},$$

where $p^* = \frac{m+2}{m-2}$. That these two problems are essentially the same when ε and μ are small has been shown by V. Benci and G. Cerami (see for instance [**8**]).

Then we can obtain a "simple" solution by using the direct methods of the calculus of variations, that is by minimizing

$$I(u) = \int_{S^m} \frac{1}{2}|\nabla u|^2 + \frac{\lambda}{2}u^2,$$

among positive functions in $H^1(S^m)$ subject to the additional constraint

$$\int_{S^m} u^{p^*+1-\mu} = 1.$$

The fact that μ is positive and small allows us to use standard techniques (for a complete treatment see [**55**]). This simple solution corresponds to the least energy nontrivial solution (a ground state in the physical terminology). More importantly, if we consider what happens when $\mu \to 0$, it is possible to show that a corresponding subsequence of solutions $\{u_\mu\}$ concentrates more and more around a point, that is a spike is formed. Once we know that least energy solutions concentrate when μ is small, it is reasonable to expect that some sort of superposition principle is valid, that is, linear combinations of solutions concentrating around different points will be approximate solutions. In fact a stronger assertion holds: if $\mu \to 0$ then ALL solutions of (3) will tend to a superposition of simple solutions. This is the content of the main result in [**54**]. In particular, it is possible to construct, given any positive integer k, a solution that has k concentrations. A natural question is how these concentrations arrange themselves. If among those solutions with k concentrations we are interested in those with least energy, we can prove that as $\mu \to 0$ the peaks of such solutions tend to be as far from each other as possible, thus solving a sphere packing problem on the sphere. The details of the proof of this assertion can be found in [**25**]. We mention that C. Gui and J. Wei [**31**] have obtained a similar result for Equation 3 on domains in R^n with Neumann boundary conditions carrying out the analysis when ε is small using implicit function techniques, rather than the variational approach based on concentration compactness methods described here.

There are many related open problems; we mention only a few. First, one finds solutions that are not necessarily positive and seek to understand how they behave when the relevant parameter tends to zero and observe what concentrations appear. By considering simple geometries, it is not difficult to convince oneself that concentration can occur in objects of higher dimension. Is it possible to characterize them in general? In other words, is there a geometric description for them? We refer to the work of A. Malchiodi and M. Montenegro for a recent contribution to this question [**40**]. We have only talked about equilibrium configurations, but what happens to the dynamics of the original reaction-diffusion equation is to a great extent an open problem. See [**17, 36**] and the references therein for a discussion of issues related to the dynamics of this problem. It has also been pointed out that boundary conditions do play an important role. For instance, if we consider the problem on a disc with zero Dirichlet boundary conditions, a well known result by Gidas, Ni and Nirenberg implies that the only positive solution is radially symmetric. However, if the boundary condition is changed to zero Neumann, then multiple (not necessarily radial) solutions can be obtained. For related issues we refer to the paper by A. Castro in this proceedings volume. Many authors have studied Dirichlet or Neumann boundary conditions. The results can be very different; an interesting problem is to consider a continuous transition from one boundary condition to another, .e.g.,

$$tu + (1-t)\frac{\partial u}{\partial \nu} = 0,$$

where t is a bifurcation parameter.

3. Closed geodesics on oval surfaces and related problems

We continue with a variational problem that for several reasons has attracted much attention. In one form or another, it is related to many interesting phenomena: phase transitions, superconductivity, alloy cooling, population genetics and many more (see [**51**] for some references on applications). In its simplest form, it can be thought of as choosing the nonlinearity

$$F(u) = (1 - u^2)^2,$$

in

$$\varepsilon^2 \Delta u - f(u) = 0,$$

where as before $F'(u) = f(u)$. What really matters is the structure of this nonlinearity, rather than its analytic expression (see also [**51**]). The reason for the wide variety of applicability of such problems is because they describe (in perhaps the simplest possible way) a system that has two equilibrium states, in fact two equally likely minima. This is a bistable system in which the only two stable steady states are global minima of the potential energy function; such a function is called a like a double-well.

From the variational point of view, this is perhaps the simplest nontrivial system (in the sense that it gives rise to a non convex functional) and is ubiquitous in Nature, from superconductivity to transmission of electrical impulses in the nervous system. The variational problem of determining the equilibrium configurations is then equivalent to finding the critical points of

$$E(u) = \int_\Omega \frac{\varepsilon}{2} |\nabla u|^2 + \frac{1}{\varepsilon} F(u)\, dx,$$

whose Euler-Lagrange equation is (3). Notice that we have rewritten the functional in such a way that the ε appears in the coefficients of both terms. This scaling turns out to be useful. Moreover, in this form, it is easier to guess what happens when ε is small. Indeed, if $\varepsilon << 1$, then, unless $u(x)$ is close to ± 1, where F is zero, the energy will become very big. In other words, if we are looking for solutions that have bounded energy (independently of ε), then, as $\varepsilon \to 0$, $u(x) \to \pm 1$ almost everywhere. If we imagine $u(x)$ to determine a black and white coloring of the domain, and associate the "white" with $u = -1$, correspondingly the "black" with $u = 1$ and the intermediate values with the possible intermediate "greys", we see that the variational problem determines a partition of the domain which in the limit $\varepsilon = 0$ corresponds to a black and white pattern. However, if we are dealing with continuous functions, say Lipschitz continuous, then the derivative of u will become very large for ε small in regions where there is a transition from white to black. We see that the presence of the gradient term in E penalizes these transitions and so there is a balance between these two effects: u trying to get closer to ± 1 and the transition regions being minimized. It turns out that for ε small, the functional is proportional to the length of the transition region. This statement makes clear the connection with a geometric problem.

Much work has been devoted to make this assertion precise. From the mathematical point of view, it was because of this property that E. De Giorgi introduced the previous functional (we refer to [**51**] and [**34**] and the corresponding bibliography). Of course the simplest situation is to have no transitions and in fact, $u \equiv \pm 1$ are indeed both global minima and solutions to the Euler-Lagrange equation. However, we are interested in nontrivial patterns and therefore, we either look for solutions that are not global minima or more generally for other critical points. One way of preventing the minimization problem from having trivial solutions is to impose a volume constraint, that is, to consider functions that satisfy

$$\int_\Omega u = m.$$

Of course in general such a constraint implies that a Lagrange multiplier term will appear in the equation.

Here we use a related but different constraint which in turn leads to a different geometric problem. In order to make the discussion as simple as possible, we restrict ourselves to the case in which Ω is a surface. Moreover, we will be considering only oval surfaces, that is, surfaces that are boundaries of strictly convex domains in R^3. The reason for this will become apparent soon. Let us for the time being just mention that we are interested in relating the qualitative properties of critical points of the functional with the existence of closed geodesics on the surface. We need first to recall some geometric results considered originally by Poincaré. What he realized was that it is possible to give a simple variational characterization of the shortest closed geodesic on an oval surface. Let $g : \Omega \to S^2$ denote the Gauss map of the surface in R^3, which by definition assigns to every $x \in \Omega$ its unit normal. If we consider a simple closed curve γ in Ω, then $g(\gamma)$ will be a simple closed curve in S^2, dividing the unit sphere in two regions; we denote by A_1 and A_2 the corresponding areas. Consider only those curves in Ω that under the Gauss map divide the unit sphere in regions of equal area. We let Γ denote this class, that is curves for which $A_1 = A_2$.

Then the assertion is that the curve that minimizes length in Γ is a closed geodesic of minimal length. We will discuss the importance of the problem of finding closed geodesics at the end of this section, but notice that this variational characterization suggests a natural constraint for the original problem we considered, namely,

$$\int_{S^2} u(g^{-1}(y))\, d\sigma = 0. \tag{4}$$

Indeed, if we think of u as a step function suitably smoothed out, it is clear that the previous constraint says that the transition region divides S^2 into two parts that have to be of the same area in order for the region where u is close to -1 to cancel the complement where u is close to 1 (recall we are assuming ε is small). We expect the following statements to be true.

Let u_ε be a minimizer of E among functions satisfying the constraint (4). Then, as $\varepsilon \to 0$ u_ε converges to a step function having a transition at a minimal closed geodesic. This is the content of the result proved in [**26**]. In order to make all the previous rigorous, we need to make precise what we mean by convergence of u, and of the transition, and so on. The natural notion is given in terms of varifold convergence. We refer to [**51, 34, 26**] for the details and proofs, but for the sake of completeness we describe the result here.

Let $\epsilon_i \to 0$ as $i \to \infty$ and $\{u_i\}$ be minima of the functional E for the corresponding value of $\epsilon = \epsilon_i$ under the constraint determined by the Gauss map discussed above. Denote by V_i the varifold associated with u_i, that is

$$V_i(A) = \int_{-1}^{1} v(w^i = t)(A)\, dt,$$

where A is a Borel set. We refer to [**26**] for the precise definition of v and w^i and we note that it is proved that the varifold is determined by the density $\|\nabla u_i\|^2$. This could be taken as an alternative definition.

Theorem:

When $\epsilon \to 0$ the varifolds V^i associated to minimal solutions of E, with the constraint 4 converge (up to a subsequence) in the sense of varifolds to a closed geodesic γ_0 of minimal length on the oval surface. In fact, for each compact $\tilde{U} \subset \Omega$ and $0 < b < 1$, $\{|u^i| \leq 1-b\} \cap \tilde{U}$ converges to $\tilde{U} \cap \mathrm{supp}\|V_{\gamma_0}\|$ in the Hausdorff distance sense, where V_{γ_0} is the varifold associated with γ_0.

We point out that from the previous discussion, the transition regions that form when ε tends to 0 have important properties; even locally they have to be minimal. So even locally, they should be geodesics or, for higher dimensions, minimal surfaces (see [**51, 34**]). It is interesting to mention that the previous approach can then be viewed as a regularization of the problem of finding minimal closed geodesics or minimal surfaces in general. This opens the possibility of applying PDE methods to this geometric problem. In particular, any multiplicity result for the PDE would imply a multiplicity result for the geometric problem. In the case of closed geodesics, a general result for surfaces that are diffeomorphic to the sphere was established by Birkhoff, who proved the existence of a minimal closed geodesic (see [**11**]). Later on, by using variational methods and studying the structure of the loop space, Ljusternik and Schirelman proved that in fact there are at least three closed geodesics. The natural question is whether these known results and other new results could be obtained using a PDE approach, which might be analytically simpler. In [**21**] some extensions to general Riemannian manifolds are considered.

The problem of studying closed geodesics is important for several reasons. We refer the reader to the book by Klingenberg for more details, which is by now a standard reference. This has been a very active area. The work of Berger, Bombieri, and Corke and the bibliography therein [**9, 19**] should also be mentioned. Other interesting connections with topology and analysis have been explored. For instance, in [**15**] E. Calabi and J. Capo establish that the shortest closed geodesic on a smooth Riemannian manifold of non-negative curvature is simple (non self-intersecting). The relationship with estimates of the first eigenvalue of the Laplacian on a manifold as well as other problems can be found in the previous work and its bibliography. We remark that the authors also discuss several important isoperimetric type inequalities; it would be interesting to see the relationship with standard elliptic estimates in the same way other analytic and geometric inequalities are related. One can consider, for example, Poincaré-Sobolev and standard isoperimetric inequalities (see for instance the contribution by P. Li in [**61**]).

Another related question is if Morse theoretical methods that are well established for elliptic operators can be useful in this context. Another area in which the existence of closed geodesics has played a central role is Hamiltonian dynamics and more generally in symplectic geometry and topology (see for instance [**33**] or [**41**]). In particular, it would be interesting to investigate if the ideas outlined here could give some understanding in the study of periodic orbits in Hamiltonian systems via a semiclassical approximation. Several works can be found on L. C. Evans web page where the semiclassical approximation in quantum mechanics and weak KAM and Aubry-Mather theory in Hamiltonian dynamics are related.

Again, we have not yet mentioned the dynamics of the problem. Extensive research has been done in this subject; the interested reader could consult [**35**]. The connection of these methods for studying the dynamic problem with the numerical methods developed by J. M. Neuberger is also to be explored (see his article in this proceedings and [**47**]). Just to end this discussion, we observe that if the oval surface we considered is "flattened out", it becomes a convex planar domain. In the limit, the transition becomes a straight line intersecting the boundary orthogonally (see [**27**] where we establish this result rigorously using variational methods). A similar result using a different approach has been obtained by M. Kowalsczyk.

4. A problem arising in phase transitions and capillarity theory

If the variational problem discussed in the previous section is prototypical of systems having two equilibria, the case when there are three or more such equilibria is also important and appears in several contexts. In particular, in physics it corresponds to a system with three stable phases, say water, ice and vapor, or in capillarity theory, where a liquid (phase one) is in contact with a gas (phase two) and the container (phase three). From the mathematical point of view there is an important qualitative difference with the bistable system. Indeed, one could think that a suitable generalization of the two phase to the three phase case could be given by a potential, just as the one given in Section 2, but with three minima, a,b, and c, instead of two. The problem with this approach is that the phases would be ordered in the sense that if the solution u to the problem (the order parameter, as it is known in this context) is a continuous function, then, assuming that $a < b < c$, there could never be an ac interface. The order parameter would necessarily take the value b and so, between phase a and phase c, there always would be a region occupied by phase b. This might be a good model in some situations, but in the applications we mentioned one important feature is the existence of the so called triple points, i.e., regimes in which the three phases coexist (pairwise) in equilibrium. In our example, ice and water, water and vapor, and ice and vapor. A way of circumventing this difficulty is to consider a vector valued order parameter $u = (U, V)$, and modify the potential to be a function of U and V taking real values, with three minima $a, b, c \in R^2$. In this way the order parameter could continuously vary from one stable phase to another, without having to take the value corresponding to the remaining one.

This problem has been studied by many authors and again we do not give an exhaustive list of references but provide as a starting point [**24**] and some useful references therein. There are several reasons why we include this problem. First, it is an example in which ideas from the calculus of variations and dynamical systems can be combined in order to obtain nontrivial solutions. Specifically, we use minimax methods from the calculus of variations to prove the existence of such solutions. It is suggested by this variational procedure that one of the qualitative features as ε tends to zero is precisely to develop a triple point, i.e. a point $x \in \Omega$ in which the three phases meet. This problem has been successfully studied using numerical methods, in particular by using the Gradient Newton Galerkin Algorithm, a version of Newton's method for elliptic equations implemented by J. Neuberger, D. Rice, and J. Swift (see [**48**]). They demonstrate that indeed the minimax solution exhibits a triple junction. Moreover they are able to use this approach to study the bifurcation structure of the solution set and its relation with symmetry and nodal properties of the solutions. The interaction between analytical and numerical methods is by itself an important issue and the scheme mentioned above provides a way of implementing abstract variational methods numerically, making them not only a powerful tool in proving the existence of solutions, but in actually computing them.

In what follows, we outline the construction presented in [**24**]. We remark that the equation and the corresponding functional are

$$\varepsilon^2 \Delta u + \nabla f(u) = 0$$

and

$$E(u) = \int_\Omega \frac{\varepsilon}{2}|\nabla u|^2 + \frac{1}{\varepsilon}F(u)\, dx,$$

respectively. That is, formally the same as in the previous section. However, the equation is really a system and $|\nabla u|^2$ is to be taken as $|\nabla U|^2 + |\nabla V|^2$. For simplicity, we consider the case of a planar domain Ω and Neumann, zero flux, boundary conditions $\frac{\partial u}{\partial \nu} = 0$, where as usual ν is the outer normal.

Now there are three trivial (spatially homogeneous solutions), $u_a \equiv a$ $u_b \equiv b$ and $u_c \equiv c$, which in function space correspond to three points. For every pair of them, a nontrivial solution can be constructed using the Mountain Pass Lemma (see [**52, 54**] for a discussion of the MPL and minimax methods in general). That is, there are solutions u_{ab}, u_{ac} and u_{bc} of the Euler-Lagrange equation that under reasonable assumptions on the potential (basically that the triple-well structure is well defined, see [**24**]) are different. These three solutions can be viewed as points in the function space $H^1(\Omega) \times H^1(\Omega)$. The next step is to construct a one dimensional family of functions, a triangle so to speak, that is invariant under minus the gradient flow of E and which provides a natural way of implementing a minimax procedure. This set is formed by constructing connecting trajectories (heteroclinic connections) from u_{ab} to u_a and u_b of the negative gradient flow

$$\frac{\partial u}{\partial t} = -\nabla E(u).$$

The same procedure can be repeated for the other solutions u_{ac} and u_{bc}. Thus, we have the one dimensional family that is formed by solutions of minus the gradient flow and so invariant by construction. The existence of these connecting trajectories is proved using Conley index techniques which have been applied to study the structure of invariant sets in many dynamical systems (see [**53**] and references therein for an introduction to Conley index theory and [**24**] for the proof in our setting).

Let τ be the invariant triangle just described. Then we can consider all two dimensional families that are spanned by τ. We denote by s one of these two dimensional families and by S the corresponding class. It can be shown that

$$c_\varepsilon = \inf_{s \in S} \max E(s)$$

is a critical value and that a corresponding critical point is different from the previous six already mentioned.

In the generic case, this solution has Morse index two (see [**42, 22**] for an introduction to Morse theory and [**53**] for applications to differential equations). This[1] and other considerations suggest that in fact this solution has a triple point structure. However, this is still to be rigorously proved.

5. An example of localized structures in particle physics

We finish this series of examples of singularly perturbed equations with an apparently different problem, namely, consider the sine-Gordon equation in $2+1$ dimensions:

$$u_{tt} - \varepsilon \Delta u + \frac{1}{\varepsilon} \sin u = 0.$$

This equation appears in relativistic field theory and is related to the simplest relativistic treatment of quantum fields, which leads to the related Klein-Gordon equations (see [**38**]). In the context of particle physics, it is the simplest case of the so called sigma models which were proposed by Skyrme (see also [**38, 60**] for a discussion of its relation with solitons and

[1]See [**6**] for a rigorous result that relates the nodal properties to the Morse index and some a priori estimates of solutions to some semilinear elliptic equations, or [**14**] where both variational ideas and Morse index techniques are used.

instantons in field theory) . In one spatial dimension, the sine-Gordon equation is a completely integrable system and has soliton solutions, which look like bumps. These localized structures are then identified with particles and are referred to as quasiparticles. In $2+1$ dimensions it is well known that the analogous solutions don't exist. It was proposed by J. Neu that different localized structures might exist, provided that they rotate, i.e. support some kind of angular momentum. These solutions were studied asymptotically in [**44**]. In mathematical terms, this might be achieved by establishing the existence of solutions that are not radially symmetric and that carry a traveling wave in the angular direction. That is, we look for solutions of the form

$$u(\theta - \sqrt{\varepsilon V} t, r),$$

where θ and r are polar coordinates. Therefore, they do not exhibit an explicit time dependence, but only through the angular variable. If we substitute u into the original equation, we have

$$\varepsilon V^2 u_{\theta\theta} - \varepsilon \Delta u + \frac{1}{\varepsilon} \sin u = 0,$$

which is formally the Euler-Lagrange equation of

$$I(u) = \int \frac{\varepsilon}{2} \left(u_r^2 + (\frac{1}{r^2} - V^2) u_\theta^2 \right) + \frac{1}{\varepsilon} W(u).$$

We have written the gradient in polar coordinates and denoted by W a primitive of $\sin u$. There are still several difficulties in handling this functional; the main one being that it is not elliptic. However, by considering the problem in a ball and choosing V sufficiently small, the functional becomes elliptic and well defined. In particular, standard variational tools can be applied to obtain nontrivial solutions, e.g., the MPL. However, as pointed out before, we are interested in nonradial solutions. This is accomplished in two steps. First, an estimate for the energy of nontrivial radial solutions is deduced, which guarantees that $I(u) > c$ for some constant c. Then, by constructing asymptotic solutions of the original equation, it is possible to exhibit an admissible path for the mountain pass minimax procedure for which the maximum of the energy is strictly less than c. This implies that the solution cannot be radial. We refer to [**46**] for the details and also for a discussion of the numerics. The limit when $\varepsilon \to 0$ is also important in this context. Both numerically and asymptotically, it is observed that the solutions develop transition layers. This is not surprising taking into account what has been discussed already. As it was pointed out in the paper by Neu, the region where the transition occurs should collapse to a minimal surface in Minkowski space as $\varepsilon \to 0$. Notice that they are very unstable and so disappear as time increases.

As with any of the previous examples, there are many open problems. In this case the study of solutions that have more than one transition layer is particularly interesting. From numerical simulations it is observed that the way in which these solutions collapse is far from trivial; they develop spiral arms and other localized structures typical of hyperbolic equations, e.g. cusps singularities. From a more general point of view, the study of hyperbolic equations from this perspective poses many interesting questions (see for instance [**7**]). In particular it would be interesting to study the relationship of this approach with other localized structures is, e.g., shock profiles or solitons (see [**29**] for a recent reference in a discussion of shocks is given in terms that might be naturally linked to the concepts outlined here).

6. Other problems

We conclude with a series of general questions more or less related to the material discussed so far. Before referring to specific problems we mention two specific areas that have not yet been explicitely discussed. The first one is the study of genetic control systems or more generally of genetic control networks. This is indeed a fascinating subject and its connections with PDE's are just beginning to be exploited. In particular in the recent work *Gradients, Diffusion, and Genes in Pattern Formation*, Nijihout presents some important open questions related to reaction-diffusion models and genetics. In considering several of these questions, a variational perspective would be very useful. We refer to Nijihout's paper [**45**]. The second new idea refers to optimization problems in material science and optimal shape design issues. The study of composite materials and related applications is also a very active area where the calculus of variations plays a central role. We refer for instance to the book on variational methods and structural optimization by A. Cherkaev for a recent reference [**18**].

a: In all of our featured problems, the Laplacian plays a major role. To understand what happens when another general second order elliptic operator is substituted is not merely an academic question. Such operators appear in a natural way in many examples, ranging from anisotropic crystal growth, chemical reactions, etc., to population dynamics and Riemannian geometry. For the Laplacian, spots and stripes are natural. To understand the geometry of the problem when a general elliptic operator is used would also be interesting. Along the same lines, extensions of these results to systems would be very useful.

b: Is it possible to extend the results on the characterization of the transition as a minimal surface (see [**34**]) to the hyperbolic case?

c: The problem of pattern formation in growing domains or non planar domains has become more and more a subject of study, because many problems in developmental biology and morphogenesis are connected to it. A clarification and a better understanding of the underlying geometric mechanisms is still needed.

d: As it was mentioned in several of the examples, the study of the dynamics of the localized structures is interesting in itself and although there has been lots of work in this direction, there is still much to be done (see for instance [**30, 23, 13, 12, 56, 57**]). We have yet to consider problems in fluid dynamics, superconductivity, etc., where other important localized structures form.

e: Introducing a small parameter and letting the parameter go to zero is precisely what is done in studying the semiclassical approximation in quantum mechanics. Using this approach to understand both the classical system from what is known of the quantum and the other way around is the subject of current research and the source of many interesting questions. In particular, to understand some important special solutions such as heteroclinic or homoclinic orbits, quasiperiodic solutions and invariant tori or certain behaviors such as Arnold's diffusion or turbulence from a variational perspective is proving very fruitful (see [**3, 2, 10, 37, 32, 50**]and also [**20**] for the study of localized solutions of the Schroedinger equation).

f: The numerical implementation of many of these ideas and of other variational methods is crucial if they are going to be of any practical use.

g: Instead of focusing on two specific nonlinearities, one could ask what happens with a generic nonlinearity. Besides stripes and spots some other structures are formed, e.g. laberynthic and spiral patterns. It would be very useful to have a more global picture of the structures.

7. Conclusion

We have given an idea of how variational problems in which a small parameter is present lead to an underlying geometric or dynamical question that can be used to understand the structure of solutions. We hope that some of the questions and problems given are of interest to the readers and that they find the bibliography a useful starting point. However, we stress that this is only the very tip of the iceberg; we make no attempt to incorporate all the references we were aware of. For brevity, we have not touched upon other relevant subjects such as optimization problems in Economics, Finance, and control theory.

References

[1] Adimurthi, Struwe, M. *Global Compactness properties of semilinear elliptic equations with critical exponential growth.* J. Func. Anal, 175, 125-167, (2000).

[2] Ambrosetti, A. and Badiale, M. *Homoclinics: Poincaré-Melnikov type results via a variational approach.* Ann. Inst. H. Poincaré, Analyse non lin.,15, 233-252, (1998)

[3] Ambrosetti, A. and Berti, M. *Homoclinics and complex dynamics in slowly oscilating dynamical systems.* Disc. cont. dyn. sys., 4 (3), 393-403, 1998.

[4] Bahri, A. *Critical points at infinity in some variational problems.* Pitman Res. Notes Math., 182.

[5] Bahri, A. and Coron, J.M. *On a nonlinear elliptic equation involving the critical Sobolev exponent: the effect of the topology of the domain.* Comm. Pure Appl. Math.

[6] Bahri, A and Lions, P.L. *Solutions of superlinear elliptic equations and their Morse Indices.* Comm. Pure Appl. Math., 41, 1027-1037, (1988).

[7] Benci, V. and Fortunato, D. *On the existence of the impossible pilot wave* . Rapporto del Dipartamento Interuniversitario di Matematica, n. 14, Bari, 1988.

[8] Benci, V. and Cerami, G. *Positive solutions of some nonlinear elliptic problems in exterior domains.* Arch. Rat. Mech. Anal. 99, 4, 283-300, 1987.

[9] Berger, M.S and Bombieri, E. *On Poincaré's isoperimetric problem for simple closed geodesics.* J. Func. Anal., 42, 274-298, 1981.

[10] Bessi, U. Aubry-Mather theory and Hamilton-Jacobi equations. Comm. Math. Phys., 235, 495-511, 2003.

[11] Birkhoff, G. *Dynamical systems with two degrees of freedom.* Trans. Amer. Math. Soc., 18, 199-300, 1917.

[12] Bronsard, L. and Reitich, F.*On three-phase boundary motion and the singular limit of a vector-valued Ginzburg-Landau equation.* Arch. Rat. Mech. and Anal., vol. 124, no. 4, 355-379, 1993.

[13] Bronsard, L. and Wetton, B. *A numerical method for tracking curve networks moving with curvature motion.* J. Comp. Phys., 120, 66-87, 1995.

[14] Castro, A. Cossio,J. and Neuberger, J.M.*A sign-changing solution for a superlinear Dirichlet problem.* Rocky Moun. J. Math., Vol. 27, No. 4., 1001-1053.

[15] Calabi, E. and Cao, J. *Simple closed geodesics on convex surfaces.* J. Diff. Geom., 36, 517-549, 1992.

[16] Chaplain, M.A.J., Ganesh, M. and Graham, I.G. *Spatio-temporal pattern formation on spherical surfaces: Numerical simulation and application to solid tumour growth.* J. Math. Biol., 42 (5), 387-423, 2001.

[17] Chen, X. and Kowalczyk, M. *Dynamics of an interior spike in the Gierer-Meinhardt system.* SIAM J. Math. Anal., 33, 1, 172-193, 2001.

[18] Cherkaev, A. *Vatiational Methods and Structural Optimization.* Springer-Verlag, (2002) .

[19] Croke, C. *Poincare's problem and the length of the shortest closed geodesic on a convex hypersurface.* J. Diff. Geom, 17, 595-634, 1980.

[20] Del Pino, M. and Felmer, P.

[21] Del Río, H., Garza, C. and Padilla, P. *Geodesics, soap bubbles and pattern formation in Riemannian manifolds.* J. Geom. Anal., Vol. 13, No. 4, 2003.

[22] Dubrovin, Fomenko, Novikov. *Modern Geometry.* Springer-Verlag.

[23] Lin, F. *Complex Ginzburg-Landau equations and dynamics of vortices, filaments, and codimension 2 submanifolds.* Comm. Pure Appl. Math., 51, 385-441 (1998).

[24] Flores, G., Padilla, P. and Tonegawa, Y. Higher energy solutions in the theory of phase transitions. J. Diff. Eq. , 2001.

[25] Garza, C. and Padilla, P. *A pattern formation problem on the sphere.* Preprint.

[26] Garza, C. and Padilla, P. *Closed geodesic on oval surfaces and pattern formation.* J. Geom. Anal, Vol 11, No. 2, 223-233, 2003.

[27] Garza, C. and Padilla, P. *Location of interfaces on canvex domains for the Allen-Cahn equations.* Elec. J. Diff. Equations.

[28] Gierer, A. and Meinhardt, H. A theory of biological pattern formation. Kibernetik, 12, 30-39, 1972.

[29] Giga, *Viscosity solutions with shocks* . Comm. Pure Appl. Math., 55, 431-480, 2002.

[30] Girao, P.M. and Kohn, R.V. *Convergence of a crystalline algorithm for the heat equation in one dimension and for the motion of a graph.* Numer. Math., 67, 41-70, 1994.

[31] Gui, C. and Wei, J. *Multiple interior peaks solutions for some singularly perturbed Neumann problems.* J. Diff. Eq., 158, 1, 1-27, 1999.

[32] Gomes, D. Iturriaga, R. Khanin, K and Padilla, P. *Convergence of viscosity solutions in Burger's equations.* Preprint.

[33] Hofer, H and Zehnder, E. *Symplectic Capacities.* Birkhauser, Basel, (1994).

[34] Hutchinson, J. and Tonegawa, Y. *Convergence of phase interfaces in the van del Waals-Cahn-Hilliard theory.* Calc. Var. Partial Diff. Eq., 10 (1), 49-84, 2000.

[35] Ilmanen, T. *Convergence of the Allen-Cahn Equation to Brakke's Motion by Mean Curvature.* J. Diff. Geom 38 (1993) 417-461.

[36] Iron, D. and Ward, M. *The dynamics of multispikes solutions to the one-dimensional Gierer-Meinhardt model* SIAM J. Appl. Math, 62, 6, 1924-1951, 2002.

[37] Iturriaga, Renato and Khanin, K.*Burgers turbulence and random Lagrangian systems.* Comm. Math. Phys. 232, 377-428, 2003.

[38] Kaku, M. *Quantum Field Theory.* Oxford Univ. Press, 1993.

[39] Lions, P.L. *The concentration compactness method in the calculus of variations, the limit case.* Rev. Mat. Iberoamericana 1,2, 145-201, 1985.

[40] Malchiodi, A. and Montenegro, M. *Boundary concentration phenomena for a singularly perturbed elliptic problem.* Comm. Pure Appl. Math. Vol LV, 1507-1568, 2002.

[41] McDuff, D. and Salamon, D. *Introduction to Symplectic Topology.* Clarendon Press, Oxford, (1995).

[42] Milnor, J. *Morse Theory.* Princeton University press, (1974).

[43] Murray, J.D. *Mathematical Biology.* Springer-Verlag, Berlin, 2nd edition, (1993).

[44] Minzoni, A. Smyth, N. and Worthy, A. *Pulse evolution for a two-dimensional sine-Gordon equation.* Physica D, Volume: 159, Issue: 1-2, November 1, 101-123 (2001).

[45] Müeller, G.B and Newman, S.A., *Origination of Organismal Form.* The Vienna Series in Theoretical Biology. MIT press. Cambridge, MS, (2003).

[46] Existence of travelling wave solutions for the sine-Gordon equation. Preprint.

[47] Neuberger, John M.; Swift, James W. , *Newton's method and Morse index for semilinear elliptic PDEs*, Internat. J. Bifur. Chaos Appl. Sci. Engrg. **11** (2001), no. 3, p801-820.

[48] Neuberger, John M.; Rice, Dennis; Swift, James W. , *Numerical Solutions to a Vector Ginzburg-Landau Equation with Triple-Well Potential*, Int. J. Biff. Chaos. **13**, No. 11 (2003) p3295-3306.

[49] Ni, W.M. *Diffusion, cross-diffusion and their spike-layer steady state.* Notices of the AMS, 45 (1), 9-18, 1998.

[50] Padilla, P. *Nonminimal quasiperiodic solutions of some second order equations.* To appear in TMNA.

[51] Padilla, P. and Tonegawa, Y. *On the convergence of stable phase transitions.* Comm. Pure Appl. Math. 51 (6), 551-579, (1998).

[52] Rabinowitz, P. *Minimax methods in critical point theory*

[53] Smoller, J. *Shock waves and reaction diffusion equations.* Sprincger-Verlag, Berlin, 1992.

[54] Struwe, M. *A global compactness result for elliptic boundary value problems involving nonlinearities.* Math. Z., 187,511-517, 1984

[55] Struwe, M. *Calculus of variations.* Springer-Verlag, Berlin, Heidelberg, 1990.

[56] Taylor, J.*Some mathematical challenges in material sciences.* Bull. AMS, 40, 69-87, 2003.

[57] Tonegawa, Y. *On the regularity of a chemical reacion interface.* Comm. PDE's. 23,1181-1207 ,1998.

[58] Turing, A. M. *The chemical basis of morphogenesis.* Phil. Trans. Roy. Soc. B (London), 237, 37-72, 1952.

[59] Varea, C. Aragón, J.L and Barrio, R.A. *Turing patterns on the sphere.* Phys Rev. E, 60, 4, 4588-4592, (1999).

[60] Yang, Y. *Solitons in Field Theory and Nonlinear Analysis.* Springer-Verlag, New York, Berlin, Heidelberg, 2001.

[61] Yau, S.T., editor. *Seminar on differential geometry.* Annals of Mathematics Studies 102, Princeton University Press, 1982.

Mathematical Institute, Oxford University, 24-29 St. Giles, OX1 3LB

Permanent address: IIMAS-FENOMEC, Universidad Nacional Autónoma de México, IIMAS, UNAM, Circuito Escolar, Cd. Universitaria. México D.F. 04510, México

E-mail address: pablo@mym.iimas.unam.mx

Contemporary Mathematics
Volume **357**, 2004

The Perturbed p-Laplacian and Quadratic Growth

Victor L. Shapiro

ABSTRACT. In the space $W^{1,p}(\Omega)$, where Ω is the N-torus, N$\geq$ 2, an existence theorem for periodic solutions of the perturbed p-Laplacian plus lower order terms with quadratic growth conditions is established. Galerkin techniques involving various inequalities are used. Also, use is made of the solenoidal condition involving the lower order terms. The exposition is essentially self- contained.

1.Introduction.

In this paper, we establish an existence theorem for periodic solutions of the perturbed p-Laplacian plus lower order terms with quadratic growth conditions. We use Galerkin techniques to establish our result. What we do is solve the equation in (1.8) below in the space $W^{1,p}(\Omega)$ by introducing finite-dimensional subspaces S_n such that $\cup_{n=0}^{\infty} S_n = W^{1,p}(\Omega)$ and find approximating solutions in the space S_n. We then show that these approximating solutions give rise to the solution for the equation in (1.8). The techniques are noncoercive, but nevertheless make use of the monotonicity condition for the perturbed p-Laplacian (see (3.20) below) coupled with the idea of Leray and Lion in their paper, [Le-Li]. Also, use is made of some of the ideas in [Sh1].

To state our results, we let Ω be the N-torus, N$\geq$ 2, so that

$$\Omega = \{x : -\pi \leq x_j < \pi, j = 1, ..., N\}.$$

$\phi \in C^{\infty}(\Omega)$ means that $\phi \in C^{\infty}(\mathbf{R}^N)$ and is periodic of period 2π in each variable. $W^{1,p}(\Omega)$, 1<p<∞, will designate the familiar closure of $C^{\infty}(\Omega)$, i.e., we set for $\phi \in C^{\infty}(\Omega)$

$$(1.1) \qquad \|\phi\|_p = \{\int_{\Omega} |\phi|^p\}^{1/p} \text{ and } \|\phi\|_{1,p} = \|\phi\|_p + \| \ |D\phi| \ \|_p$$

1991 *Mathematics Subject Classification.* Primary 35J60; Secondary 35D05, 34C25, 58J05.

Key words and phrases. Galerkin methods, p-Laplcian, periodic solutions.

where $D\phi = (D_1\phi, ..., D_N\phi)$ and $D_j\phi = \partial\phi/\partial x_j$. Then $L^p(\Omega)$ and $W^{1,p}(\Omega)$ will be the Banach spaces obtained when $C^\infty(\Omega)$ is closed with respect to the norms $\|\cdot\|_p$ and $\|\cdot\|_{1,p}$, respectively.

We shall deal with the second order, quasilinear elliptic operator Q, operating on $W^{1,p}(\Omega)$, of the form

$$Qu = -\sum_{j=1}^{N} D_j[a(x,u)\,|Du|^{p-2}\,D_j u] + \sum_{j=1}^{N} b^j(x,u,Du)D_j u \tag{1.2}$$

The coefficients of Q, namely $a(x,s)$ and $b^j(x,s,q)$ are assumed to be defined for all values of $(x,s) \in (\Omega\backslash E) \times \mathbf{R}$ and $(x,s,q) \in (\Omega\backslash E) \times \mathbf{R} \times \mathbf{R}^N$, respectively, where E$\subset \Omega$ is of N-dimensional Lebesgue measure zero. Furthermore, we shall assume the following throughout the paper.

(Q-1) The coefficients $a(x,s)$ and $b^j(x,s,q)$ satisfy the Caratheodory conditions: For each fixed $s \in \mathbf{R}$ and each fixed $q \in \mathbf{R}^N$, the functions $a^j(x,s)$ and $b^j(x,s,q)$ are measurable; for a.e. $x \in \Omega$, the functions $a(x,s)$ and $b^j(x,s,q)$ are respectively continuous in $\mathbf{R}$ and $\mathbf{R} \times \mathbf{R}^N$, $j = 1, ..., N$.

(Q-2) $\exists$ two positive constants η_0, η_1 such that $\eta_0 \leq a(x,s) \leq \eta_1$ for $s \in \mathbf{R}$ and a.e. $x \in \Omega$.

(Q-3) $\exists$ a nonnegative function $b(x) \in L^p(\Omega)$ and positive constants η_2 and η_3 such that

$$\left|b^j(x,s,q)\right| \leq b(x) + \eta_2\,|s| + \eta_3\,|q|$$

for $s \in \mathbf{R}$, $q \in \mathbf{R}^N$, and a.e. $x \in \Omega$, $j = 1, ..., N$.

(Q-4) For every $u \in W^{1,p}(\Omega)$, the vector-valued function

$$\mathbf{b}(x,u,Du) = [b^1(x,u,Du), ..., b^N(x,u,Du)]$$

is weakly solenoidal, i.e.,

$$\sum_{j=1}^{N} \int_\Omega b^j(x,u,Du)D_j\phi = 0 \tag{1.3}$$

for every $u \in W^{1,p}(\Omega)$ and every $\phi \in C^\infty(\Omega)$.

Before proceeding, we give an example of a vector-valued function

$$\mathbf{b}(x,s,q) = [b^1(x,s,q), ..., b^N(x,s,q)]$$

which meets (Q-1), (Q-3),and (Q-4). We set

$$b^1(x,s,q) = q_2 \sin x_1\ ,\quad b^2(x,s,q) = -q_1 \sin x_1 - s\cos x_1,$$

and $b^j(x,s,q) \equiv 0$ for $j = 3, ..., N$. $\mathbf{b}(x,s,q)$ clearly meets (Q-1) and (Q-3). If $\psi \in C^\infty(\Omega)$, then, as a computation easily shows, $\sum_{j=1}^N D_j b^j(x,\psi,D\psi) = 0\ \forall x \in \Omega$. Therefore, using integration by parts, we see that (1.3) holds when $u \in C^\infty(\Omega)$. Then it follows from (Q-1) and (Q-3), and from the very definition of $W^{1,p}(\Omega)$, that (1.3) holds when $u \in W^{1,p}(\Omega)$ and $\phi \in C^\infty(\Omega)$. Consequently, (Q-4) holds.

In this paper, we shall always assume that $p > 3N/(N+1)$ and that $N \geq 2$. With this in mind, we set for $u, v \in W^{1,p}(\Omega)$,

$$Q(u,v) = \sum_{j=1}^{N} \int_{\Omega} [a(x,u)\,|Du|^{p-2}\, D_j u D_j v + b^j(x,u,Du) D_j u v] \tag{1.4}$$

and observe from (Q-2) and (Q-3) that $Q(u,v)$ is well-defined for $u, v \in W^{1,p}(\Omega)$. (Note, from [Ad, p. 97], $W^{1,p}(\Omega)$ is, in particular, continuously imbedded in $L^{p^\diamond}(\Omega)$, where $p^\diamond = p/(p-2)$. Hence, because the expression $b^j(x,u,Du)D_j u$ has quadratic growth in $|Du|$, the integral involving this term and v is indeed well-defined.)

Next, we observe from [Ne, p. 32] that there exists a constant $\eta^\diamond > 0$ such that

$$\eta^\diamond \, \|u\|_p^p \leq \| \ |Du| \ \|_p^p \text{ for } u \in C^\infty(\Omega) \text{ with } \int_\Omega u = 0. \tag{1.5}$$

In the sequel, we shall set

$$u = u_a + u_b \quad \text{where } u_a = (2\pi)^{-N} \int_\Omega u \text{ for } u \in W^{1,p}(\Omega). \tag{1.6}$$

Observing that for $\phi \in C^\infty(\Omega)$, $\phi D_j \phi = 2^{-1} D_j \phi^2$ and that $D_j \phi = D_j \phi_b$, we see from (Q-2), (Q-4), (1.4), and (1.5) that

$$Q(\phi, \phi_b) \geq \eta_0 \, \| \ |D\phi_b| \ \|_p^p \geq \eta_0 \eta^\diamond \, \|\phi_b\|_p^p \quad \forall \phi \in C^\infty(\Omega).$$

We obtain from this last computation that

$$\eta^* \, \|\phi_b\|_p^p \leq Q(\phi, \phi_b) \quad \forall \phi \in C^\infty(\Omega), \tag{1.7}$$

where $\eta^* > 0$, and $\eta^* = \eta_0 \eta^\diamond$. Also, ϕ_b is defined in (1.6).

In this paper, we study the equation

$$Qu = \lambda\, |u|^{p-2}\, u + f(x,u) + G, \tag{1.8}$$

where $\lambda \in \mathbf{R}^+$, $G \in [W^{1,p}(\Omega)]^*$, the dual of $W^{1,p}(\Omega)$, and $f(x,s)$ is such that

$$f : \Omega \times \mathbf{R} \to \mathbf{R}$$

and meets the following three conditions:

(f-1) $f(x,s)$ meets the Caratheodory conditions;

(f-2) $\exists\ f_0 \in L^{p'}(\Omega)$ such that $|f(x,s)| \leq \lambda\, |s|^{p-1} + f_0(x)$ $\forall s \in \mathbf{R}$ and a.e. $x \in \Omega$;

(f-3) $\exists\ f_1, f_2 \in L^{p'}(\Omega)$ with $f_1, f_2 \geq 0$ such that $sf(x,s) \geq -f_1(x)\,|s| - f_2(x)$ $\forall s \in \mathbf{R}$ and a.e. $x \in \Omega$.

In the above, $p' = p/(p-1)$. Also, $\lambda > 0$ and will satisfy an upper bound depending on η^*, defined in (1.7). We shall say that $u \in W^{1,p}(\Omega)$ is a weak solution of (1.8) if the following obtains:

$$Q(u,v) = \int_\Omega [\lambda\, |u|^{p-2}\, u + f(x,u)]v + G(v) \quad \forall v \in W^{1,p}(\Omega), \tag{1.9}$$

where $Q(u,v)$ is defined in (1.4).

The existence theorem dealing with weak solutions of (1.8) that we shall establish here is the following:

Theorem. *Let Ω be the N-torus, $N\geq 2$, and let $p>3N/(N+1)$. Assume that Qu, given by (1.2), meets (Q-1)-(Q-4). Suppose, also, that f(x,s) meets (f-1)-(f-3) and $G\in [W^{1,p}(\Omega)]^*$. Furthermore, suppose that*

$$0 < \lambda < p\eta^*(p')^{p-1}/4^p \tag{1.10}$$

where $p' = p/(p-1)$ and η^* is defined in (1.7). Then there exists $u\in W^{1,p}(\Omega)$ which is a weak solution of (1.8).

We observe by (Q-3), the lower order terms of Qu, namely $\sum_{j=1}^{N} b^j(x,u,Du)D_j u$, do indeed have quadratic growth in $|Du|$. Also, we see from [Dr-Ku-Ni, p.188] that the principal order terms of our operator Qu is in the form of the perturbed $p-$ *Laplacian.*

The existence theorem above was motivated by our previous paper, [Sh1]. The Galerkin techniques that we shall use here, however, are considerably different from those used in [Sh1].

2. Fundamental Lemmas.

With CONS standing for a complete orthonormal system, the first lemma that we establish is the following:

Lemma 1. *Let Ω be the N-torus, $N\geq 2$, and let $p>3N/(N+1)$. Then, with $\phi_0(x) = (2\pi)^{-N}$,*

$$\exists\{\phi_n\}_{n=0}^{\infty} \text{which is a CONS in } L^2(\Omega) \text{with } \phi_n \in C^\infty(\Omega)\ \forall n. \tag{2.1}$$

Let S_J be the subspace spanned by $\{\phi_0,...,\phi_J\}$. Then, given $v\in W^{1,p}(\Omega)$, $\exists\{v_J\}_{J=0}^{\infty}$ with $v_J\in S_J$ such that

$$\lim_{J\to\infty} \|v - v_J\|_{1,p} \to 0. \tag{2.2}$$

To establish this lemma, we take it as well-known that $\{\cos jt\}_{j=0}^{\infty} \cup \{\sin jt\}_{j=1}^{\infty}$ properly normalized gives rise to a CONS in $L^2(T_1)$, where T_1 is the 1-torus. Likewise, the union of $\{\cos j_1 t \cos j_2 s\}_{j_1=0,j_2=0}^{\infty,\infty}$, and $\{\cos j_1 t \sin j_2 s\}_{j_1=0,j_2=1}^{\infty,\infty}$, and $\{\sin j_1 t \cos j_2 s\}_{j_1=1,j_2=0}^{\infty,\infty}$, and $\{\sin j_1 t \sin j_2 s\}_{j_1=1,j_2=1}^{\infty,\infty}$ gives rise to a CONS in $L^2(\Omega)$ where Ω is now the 2-torus. Continuing in this vein, we see that with Ω now the N-torus, N$\geq$ 2, there exists $\{\phi_n\}_{n=0}^{\infty}$which is a CONS in $L^2(\Omega)$ with $\phi_0(x) = (2\pi)^{-N}$ and

$$\phi_n(x) = \tau_{j_1}(x_1)\cdots\tau_{j_N}(x_N) \text{ where } \tau_{j_k}(x_k) = \cos j_k x_k \text{ or } \sin j_k x_k.$$

Consequently, (2.1) does indeed hold.

To show that (2.2) is valid, given $\psi \in C^\infty(\Omega)$, we observe from well-known facts about multiple Fourier series (see [St-We] or [Sh2]) that there exists $\psi_J \in S_J$ such that $\psi_J \to \psi$ and $D_k\psi_J \to D_k\psi$ as $J \to \infty$ uniformly in Ω for k=1,...,N. Hence, (2.2) holds for $\psi \in C^\infty(\Omega)$.

Next, given $v \in W^{1,p}(\Omega)$ and a positive integer n, from the very definition of $W^{1,p}(\Omega)$, it follows that there exists $\psi^n \in C^\infty(\Omega)$ such that $\|v - \psi^n\|_{1,p} \leq$

$1/2n$. Also, from the above paragraph, it follows that there exists $\psi^n_{J_n} \in S_{J_n}$ with $\|\psi^n_{J_n} - \psi^n\|_{1,p} \leq 1/2n$. Consequently, setting $v_{J_n} = \psi^n_{J_n}$, we see that $\|v - v_{J_n}\|_{1,p} \leq 1/n$. Hence, we obtain that

$$\text{there exists } J_1 < J_2 < \cdots < J_n \to \infty \text{ and } \{v_{J_n}\}_{n=1}^{\infty} \text{ with } v_{J_n} \in S_{J_n}$$

such that $\|v - v_{J_n}\|_{1,p} \to 0$ as $n \to \infty$. But this fact implies that (2.2) holds (see [Sh3, p. 1854]), completing the proof of the lemma.

The next lemma we establish is the following:

Lemma 2. *Suppose* $0 < \lambda < p\eta^*(p')^{p-1}/4^p$ *where* η^*, p, *and* p' *are as in the Theorem. Then* $\exists \varepsilon_0 > 0$ with $\varepsilon_0 < \eta^*, \lambda$ *such that*

$$(2.3) \qquad (\eta^* - \varepsilon_0)A^p + (\lambda - \varepsilon_0)B^{p'} > 4\lambda AB \quad \forall A, B \in \mathbf{R}^+.$$

To establish the lemma, we observe that $\eta^*\lambda^{p-1}(p')^{p-1}/(4\lambda)^p > 1/p$. Hence, there exists $\varepsilon_0 > 0$ with $\varepsilon_0 < \eta^*, \lambda$ such that

$$(2.4) \qquad (\eta^* - \varepsilon_0)(\lambda - \varepsilon_0)^{p-1}(p')^{p-1}/(4\lambda)^p > 1/p.$$

Next, with A and B two given positive real numbers, we set $C = 4\lambda A/(\lambda - \varepsilon_0)p'$. Then it follows from Young's inequality, [Ad, (5), p. 229], that

$$C^p/p + B^{p'}/p' \geq CB.$$

Replacing C by the corresponding value involving A, we obtain from this last inequality that

$$(4\lambda A)^p/p[(\lambda - \varepsilon_0)p']^p + B^{p'}/p' \geq 4\lambda AB/(\lambda - \varepsilon_0)p'.$$

Consequently,

$$(2.5) \qquad \frac{(4\lambda)^p}{p(\lambda - \varepsilon_0)^{p-1}(p')^{p-1}}A^p + (\lambda - \varepsilon_0)B^{p'} \geq 4\lambda AB.$$

But, it follows from the inequality in (2.4) that $(\eta^* - \varepsilon_0)$ is strictly greater than the coefficient of A^p in the inequality in (2.5). The inequality in (2.3) follows immediately from this fact, and the lemma is established.

Next, let g: $\Omega \times \mathbf{R} \to \mathbf{R}$, and g(x,s) satisfy the following two conditions:

(g-1) g(x,s) satisfies the Caratheodory conditions;

(g-2) $\exists K > 0$ such that $|g(x,s)| \leq K \quad \forall s \in \mathbf{R}$ and a.e. x$\in \Omega$.

For such a g(x,s) as above, we establish the following lemma:

Lemma 3. *Suppose* λ, η^*, p, *and* G *are as in the Theorem. Suppose also that g(x,s) meets (g-1) and (g-2). Let* J *be a positive integer, and let* S_J *be the subspace given in Lemma 1. Then there exists* $u^* \in S_J$ *such that*

$$(2.6) \qquad Q(u^*, v) = \int_\Omega [\lambda\, |u^*|^{p-2}\, u^* + g(x, u^*)]v + G(v) \quad \forall v \in S_J.$$

To establish the lemma, we set $\alpha = (\alpha_0, \alpha_1, ..., \alpha_J) \in \mathbf{R}^{J+1}$, and

$$(2.7) \qquad u = \sum_{j=0}^{J} \alpha_j \phi_j \quad \text{and} \quad \widetilde{u} = -\alpha_0\phi_0 + \sum_{j=1}^{J} \alpha_j \phi_j$$

where ϕ_j is given in Lemma 1. Next, set

$$F_0(\alpha) = -Q(u,\phi_0) + \int_\Omega [\lambda\,|u|^{p-2}\,u + g(x,u)]\phi_0 + G(\phi_0), \tag{2.8}$$

$$F_j(\alpha) = Q(u,\phi_j) - \int_\Omega [\lambda\,|u|^{p-2}\,u + g(x,u)]\phi_j - G(\phi_j) \tag{2.9}$$

for j=1,...,J where u is given in (2.7). It follows from the fact that each $\phi_j \in C^\infty(\Omega)$, from (g-1), (g-2), and from(Q-1)-(Q-3) that $F = (F_0, ..., F_J)$ is such that

$$F : \mathbf{R}^{J+1} \to \mathbf{R}^{J+1} \text{ and } F \in C^0(\mathbf{R}^{J+1}, \mathbf{R}^{J+1}). \tag{2.10}$$

It also follows from (2.7)-(2.10), that

$$F(\alpha)\cdot\alpha = Q(u,\ \widetilde{u}) - \int_\Omega [\lambda\,|u|^{p-2}\,u + g(x,u)]\ \widetilde{u} - G(\ \widetilde{u}). \tag{2.11}$$

Using the fact that ϕ_0 is a constant, we see from (2.1), (1.6), and (2.7) that

$$u_a = \alpha_0\phi_0,\ u_b = u - u_a, \text{ and } \widetilde{u} = -u_a + u_b = -u + 2u_b. \tag{2.12}$$

Also, we see from(1.4) that $Q(u,\widetilde{u}) = Q(u,u_b) = Q(u,u)$. Consequently, we obtain from (Q-2), (Q-4), (1.4), and (2.12) that

$$Q(u,\widetilde{u}) \geq \eta_0 \sum_{j=1}^{N} [\int_\Omega |Du|^{p-2}\,|D_j u|^2 + \int_\Omega b^j(x,u,Du)D_j u_b^2/2].$$

Hence, from (1.3) in (Q-4), we have $Q(u,\widetilde{u}) \geq \eta_0 \|\ |Du|\ \|_p^p$, and we conclude from (2.11) and this last inequality that

$$F(\alpha)\cdot\alpha \geq \eta_0 \|\ |Du_b|\ \|_p^p - \int_\Omega \lambda\,|u|^{p-2}\,u\widetilde{u} - \int_\Omega g(x,u)\widetilde{u} - G(\widetilde{u}). \tag{2.13}$$

From (2.12), we see that $\widetilde{u} = -u + 2u_b$ and obtain that
(2.14)

$$-\int_\Omega \lambda\,|u|^{p-2}\,u\ \widetilde{u} = \lambda\,\|u\|_p^p - 2\lambda \int_\Omega |u|^{p-2}\,uu_b \geq \lambda\,\|u\|_p^p - 2\lambda\,\|u\|_p^{p-1}\,\|u_b\|_p\,.$$

Next, we observe from (1.7) that $\eta^* = \eta_0\eta^\diamond$, and we choose $\varepsilon_1 > 0$, so that

$$(\eta_0 - \varepsilon_1)\eta^\diamond \geq \eta^* - \varepsilon_0 > 0,$$

where ε_0 is defined in Lemma 2. Consequently, from (1.5), (2.14), and this last inequality, we see that

$$(\eta_0 - \varepsilon_1)\,\|\ |Du_b|\ \|_p^p - \int_\Omega \lambda\,|u|^{p-2}\,u\widetilde{u} \geq \varepsilon_0\,\|u\|_p^p$$

where we also have made use of Lemma 2 with $A = \|u_b\|_p$ and $B = \|u\|_p^{p-1}$. Using this last inequality in conjunction with the inequality in (2.13), we obtain that

$$F(\alpha)\cdot\alpha \geq \varepsilon_1 \|\ |Du_b|\ \|_p^p + \varepsilon_0 \|u\|_p^p - \int_\Omega g(x,u)\widetilde{u} - G(\widetilde{u}), \tag{2.15}$$

where $\varepsilon_0, \varepsilon_1 > 0$.

It is easy to see from (g-2), (2.12), and the fact that G$\in [W^{1,p}(\Omega)]^*$ that $\exists K_1 > 0$ such that

$$\left|\int_\Omega g(x,u)\widetilde{u} + G(\widetilde{u})\right| \leq K_1[\|\ |Du_b|\ \|_p + \|u_b\|_p + \|u\|_p\ ] \tag{2.16}$$

$\forall u \in S_J$. Since $p > 2$, we have from (1.5), (2.15), and (2.16) that there is a constant $K_2 > 0$ such that

$$F(\alpha)\cdot\alpha \geq \varepsilon_0 \|u\|_p^p/2 - K_2 \qquad \forall\alpha \in \mathbf{R}^{J+1} \tag{2.17}$$

where $u = \sum_{j=0}^J \alpha_j\phi_j$. But $|\alpha|^2 = \int_\Omega u^2$, and we conclude from Holder's inequality and the fact that $p > 2$, that $\lim_{|\alpha|\to\infty} \|u\|_p = \infty$. Hence it follows from (2.17),

$$\exists s_0 > 0 \text{ such that } F(\alpha)\cdot\alpha > 0 \qquad \text{for } |\alpha| \geq s_0.$$

But then from (2.10) and [Nir, p.11], we obtain that

$$\exists\alpha^* = (\alpha_0^*, \alpha_1^*, ..., \alpha_J^*) \text{ with } |\alpha^*| < s_0 \text{ such that } F_j(\alpha^*) = 0, \tag{2.18}$$

for $j = 0, 1, ..., J$.

With α^* as in (2.18), we set $u^* = \sum_{j=0}^J \alpha_j^*\phi_j$ and obtain from (2.8) and (2.9) that

(2.19)

$$Q(u^*, \phi_j) = \int_\Omega [\lambda |u^*|^{p-2} u^* + g(x,u^*)]\phi_j + G(\phi_j) \quad for \quad j = 0, 1, ..., J.$$

Given $v \in S_J$, $\exists\beta = (\beta_0, \beta_1, ..., \beta_J)$ such that $v = \sum_{j=0}^J \beta_j\phi_j$. The conclusion to Lemma 3 follows immediately from this fact and (2.19).

3. Proof of Theorem.

3. Proof of Theorem. To prove the theorem, we set

(3.1)

$$f_n(x,s) = f(x,s) \quad \text{if} \quad |f(x,s)| \leq n, \text{ and } \ f_n(x,s) = \begin{cases} n & \text{if } f_n(x,s) > n \\ -n & \text{if } f_n(x,s) < -n \end{cases}.$$

Since $f_n(x,s)$ meets (g-1) and (g-2), it qualifies for Lemma 3. Hence, there exists a sequence $\{u_n\}_{n=1}^\infty$ with each $u_n \in S_n$ such that

$$Q(u_n, v) = \int_\Omega [\lambda |u_n|^{p-2} u_n + f_n(x,u_n)]v + G(v) \quad \forall v \in S_n \tag{3.2}$$

where S_n is the subspace of $C^\infty(\Omega)$ spanned by $\{\phi_0, ..., \phi_n\}$ and $Q(u_n, v)$ is defined by (1.4).

We first propose to show that

$$\text{the sequence } \left\{\|u_n\|_{1,p}\right\}_{n=1}^\infty \text{ is uniformly bounded.} \tag{3.3}$$

In order to accomplish this, we set

(3.4)
$$u_{na} = (2\pi)^{-N} \int_{\Omega} u_n, \; u_{nb} = u_n - u_{na}, \; \widetilde{u}_n = -u_{na} + u_{nb} = -u_n + 2u_{nb},$$

and observe from (1.4), (Q-2), and (Q-4) that

$$Q(u_n, \widetilde{u}_n) = Q(u_n, u_{nb}) \geq \eta_0 \| \; |Du_{nb}| \; \|_p^p \quad \forall n. \tag{3.5}$$

Also, we see from (3.4) that

$$\int_{\Omega} |u_n|^{p-2} u_n \widetilde{u}_n = - \int_{\Omega} |u_n|^p + 2 \int_{\Omega} |u_n|^{p-2} u_n u_{nb}, \tag{3.6}$$

and from (f-2) and (f-3) (we note that f_n meets (f-1) and (f-2) with the same f_0, f_1, and f_2) that

$$\int_{\Omega} f_n(x, u_n)\widetilde{u}_n \leq \int_{\Omega} [f_1(x) |u_n| + f_2(x)] + 2 \int_{\Omega} [\lambda |u_n|^{p-1} + f_0(x)] |u_{nb}| \tag{3.7}$$

where f_0, f_1, and f_2 are nonnegative functions in $L^{p'}(\Omega)$.

Next, with $\eta^* = \eta_0 \eta^\diamond$ as defined in (1.7), we choose $\varepsilon_1 > 0$ so that

$$(\eta_0 - \varepsilon_1)\eta^\diamond \geq \eta^* - \varepsilon_0 > 0 \text{ where } \varepsilon_0 \text{ is given in Lemma 2.}$$

We then observe from (1.5), (3.4), (3.5), and this last inequality that

$$\varepsilon_1 \| \; |Du_{nb}| \; \|_p^p + (\eta^* - \varepsilon_0) \|u_{nb}\|_p^p \leq Q(u_n, \widetilde{u}_n). \tag{3.8}$$

Then it follows from (3.2) and (3.6)-(3.8) that

(3.9)
$$\varepsilon_1 \| \; |Du_{nb}| \; \|_p^p + \varepsilon_0 \|u_n\|_p^p \leq \int_{\Omega} [f_1(x) |u_n| + f_2(x) + 2f_0(x) |u_{nb}|] + G(\widetilde{u}_n)$$

for every n, where we have also made use of Holder's inequality as well as Lemma 2 with $A = \|u_{nb}\|_p$ and $B = \|u_n\|_p^{p-1}$.

Since the constant $|u_{na}| \leq (2\pi)^{-N+N/p'} \|u_n\|_p$ and since $G \in [W^{1,p}(\Omega)]^*$, we see from (1.5) that there are two constants K_1 and K_2 such that the right-hand side of the inequality in (3.9) is majorized by

$$K_1(\| \; |Du_{nb}| \; \|_p + \|u_n\|_p) + K_2 \quad \forall n.$$

Consequently, we have from (3.9) that

$$\varepsilon_1 \| \; |Du_{nb}| \; \|_p^p + \varepsilon_0 \|u_n\|_p^p \leq K_1(\| \; |Du_{nb}| \; \|_p + \|u_n\|_p) + K_2 \quad \forall n.$$

But ε_0 and ε_1 are positive real numbers and $p > 2$; so (3.3) follows immediately from (1.1) and this last inequality.

Now, as is well-known, $W^{1,p}(\Omega)$ is a separable reflexive Banach space. As a consequence, we obtain from (3.3) and the compact imbedding theorem for Sobolev spaces, [Ad, p. 144], the existence of a subsequence and a $u \in W^{1,p}(\Omega)$ for which the following five statements hold (for ease of notation, we use the full sequence):

$$\exists K_3 \text{ such that } \|u_n\|_{1,p} \leq K_3 \quad \forall n, \tag{3.10}$$

$$(3.11) \qquad \lim_{n\to\infty} \|u_n - u\|_{\widetilde{p}} = 0 \quad \text{for } \widetilde{p} < p^*,$$

$$(3.12) \qquad \lim_{n\to\infty} \int_\Omega D_j u_n w = \int_\Omega D_j u w \quad \forall w \in L^{p'}(\Omega) \text{ and for } j = 1, ..., N,$$

$$(3..13) \qquad \lim_{n\to\infty} T(u_n) = T(u) \quad \forall T \in [W^{1,p}(\Omega)]^*,$$

$$(3.14) \qquad \lim_{n\to\infty} u_n(x) = u(x) \quad \text{a.e. in } \Omega,$$

where $p^* = Np/(N-p)$ for $p < N$ and $p^* = \infty$ for $p \geq N$.

From (3.10), we see that if $\exists J$ such that $v \in S_J$, then $\{|u_n|^{p-1} |v|\}_{n=0}^\infty$ is absolutely equi-integrable, i.e., *given* $\varepsilon > 0$, $\exists \delta > 0$ *such that if* $E \subset \Omega$ *and* $meas(E) < \delta$, *then* $\int_E |u_n|^{p-1} |v| < \varepsilon$. Consequently, we obtain from (3.13), (3.14), (f-1), (f-2), and Egoroff's theorem applied to the right-hand side of (3.2) that

$$(3.15) \quad \lim_{n\to\infty} Q(u_n, v) = \int_\Omega [\lambda |u|^{p-2} u + f(x,u)]v + G(v) \quad \forall v \in S_J \text{ and } \forall J.$$

We also claim that there exist a subsequence $\{u_{n_k}\}_{k=1}^\infty$ such that

$$(3.16) \qquad \lim_{k\to\infty} Q(u_{n_k}, v) = Q(u,v) \quad \forall v \in S_J \text{ and } \forall J.$$

In order to establish (3.16), we see that it is sufficient to show that there exists subsequences $\{D_j u_{n_k}\}_{k=1}^\infty$, j=1,...,N, such that

$$(3.17) \qquad \lim_{k\to\infty} D_j u_{n_k} = D_j u \quad \text{a.e. in } \Omega \quad \text{for j} = 1, ..., N.$$

For if (3.17) is established, then we see from (Q-1), (3.14) and (3.17) that both

$$\lim_{k\to\infty} a(x, u_{n_k}) |Du_{n_k}|^{p-2} D_j u_{n_k} = a(x,u) |Du|^{p-2} D_j u \qquad \text{a.e. in } \Omega,$$

and

$$\lim_{k\to\infty} b^j(x, u_{n_k}, Du_{n_k}) D_j u_{n_k} = b^j(x,u,Du) D_j u \quad \text{a.e. in } \Omega$$

for j=1,...,N. Likewise, it follows from (Q-2), (Q-3), and (3.10) that

$$\left\{a(x, u_{n_k}) |Du_{n_k}|^{p-2} D_j u_{n_k} D_j v\right\}_{k=1}^\infty \quad \text{and} \quad \{b^j(x, u_{n_k}, Du_{n_k}) D_j u_{n_k} v\}_{k=1}^\infty$$

are absolutely equi-integrable for j=1,...,N where we have also made use of the fact that p>2 and $v \in C^\infty(\Omega)$,and therefore, in $L^\infty(\Omega)$. Consequently, it follows from Egoroff's theorem and (1.4) that

$$\lim_{k\to\infty} \int_\Omega [a(x, u_{n_k}) |Du_{n_k}|^{p-2} D_j u_{n_k} D_j v + b^j(x, u_{n_k}, Du_{n_k}) D_j u_{n_k} v] = Q(u,v).$$

Hence, to establish (3.16) it is sufficient to show that (3.17) holds.

In order to establish this (3.17), we make the observation from Lemma 1 that

$$(3.18) \quad \exists \{P_n u\}_{n=0}^\infty \text{ with } P_n u \in S_n \text{ such that } \lim_{n\to\infty} \|P_n u - u\|_{1,p} = 0.$$

Also, since p>3N/(N+1), we obtain from [Ad, p.97] and this last fact that

$$(3.19) \qquad \lim_{n\to\infty} \|P_n u - u\|_{p/(p-2)} = 0.$$

Next, we see that for $q \in \mathbf{R}^N$ with $q = (q_1, ..., q_N)$ that

$$(3.20) \quad \sum_{j=1}^{N} a(x,s)[|q|^{p-2} q_j - |q'|^{p-2} q'_j](q_j - q'_j) > 0 \quad \forall q \neq q', \text{ and } \forall s \in \mathbf{R},$$

and for a.e. x$\in \Omega$ (see [Ki-St, p. 16]).

To establish (3.17), besides (3.20), we also need

$$(3.21) \qquad \lim_{n\to\infty} Q(u_n, u_n - u) = 0.$$

To show (3.21) holds, we see that (Q-3), (1.4), (3.10), (3.18), and (3.19) imply that $\lim_{n\to\infty} Q(u_n, P_n u - u) = 0$. Consequently, to obtain (3.21), we infer from the linearity of Q$(\cdot,\cdot)$ in the second variable, that it is sufficient to show that

$$(3.22) \qquad \lim_{n\to\infty} Q(u_n, u_n - P_n u) = 0.$$

To establish (3.22), we observe from (3.2) that

(3.23)

$$Q(u_n, u_n - P_n u) = \int_\Omega [\lambda |u_n|^{p-2} u_n + f_n(x, u_n)](u_n - P_n u) + G(u_n - P_n u).$$

It is clear from (3.13) and (3.18) that $G(u_n - P_n u) \to 0$ as $n \to \infty$. Likewise, it is clear from (3.10), (3.11), and (3.18) that $\int_\Omega |u_n|^{p-1} |u_n - P_n u| \to 0$ as $n \to \infty$. Hence, using (f-2), we see that the integral on the right-hand side of (3.23) also tends to zero as $n \to \infty$. We conclude that (3.22) does indeed hold, and (3.21) is therefore established.

We now proceed along the lines of [Sh4, pp. 371-6] and use (3.20), (3.21), along with (3.10)-(3.14) to show that (3.17) holds. To accomplish this, we will establish the existence of a sequence $\{u_{n_k}\}_{k=1}^{\infty}$ such that the following two facts hold:

(3.24)

$$\lim_{k\to\infty} \sum_{j=1}^{N} a(x, u_{n_k})(|Du_{n_k}|^{p-2} D_j u_{n_k} - |Du|^{p-2} D_j u)(D_j u_{n_k} - D_j u) = 0 \text{ a.e. in } \Omega;$$

$$(3.25) \qquad \{|Du_{n_k}|\}_{k=1}^{\infty} \text{ is pointwise bounded for a.e. } x \in \Omega.$$

Using the analog of the argument in [Sh4, p. 372], we first show that (3.24) and (3.25) together do indeed imply (3.17). So assuming (3.24) and (3.25), let $\Omega_1 \subset \Omega$ with $meas(\Omega_1) = meas(\Omega)$ be the set where a.e. is replaced by everywhere in (3.14), (3.20), (3.24), (3.25), in(Q-1) with respect to a(x,s), and in (Q-2). Also, we assume that both $u(x)$ and $D_j u(x)$ are finite for $x \in \Omega_1$ and j=1,...N. Suppose there is an $x^* \in \Omega_1$ and

$$(3.26) \quad \exists j_0 \text{ with the property that } lim_{k\to\infty} D_{j_0} u_{n_k}(x^*) = D_{j_0} u(x^*) \text{ is false.}$$

If we can show that this fact leads to a contradiction, then our assertion that (3.24) and (3.25) imply (3.17) will be established.

It follows from (3.25) and (3.26) that $\exists\{u_{n_{km}}(x^*)\}_{m=1}^{\infty}$ such that

$$\lim_{m\to\infty} D_j u_{n_{km}}(x^*) = \zeta_j,\ j=1,...N, \text{ and } \zeta \neq Du(x^*) \text{ where } \zeta = (\zeta_1,...,\zeta_N).$$

Since $x^* \in \Omega_1$, we obtain from this last set of equalities in conjunction with (3.24) that

$$\sum_{j=1}^{N} a(x^*,u)[|\zeta|^{p-2}\zeta_j - |Du(x^*)|^{p-2} D_j u(x^*)][\zeta_j - D_j u(x^*)] = 0 \tag{3.27}$$

On the other hand, since $\zeta \neq Du(x^*)$ and $x^* \in \Omega_1$, we have from (3.20) that

$$\sum_{j=1}^{N} a(x^*,u)[|\zeta|^{p-2}\zeta_j - |Du(x^*)|^{p-2} D_j u(x^*)][\zeta_j - D_j u(x^*)] > 0.$$

But this is a contradiction to (3.27). Hence, (3.24) and (3.25) do indeed imply (3.17).

We now show that (3.24) implies that (3.25) holds. Suppose then, assuming (3.24), that there is an $E \subset \Omega$ with $meas(E) > 0$ such that $|Du(x)|$ is finite for $x \in E$, $\eta_0 \leq a(x,s) \leq \eta_1$ for $s \in \mathbf{R}$ and $x \in E$, and $\{|Du_{n_k}(x)|\}_{k=1}^{\infty}$ is unbounded for $x \in E$. Let $x_0 \in E$. Then

$$\exists\left\{\left|Du_{n_{km}}(x_0)\right|\right\}_{m=1}^{\infty} \text{ such that } \left|Du_{n_{km}}(x_0)\right| \to \infty \text{ as } m \to \infty. \tag{3.28}$$

If we can show that this fact leads to a contradiction, then (3.25) does indeed hold.

To establish this contradiction, we infer from (3.24) and (3.28) that

(3.29)

$$\lim_{m\to\infty}\sum_{j=1}^{N} a(x_0,u_{n_{km}})(\left|Du_{n_{km}}(x_0)\right|^{p-2}\left|D_j u_{n_{km}}(x_0)\right|^2 / \left|Du_{n_{km}}(x_0)\right|^{p-\frac{1}{2}} = 0.$$

From the fact that $a(x_0,u_{n_{km}}) \geq \eta_0$, it follows that $\eta_0\left|Du_{n_{km}}(x_0)\right|^{\frac{1}{2}}$ is majorized by

$$\sum_{j=1}^{N} a(x_0,u_{n_{km}})(\left|Du_{n_{km}}(x_0)\right|^{p-2}\left|D_j u_{n_{km}}(x_0)\right|^2 / \left|Du_{n_{km}}(x_0)\right|^{p-\frac{1}{2}}.$$

Therefore, we see from (3.29) that $\eta_0\left|Du_{n_{km}}(x_0)\right|^{\frac{1}{2}} \to 0$ as m$\to\infty$. But $\eta_0 > 0$. Therefore $\left|Du_{n_{km}}(x_0)\right| \to 0$ as m$\to\infty$. This is a direct contradiction of (3.28). Consequently, $\{|Du_{n_k}(x)|\}_{k=1}^{\infty}$ is not unbounded for $x \in E$, and (3.25) does indeed hold.

To show that the assertion in (3.17) is true, it remains to prove that (3.10)- (3.14) along with (3.20) and (3.21) imply that (3.24) holds. In order

to do this, we set
(3.30)

$$I(x, u_n, u) = \sum_{j=1}^{N} a(x, u_n)(|Du_n|^{p-2} D_j u_n - |Du|^{p-2} D_j u)(D_j u_n - D_j u),$$

and observe that (3.24) will be established when we show that the following holds:

$$\text{(3.31)} \qquad \exists \{u_{n_k}\}_{k=1}^{\infty} \text{ such that } \lim_{k\to\infty} I(x, u_{n_k}, u) = 0 \text{ a.e. in } \Omega.$$

To show that (3.31) does indeed hold, from (3.10) and (3.11), we observe that

$$\lim_{n\to\infty} \| |Du_n| \|_p^2 \|u_n - u\|_{p/(p-2)} = 0.$$

Using this fact in conjunction with (Q-3) and Holder's inequality, we then obtain

$$\lim_{n\to\infty} \sum_{j=1}^{N} \int_\Omega [b^j(x, u_n, Du_n) D_j u_n](u_n - u) = 0.$$

But then from (1.4), (3.21), and this last limit, we have

$$\text{(3.32)} \qquad \lim_{n\to\infty} \sum_{j=1}^{N} \int_\Omega [a(x, u_n) |Du_n|^{p-2} D_j u_n](D_j u_n - D_j u) = 0.$$

Next, we observe from (Q-1), (Q-2), and (3.14) that

$$\lim_{n\to\infty} \int_\Omega |a(x, u_n) - a(x, u)|^{p/(p-1)} |Du|^p = 0.$$

But then (3.10), Holder's inequality, and this last fact imply that

$$\text{(3.33)} \qquad \lim_{n\to\infty} \int_\Omega \sum_{j=1}^{N} |a(x, u_n) - a(x, u)| |Du|^{p-2} |D_j u| |D_j u_n - D_j u| = 0.$$

Also, we see that $T \in [W^{1,p}(\Omega)]^*$ where for $w \in W^{1,p}(\Omega)$,

$$T(w) = \int_\Omega \sum_{j=1}^{N} a(x, u) |Du|^{p-2} D_j u D_j w.$$

Consequently, from (3.13), we have that $\lim_{n\to\infty} T(u_n - u) = 0$, and we infer from (3.33) that

$$\lim_{n\to\infty} \int_\Omega \sum_{j=1}^{N} a(x, u_n) |Du|^{p-2} D_j u (D_j u_n - D_j u) = 0.$$

Subtracting this last limit from the one in (3.32), we see from the definition of $I(x, u_n, u)$ in (3.30) that

$$\lim_{n\to\infty} \int_\Omega I(x, u_n, u) = 0.$$

But from (3.20), we see that $I(x, u_n, u) \geq 0$ for a.e. $x \in \Omega$ and $\forall n$. Consequently, we obtain from this last limit that $|I(x, u_n, u)|$ converges to zero in $L^1(\Omega)$. Hence, as is well-known, (see [Ru, p. 70]), this fact implies that $\exists \{u_{n_k}\}_{k=1}^{\infty}$ such that $\lim_{k\to\infty} I(x, u_{n_k}, u) = 0$ a.e. in Ω. This is exactly the statement in (3.31), and our proof that (3.24) does indeed hold is complete. As a consequence, we have, as was observed earlier, that the limit in (3.17) is valid. But this implies that the limit in (3.16) is also valid, which was our desired result.

Comparing (3.15) with (3.16), we see that

$$Q(u,v) = \int_{\Omega} [\lambda |u|^{p-2} u + f(x,u)]v + G(v) \quad \forall v \in \cup_{J=0}^{\infty} S_J. \tag{3.34}$$

Next, given v$\in W^{1,p}(\Omega)$, we let $\{v_J\}_{J=0}^{\infty}$ be the sequence of functions described in Lemma 1 where $v_J \in S_J$ and the limit in (2.2) holds. Since it is also true that $\lim_{J\to\infty} \|v_J - v\|_{p/(p-2)} = 0$, [Ad, p. 97], it is easy to see from (1.4) and the fact that $u \in W^{1,p}(\Omega)$ that

$$\lim_{J\to\infty} Q(u, v_J) = Q(u, v)$$

Likewise, it follows from (f-2) that

$$\lim_{J\to\infty} \{\int_{\Omega} [\lambda |u|^{p-2} u + f(x,u)]v_J + G(v_J)\} = \int_{\Omega} [\lambda |u|^{p-2} u + f(x,u)]v + G(v).$$

We conclude from these last two limits and (3.34) that

$$Q(u,v) = \int_{\Omega} [\lambda |u|^{p-2} u + f(x,u)]v + G(v) \;\; \forall v \in W^{1,p}(\Omega).$$

But this is exactly the statement in (1.9), and the proof of the theorem is complete.

4. Open Problems.

It is clear for the perturbed p-Laplacian under periodic boundary conditions that the role of of the first eigenvalue is played by $\lambda_1 = 0$. For p=2, there are several papers in the literature, namely, [Ka-Kr] and [Sh1], that deal with quadratic growth conditions for Du in the lower order terms. However, they deal only with problems below or at the first eigenvalue. For the perturbed p-Laplacian dealing with quadratic growth conditions for Du in the lower order terms, the current paper appears to be the *first* that handles problems *strictly above the first eigenvalue.* So what really counts here is that results are obtained strictly above the first eigenvalue for $p > \frac{3N}{N+1}$.

An open problem is: *How far above the first eigenvalue can one go*? To put this question in perspective, following the reasoning that occurs in [Sh5, p.83], we see that for Qu given in (1.2), the generalized eigenvalue λ_2 under periodic boundary values is such that $\lambda_2 \geq \eta^*$ where η^* is given in (1.7). If one replaces (f-2) and (f-3) which occur in the hypothesis of the Theorem by the condition

(4.1) $\qquad f(x,s)$ is uniformly bounded for $x \in \Omega$ and $s \in \mathbf{R}$,

then it it is easy to see that the Theorem is valid for λ in the following range:

$$0 < \lambda < p\eta^*(p')^{p-1}/2^p. \tag{4.2}$$

We see that the inequality in (4.2) is the same as the one in (1.10) except that 4^p is replaced by 2^p. But for $p > 2$, $p(p')^{p-1}/2^p < 1$, as an easy calculation shows. An open problem is to show that the Theorem is still true under assumption (4.1) on $f(x,s)$ and λ in the range

$$p\eta^*(p')^{p-1}/2^p \leq \lambda < \eta^*,$$

with $p > \frac{3N}{N+1}$. If this is not the case, provide a counter-example.

Another open problem occurs if we suppose, as in [Sh1], that $u \in W^{1,p}(\Omega)$ is a distribution solution of the equation in (1.8) if the following is true:

$$Q(u,\phi) = \int_\Omega [\lambda \, |u|^{p-2} \, u + f(x,u)]\phi + G(\phi) \quad \forall \phi \in C^\infty(\Omega)$$

where $Q(u,\phi)$ is defined by (1.4). Under the quadratic growth conditions on Du that we have assumed in the Theorem, we see that $Q(u,\phi)$ is now well-defined for $u \in W^{1,p}(\Omega)$ and $\phi \in C^\infty(\Omega)$ when $p \geq 2$. An open problem would be to remove the restrictions in the Theorem that $p > \frac{3N}{N+1}$ and obtain the result in the Theorem for *distribution solutions* but now under the wider condition that $p \geq 2$.

Another interesting problem would be to establish a result similar to that obtained in the Theorem for the perturbed p-Laplacian under quadratic growth conditions on Du in the lower order terms without assuming the weak solenoidal condition (1.3) on $b(x,u,Du)$, i.e., prove the Theorem without assuming condition (Q-4).

Any solution to one of the above three problems would be a welcome addition to the mathematical literature.

References

[Ad] R. A. Adams, *Sobolev Spaces,* Academic Press, New York, 1975.

[Dr-Ku-Ni] P. Drabek, A. Kufner, and F. Nicolosi, *Quasilinear Elliptic Equations With Degenerations And Singularities,* Walter de Gruyter, New York, 1997.

[Ka-Kr] J. L. Kazdan and R. J. Kramer, *Invariant Criteria for Existence of Solutions to Second-Order Quasilinear Elliptic Equations,* Comm. Pure and Appl. Math, 31(1978), 619-645.

[Ki-St] D. Kinderlehrer and G. Stampachia, *An Introduction To Variational Inequalities And Their Application,* New York, *1980.*

[Le-Li] J. Leray and J.L. Lions, *Quelques resultats de Visik sur les problemes elliptique nonlineaires par les methodes Minty-Browder,* Bull. Soc. math. France 93 (1965), 97-107.

[Ne] J. Necas, *Introduction To The Theory Of Nonlinear Elliptic Equations,* John Wiley & Sons, Chichester,1986.

[Nir] L. Nirenberg, *Topics In Nonlinear Analysis,* American Mathematical Society, Providence, 2001.

[Ru] W. Rudin, *Real And Complex Analysis,* Second ed., McGraw-Hill, New York, 1974.

[Sh1] V.L. Shapiro, *Resonance and quasilinear ellipticity,* Trans. American Math. Soc. 294 (1986), 567-584.

[Sh2] V.L. Shapiro, *Fourier series in several variables,* Bull. American Soc. 70 (1964), 48-93.

[Sh3] V.L. Shapiro, *Quasilinear ellipticity and the 1st eigenvalue,* Comm. Partial Diff. Eqns. 16 (1991), 1819-1855.

[Sh4] V.L. Shapiro, *Resonance and the 2nd BVP,* Trans. American Math. Soc. 325 (1991), 363-387.

[Sh5] V.L. Shapiro, *Superlinear quasilinearity and the second eigenvalue,* Nonlinear Analysis 44 (2001), 81-96.

[St-We] E. Stein and G. Weiss, *Introduction To Fourier Analysis On Euclidean Spaces,* Princeton University Press, Princeton, 1975.

UNIVERSITY OF CALIFORNIA, RIVERSIDE,CA 92521
E-mail address: shapiro@math.ucr.edu

Contemporary Mathematics
Volume **357**, 2004

Variational Methods for Ill-posed Problems

Ian Knowles

ABSTRACT. The use of variational methods in the solution of ill-posed problems by minimization of functionals with unique critical points is discussed in the context of the Cauchy problem for elliptic equations and electrical impedance tomography.

1. Introduction

According to Hadamard [**9, 10**], a problem is *well-posed* if three requirements are met:

- a solution exists;
- that solution is unique, and
- depends continuously on the data.

Otherwise the problem is *ill-posed.* The clear intention of Hadamard here is that "ill-posed" means "of no physical interest". Armed with a hundred years of hindsight and a science and engineering literature of staggering proportions, we know today that such problems arise in a fundamental way in the modelling of complex physical systems.

Consider the following model Cauchy problem on the square domain[1] $\Omega = (0,1) \times (0,1)$: for $F \in \mathcal{L}^2(\Omega)$ and functions $a_1, b_1, b_2 \in \mathcal{H}^{\frac{3}{2}}(0,1)$, and $a_2 \in \mathcal{H}^{\frac{1}{2}}(0,1)$, find $u \in \mathcal{H}^2(\bar{\Omega})$ satisfying

$$\Delta u = F(x), \quad x \in \Omega, \tag{1.1}$$

together with the boundary conditions

$$u(0,y) = b_1(y), \quad 0 < y < 1, \tag{1.2}$$

$$u(1,y) = b_2(y), \quad 0 < y < 1, \tag{1.3}$$

1991 *Mathematics Subject Classification.* Primary: 65N21; Secondary 35J20.
Supported in part by US National Science Foundation grant DMS-0107492.
[1]As can be seen, the method generalizes to more general domains and equations.

and the initial conditions

$$u(x,0) = a_1(x) \quad , \quad 0 < x < 1, \tag{1.4}$$
$$\tfrac{\partial u}{\partial n}(x,0) = a_2(x) \quad , \quad 0 < x < 1. \tag{1.5}$$

We assume that the solution u exists uniquely. The problem is known to be ill-posed in that the solution does not depend continuously on the data. One can also show that the problem can become well posed with the imposition of additional constraints ("conditionally well posed" in the language of [**11, 22**]).

This problem has an extensive literature (see for example [**5, 11, 15, 19, 24**] and the references therein) deriving mainly from its relevance to practical applications in electrocardiography [**6**], which is a special case of electrical source imaging [**1**, p. 214]. Here, typically, one wishes to find the

FIGURE 1. Electrical Source Imaging

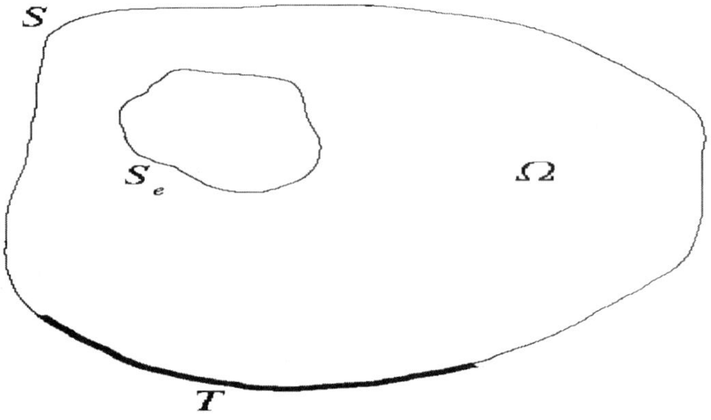

electrical cardiac (epicardial) potential generated by electrical sources inside the body (for example, the heart) on a surface S_e (known as the epicardial surface) close to the source, from the known potential on a part, T, of the surface S of the body. If we denote the region between S and S_e by Ω, then the potential u satisfies

$$\Delta u(x) = 0, \quad x \in \Omega, \tag{1.6}$$
$$\frac{\partial u}{\partial n} = 0, \text{ on } S, \tag{1.7}$$

and the problem is to recover u on S_e from a knowledge of $\phi = u|_T$.

The model problem may be solved variationally as follows. For a boundary function $f \in \mathcal{H}^{\frac{3}{2}}[0,1]$ let v_f be the solution of the Dirichlet problem formed from (1.1) with the boundary conditions (1.2), (1.3), (1.4), together with the Dirichlet condition

$$u(x,1) = f(x), \quad 0 < x < 1; \tag{1.8}$$

let w_f be the solution of the mixed problem formed from the same equation together with the conditions (1.2), (1.3), (1.5), and (1.8); these solutions exist by [**8**, Ch. 5]. For $f \in \mathcal{H}^{\frac{3}{2}}[0,1]$ define

$$G(f) = \int_\Omega |\nabla(v_f - w_f)|^2. \tag{1.9}$$

Clearly $G(f) \geq 0$. We show below that $G(f) = 0$ if and only if $v_f = w_f = u$ and $f = u|_{\{(x,1):0<x<1\}}$, where u is the solution of the Cauchy problem that we seek. Also, for $h \in \mathcal{H}^{\frac{3}{2}}[0,1]$ with $h(0) = h(1) = 0$, the Fréchet differential of G is given by

$$G'(f)[h] = 2\int_0^1 h(x)\frac{\partial}{\partial y}(w_f - v_f)(x,1)\,dx, \tag{1.10}$$

and $G'(f) = 0$ if and only if $G(f) = 0$, so that G also has a unique stationary point at the $y = 1$ boundary value of the Cauchy solution. We present later a preconditioned conjugate gradient implementation of this minimization, as well as some test examples.

While this approach has not yet been tested specifically on the electrocardiography problem the method is in principle the same. Here one is attempting to determine a Cauchy solution u in (1.6), and in particular the boundary function $f = u|_{(S-T)\cup S_e}$. For a given boundary function f in $\mathcal{H}^{\frac{3}{2}}((S-T)\cup S_e)$ one defines v_f to be the solution of (1.6) such that

$$v_f = \begin{cases} f \text{ on } (S-T)\cup S_e, \\ \phi \text{ on } T, \end{cases}$$

and w_f to be the solution of (1.6) such that

$$\begin{aligned} w_f &= f \text{ on } (S-T)\cup S_e, \\ \partial_n w_f &= 0 \text{ on } T. \end{aligned}$$

The functional $G(f)$ is also given by the formula (1.9), and the Fréchet derivative formula is analogous to (1.10).

This approach to the Cauchy problem has much in common with earlier work on the electrical impedance tomography (EIT) problem [**16**], which is in turn connected with the general class of inverse problems known as Dirichlet-Neumann map problems; the latter may be seen to include applications as diverse as reflection seismology imaging, landmine detection, optical tomography, ultrasound imaging, and inverse problems in elasticity, including the intriguing possibility of continuously monitoring subsurface material properties using natural seismic wave data with the aim of isolating earthquake precursors.

We discuss in §4 some recent work on the EIT problem, and some open questions.

2. The Functional G

We gather here some of the properties of G that will prove useful in the sequel.

First, as noted above, $G(f) \geq 0$, and, if $G(f) = 0$ then $\nabla(v_f - w_f) = 0$ on Ω, so that $v_f - w_f$ is constant. As $v_f(0, y) = w_f(0, y)$ for $0 < y < 1$, this constant must be zero, and it is therefore true that $v_f = w_f = u$. Conversely, it is immediate that $v_f = w_f$ implies that $G(f) = 0$.

Next, for $h \in \mathcal{H}^{\frac{3}{2}}[0, 1]$ with $h(0) = h(1) = 0$ we have, using integration by parts and the fact that the functions v_{f+h}, w_{f+h}, v_f, and w_f all satisfy the differential equation (1.1),

$$
\begin{aligned}
&G(f+h) - G(f) \\
&= \int_\Omega \nabla(v_{f+h} - w_{f+h}) \cdot \nabla(v_{f+h} - w_{f+h}) - \nabla(v_f - w_f) \cdot \nabla(v_f - w_f) \\
&= \int_\Omega \nabla((v_{f+h} - v_f) - (w_{f+h} - w_f)) \cdot \nabla(v_{f+h} - w_{f+h} + v_f - w_f) \\
&= \int_\Omega \nabla(v_{f+h} - v_f) \cdot \nabla(v_{f+h} - w_{f+h} + v_f - w_f) \\
&\qquad - \int_\Omega \nabla(w_{f+h} - w_f) \cdot \nabla(v_{f+h} - w_{f+h} + v_f - w_f) \\
&= \int_{\partial\Omega} (v_{f+h} - v_f) \frac{\partial}{\partial n}(v_{f+h} - w_{f+h} + v_f - w_f) \\
&\qquad - \int_{\partial\Omega} (v_{f+h} - w_{f+h} + v_f - w_f) \frac{\partial}{\partial n}(w_{f+h} - w_f) \\
&= -\int_0^1 h(x) \frac{\partial}{\partial n}(v_{f+h} - w_{f+h} + v_f - w_f)(x, 1)\, dx, \\
&= 2\int_0^1 h(x) \frac{\partial}{\partial y}(w_f - v_f)(x, 1)\, dx \\
&\qquad - \int_0^1 h(x) \frac{\partial}{\partial y}(v_{f+h} - v_f - w_{f+h} + w_f)(x, 1)\, dx
\end{aligned}
$$

after using the boundary data defining the functions v_{f+h}, w_{f+h}, v_f, and w_f. Now,

$$
\begin{aligned}
\left|\int_0^1 h(x) \frac{\partial}{\partial y}(v_{f+h} - v_f)(x, 1)\, dx\right| &\leq \|h\|_{\mathcal{L}^2(0,1)} \left\|\frac{\partial}{\partial y}(v_{f+h} - v_f)(x, 1)\right\|_{\mathcal{L}^2(0,1)} \\
&\leq \|h\|_{\mathcal{H}^{\frac{3}{2}}(0,1)} \left\|\frac{\partial}{\partial y}(v_{f+h} - v_f)(x, 1)\right\|_{\mathcal{H}^{\frac{1}{2}}(0,1)},
\end{aligned}
$$

as $\mathcal{H}^{\frac{3}{2}}(0,1)$ and $\mathcal{H}^{\frac{1}{2}}(0,1)$ are continuously embedded in $\mathcal{L}^2(0,1)$. From standard elliptic estimates at the boundary

$$\|\frac{\partial}{\partial y}(v_{f+h}-v_f)(x,1)\|_{\mathcal{H}^{\frac{1}{2}}(0,1)} \leq K\|v_{f+h}-v_f\|_{\mathcal{H}^2(\Omega)} \leq K'\|h\|_{\mathcal{H}^{\frac{3}{2}}(0,1)},$$

for some constants K and K', and it follows that

$$|\int_0^1 h(x)\frac{\partial}{\partial}(v_{f+h}-v_f)(x,1)\,dx| = O(\|h\|^2_{\mathcal{H}^{\frac{3}{2}}(0,1)}),$$

with a similar estimate for $\int_0^1 h(x)\frac{\partial}{\partial y}(w_{f+h}-w_f)(x,1)\,dx$. So we have that the Fréchet differential for G is given by the formula (1.10).

It follows that the $\mathcal{L}^2[0,1]$ gradient of G at f, $\nabla G(f)$, is given by

$$\nabla G(f)(x) = 2\frac{\partial}{\partial y}(w_f - v_f)(x,1), \quad 0 < x < 1. \tag{2.1}$$

and that the functional is analytic in the sense of [**4**, Proposition 15.2]; in particular, the Taylor formula,

$$G(f+h) = G(f) + \sum_{n=1}^{\infty}\frac{\alpha^n}{n!}G^{(n)}(f)[h],$$

is valid for G for some $r_f > 0$ and all $|\alpha| \leq r_f$.

Notice also that if u is the Cauchy problem solution that we seek, and we set $f = u|_{\{(x,1):0<x<1\}}$ then $u = v_f = w_f$ and $G'(f) = 0$. Conversely, if $G'(f^*) = 0$, then for $0 < x < 1$,

$$\frac{\partial}{\partial y}(v_{f^*} - w_{f^*})(x,1) = 0,$$

so that $\eta = v_{f^*} - w_{f^*}$ satisfies

$$\eta(x,1) = (\partial\eta/\partial n)(x,1) = \eta(0,y) = \eta(1,y) = 0.$$

Assuming uniqueness of this Cauchy problem for Ω, it follows immediately that $v_{f^*} = w_{f^*}$, and hence that $v_{f^*} = u$. Thus $f^* = f$ and we have that $u|_{\{(x,1):0<x<1\}} = f$ is the unique stationary point for G.

One may also derive a formula for the second Fréchet differential. First for $h \in \mathcal{H}^{\frac{3}{2}}(0,1)$ let G_h denote the Dirichlet Green function for the Laplace equation with boundary data equal to h on $\{(x,1) : 0 \leq x \leq 1\}$ and zero elsewhere, and let G_h^* be the Laplace equation Green function for the mixed problem involving homogeneous Neumann data on $\{(x,0) : 0 \leq x \leq 1\}$, homogenous Dirichlet data on the vertical sides of the square, and Dirichlet data equal to h on $\{(x,1) : 0 \leq x \leq 1\}$. Then we have

$$\frac{\partial}{\partial y}(v_{f+h} - v_f)(x,1) = -\int_0^1 h(t)\frac{\partial^2}{\partial x \partial y}G_h(x,t)\,dt,$$

and

$$\frac{\partial}{\partial y}(w_{f+h} - w_f)(x,1) = -\int_0^1 h(t)\frac{\partial^2}{\partial x \partial y}G_h^*(x,t)\,dt.$$

If $k \in \mathcal{H}^{\frac{3}{2}}(0,1)$ we have

$$\begin{aligned}
&G'(f+h)[k] - G'(f)[k] \\
&= 2\int_0^1 k(x)[\frac{\partial}{\partial y}(w_{f+h} - v_{f+h})(x,1) - \frac{\partial}{\partial y}(w_f - v_f)(x,1)]\,dx \\
&= 2\int_0^1 k(x)[\frac{\partial}{\partial y}(w_{f+h} - w_f)(x,1) - \frac{\partial}{\partial y}(v_{f+h} - v_f)(x,1)]\,dx \\
&= 2\int_\Omega k(x)h(t)\frac{\partial^2}{\partial x \partial y}[G_h^*(x,t) - G_h(x,t)]\,dxdt \\
&= 2\int_\Omega k(x)h(t)\frac{\partial^2}{\partial x \partial y}[G_0^*(x,t) - G_0(x,t)]\,dxdt \\
&\quad + 2\int_\Omega k(x)h(t)\frac{\partial^2}{\partial x \partial y}[(G_h^* - G_0^*)(x,t) - (G_h - G_0)(x,t)]\,dxdt.
\end{aligned}$$

Noting that $G_h^* - G_0^*$ and $G_h - G_0$ both satisfy the Laplace equation in, say, the first variable, it is not hard to show that each of these is $O(\|h\|_{\mathcal{H}^{\frac{3}{2}}(0,1)})$, from which we infer that

$$G''(f)[h,k] = 2\int_\Omega k(x)h(y)\frac{\partial^2}{\partial x \partial y}[G_0^*(x,y) - G_0(x,y)]\,dxdy. \tag{2.2}$$

Notice that there are some difficulties with the above derivations with respect to the existence of normal derivatives at the corners of the square. So, strictly speaking, one should assume that the vertical sides of the square are rounded slightly at the top and bottom to give a $\mathcal{C}^{1,1}$ boundary.

3. Conjugate Gradient Minimization

Note first that the $\mathcal{L}^2[0,1]$ gradient of G need not be zero at boundary points $x = 0$ and $x = 1$. If we wish to retain continuity at the boundary for the solution u computed via a descent minimization of G, then this gradient is not suitable. One can circumvent this problem by using a Neuberger gradient (see [**21**]), $\nabla_N G(f)$, defined by

$$(\nabla_N G(f), h)_1 = G'(f)[h], \quad \text{for all } h \in H_0^1(0,1). \tag{3.1}$$

An integration by parts shows that $g = -\nabla G_N(f)$ can be found from

$$-g'' + g = -\nabla G(f), \tag{3.2}$$

$$g(0) = g(1) = 0. \tag{3.3}$$

As $\nabla G_N(f) = (\Delta - I)^{-1}\nabla G(f)$, it is evident that the new gradient is a preconditioned (smoothed) version of the original gradient.

While one could use this new gradient together with the method of steepest descent, in practice, one generally gets faster convergence with the following adaption of the standard Polak-Ribière conjugate gradient scheme [**23**, p. 304]. Choose as the initial f_0 the linear function satisfying

$$f_0(0) = b_1(1),$$

FIGURE 2. $u = (k\pi)^{-2}\sin(k\pi x)e^{k\pi y}$

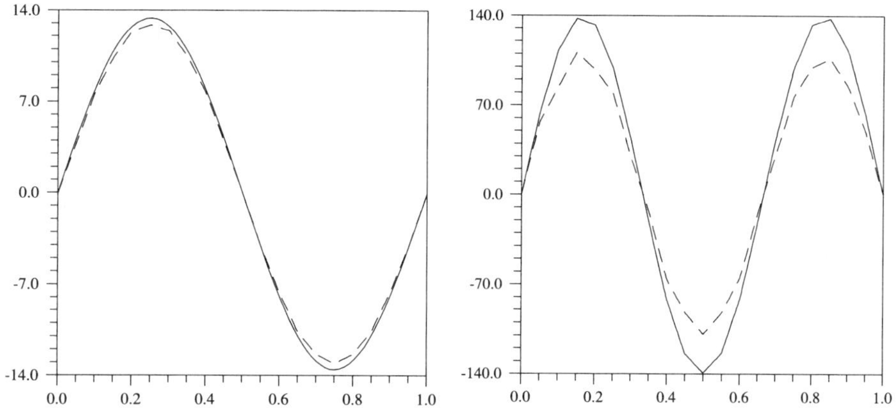

(a) $k = 2$; 2 iterations (b) $k = 3$; 200 iterations

FIGURE 3. $u = x^k + y^k$

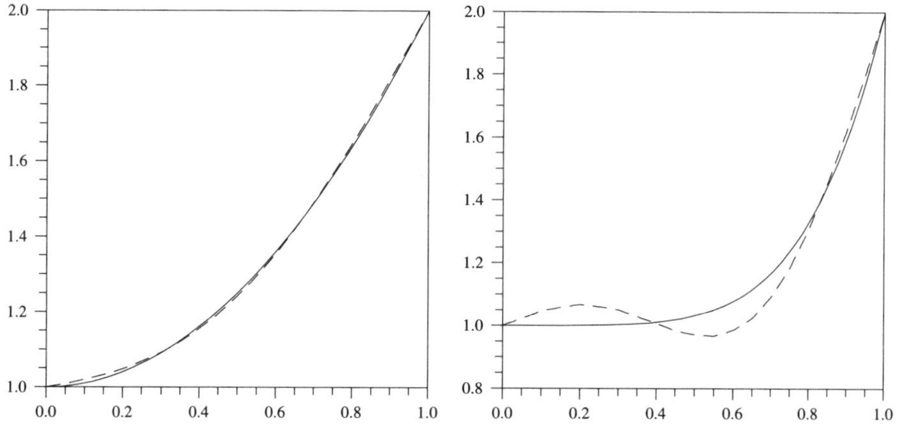

(a) $k = 2$; 2 iterations (b) $k = 5$; 100 iterations

$$f_0(1) = b_2(1) \quad .$$

The initial search direction is $h_0 = g_0 = -\nabla_N G(f_0)$. At f_i one uses the approximate line search routine to minimize $G(f)$ in the direction of h_i, resulting in $G(f_{i+1})$. Then $g_{i+1} = -\nabla_N G(f_{i+1})$ and $h_{i+1} = g_{i+1} + \gamma_i h_i$

where

$$\gamma_i = \frac{(g_{i+1} - g_i, g_{i+1})_1}{(g_i, g_i)_1} = \frac{(g_{i+1} - g_i, \nabla G(f_{i+1}))}{(g_i, \nabla G(f_i))},$$

by (3.1) and an integration by parts, where $(\cdot,\cdot)$ and $(\cdot,\cdot)_1$ denote, the usual inner products in, respectively, $\mathcal{L}^2(\Omega)$ and $\mathcal{H}^1(\Omega)$. The approximate line search procedure is implemented using bracketing and Brent minimization (see [**23**]) with all two-dimensional integrals computed as iterated one-dimensional integrals using Simpson's rule. The elliptic boundary value problems were solved using the the FIVE POINT STAR (finite difference discretization) and LINPACK SPD BAND packages within the double precision version of the ELLPACK system (see [**25**]). One bonus from this arrangement is that ELLPACK automatically calculates the derivatives of any solution by means of built-in quadratic interpolation routines; this is used to compute the boundary derivatives needed in the evaluation of the gradients.

All examples were computed using a 21×21 grid on Ω. Some results are shown in Fig. 2 for the well-known Hadamard example. The correct $f = u|_{\{(x,1):0<x<1\}}$ is shown as a solid line, and the computed f as a dashed line. The method worked well for the case $k = 2$, and for $k = 3$, although the convergence was slower; the method failed at $k = 4$. This is perhaps to be expected as the ill-posedness increases with k. In Fig. 3 the higher order polynomial case proved more difficult, but the computation was reasonably accurate.

4. Electrical Impedance Tomography

Electrical impedance tomography involves another model problem that is representative of situations in which one seeks to recover interior parameter values from measurements made at the boundary of a region. In EIT current fluxes are applied to the exterior of a body Ω and the resulting boundary voltages are measured. The voltage potential u in Ω satisfies

$$\nabla \cdot (P(x)\nabla u) = 0, \tag{4.1}$$

P representing the conductivity inside Ω. For a boundary voltage g belonging to $\mathcal{H}^{\frac{1}{2}}(\partial\Omega)$ let $\Lambda_P(g) = P\partial_\nu u$; Λ_P is the Dirichlet-Neumann map for P and represents the information obtained from all possible current-voltage pair measurements.

Of interest here is whether or not Λ_P uniquely determines P, and if so, there is then the problem of actually reconstructing P from Λ_P.

Reconstruction was considered in [**16**] using a variational approach. An appropriate functional H_{dn} is defined by way of G_{dn} defined for an f in $\mathcal{H}^{-\frac{1}{2}}(\partial\Omega)$ by

$$G_{dn}(p, f) = \int_\Omega p|\nabla(u_{p,f} - \tilde{u}_{p,f})|^2, \tag{4.2}$$

FIGURE 4. EIT Algorithm

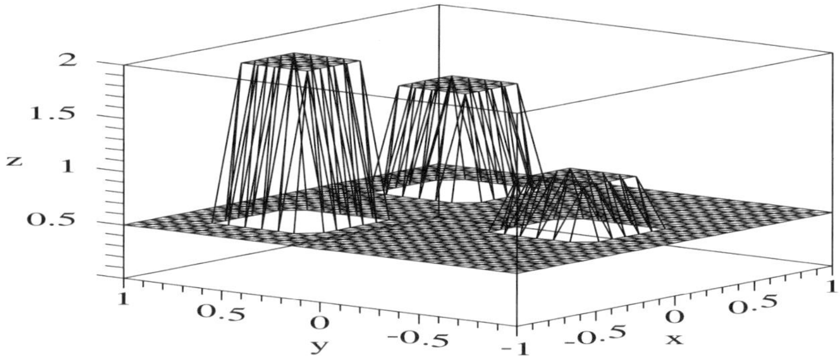

(a) True P

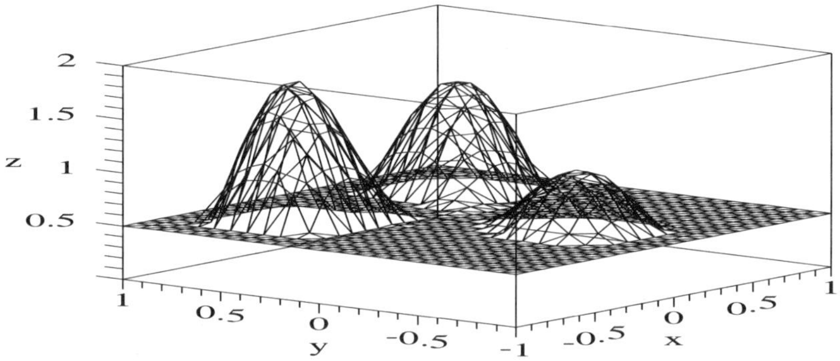

(b) Computed P; 0% noise

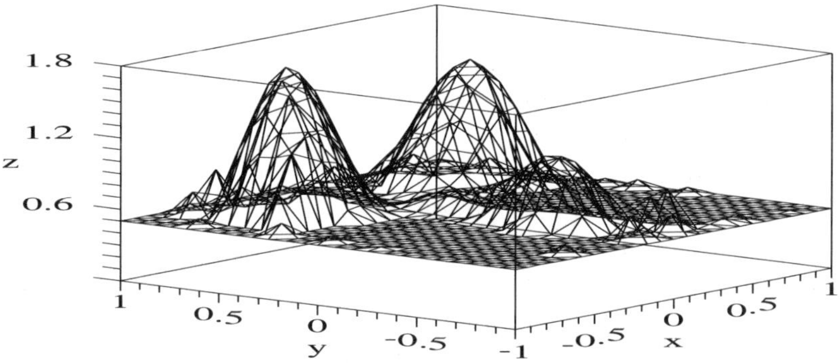

(c) Computed P; 1% noise

where $v = u_{p,f}$ is the Dirichlet solution of

$$-\nabla \cdot p\nabla v = 0, \tag{4.3}$$

satisfying $v|_{\partial\Omega} = \Lambda_P^{-1}(f)$, and $v = \tilde{u}_{p,f}$ is the solution of (4.3) satisfying the Neumann condition $p\partial_\nu v|_{\partial\Omega} = f$, as well as the condition

$$\tilde{u}_{p,f}(x_0) = u_{p,f}(x_0) \tag{4.4}$$

at some $x_0 \in \bar{\Omega}$ that we henceforth agree to keep fixed; the latter condition serves to ensure that the Neumann solution is uniquely defined. As above the Fréchet differential takes a very simple form: for h in $\mathcal{L}^\infty(\Omega)$

$$G'_{dn}(p)[h] = \int_\Omega h.(|\nabla u_{p,f}|^2 - |\nabla \tilde{u}_{p,f}|^2).$$

In [**16**] by choosing an orthonormal sequence $\{f_i\}$ from $\mathcal{H}^{-\frac{1}{2}}(\Omega)$ and then taking $H_{dn}(p)$ to be the sum of the non-negative terms $G_{dn}(p, f_i)$, we obtain a non-negative functional with the property that $H_{dn}(p) = 0$ if and only if $\Lambda_p = \Lambda_P$. As can be seen from Figure 4 minimization of this functional using steepest descent techniques proved to be an effective recovery method, even in the presence of noise in the input data. The numerical evidence indicates that H_{dn} has a unique critical point, but this has not been proven as yet.

The study of the Dirichlet-Neumann map (DN-map) uniqueness question dates back at least to the paper of Langer [**18**]. Reasonably complete survey material may be found in the papers [**20, 26**] and the book [**14**], and the references therein. In 1995 Nachman showed that uniqueness holds for $n = 2$ and P in $\mathcal{W}^{2,r}(\Omega)$ for some $r > 1$, and when $n \geq 3$ for conductivities P that lie in $\mathcal{W}^{2,r}(\Omega)$ for some $r > n/2$; this work was extended by Brown to P in $\mathcal{C}^{\frac{3}{2},\epsilon}(\bar{\Omega})$ for $n > 2$, in 1996 and by Brown and Uhlmann (1997) to P in $\mathcal{W}^{1,r}(\Omega)$ for some $r > 2$ when $n = 2$.

Uhlmann [**26**] has conjectured that uniqueness should hold for conductivities P in $\mathcal{C}^{0,1}(\bar{\Omega}) = \mathcal{W}^{1,\infty}(\Omega)$. In theory, uniqueness should in fact hold for conductivities P that (are positively bounded below and) lie in $\mathcal{L}^\infty(\Omega)$. However, the standard method, which employs complex geometrical optics and scattering theory after a transformation to the Schrödinger equation

$$-\Delta w + Q(x)w = 0, \quad x \in \Omega \subset \mathbb{R}^n, \tag{4.5}$$

by means of

$$w = P^{\frac{1}{2}}u, \quad Q = P^{-\frac{1}{2}}\Delta(P^{\frac{1}{2}}), \tag{4.6}$$

is unlikely to yield such a result, given that the smoothness on P is inevitably constrained by the transformation (4.6).

A possible path around this impasse is to try to show directly that the non-negative functional H_{dn} has a unique global minimum at P. While much of this is conjecture at present, there are some encouraging signs.

Note first that the Dirichlet-Neumann maps Λ_Q and Λ_P are related by

$$\Lambda_Q(f) = \partial_\nu w|_{\partial\Omega} = P^{-\frac{1}{2}}(\Lambda_P + \frac{1}{2}\partial_\nu P)P^{-\frac{1}{2}}.$$

Now, given Λ_Q, one can formulate an analogous functional for the equation (4.5) by setting, for an f in $\mathcal{H}^{\frac{1}{2}}(\partial\Omega)$ and $q \in \mathcal{L}^\infty(\Omega)$ with $q \geq 0$,

$$G_{dns}(q,f) = \int_\Omega |\nabla(w_{q,f} - \tilde{w}_{q,f})|^2 + q(w_{q,f} - \tilde{w}_{q,f})^2, \tag{4.7}$$

where $v = w_{q,f}$ is the Dirichlet solution of

$$-\Delta v + qv = 0 = 0, \tag{4.8}$$

satisfying $v|_{\partial\Omega} = \Lambda_Q^{-1}(f)$, and $v = \tilde{w}_{q,f}$ is the Neumann solution of (4.8) satisfying $\partial_\nu v|_{\partial\Omega} = f$. For q in $\mathcal{L}^\infty(\Omega)$ with $q \geq 0$, we define

$$H_{dns}(q) := \sup_{\|f\| \leq 1} G_{dns}(q,f). \tag{4.9}$$

There is convincing numerical evidence that in general functionals like H_{dn} and H_{dns} are not convex. However, there is the possibility of a slightly weaker condition being true. A functional H for which

$$H'(v) = 0 \Rightarrow H(v) = 0 \tag{4.10}$$

is said to be *essentially convex*. By way of example, the Cauchy problem functional G investigated above is essentially convex. We also have

PROPOSITION 4.1. *The functional H_{dns} is essentially convex.*

PROOF. Observe that from an integration by parts

$$G_{dns}(q,f) = \int_{\partial\Omega} (w_{q,f} - \tilde{w}_{q,f})\partial_\nu(w_{q,f} - \tilde{w}_{q,f}) = \int_{\partial\Omega} f.R_q f, \tag{4.11}$$

where $R_q = (\Lambda_Q^{-1} - \Lambda_q^{-1})\Lambda_q(\Lambda_Q^{-1} - \Lambda_q^{-1})$ is a compact non-negative self-adjoint operator in the Hilbert space $\mathcal{H}^{\frac{1}{2}}(\partial\Omega)$. Thus

$$H_{dns}(q) := \sup_{\|f\| \leq 1} G_{dns}(q,f) = G_{dns}(q,f_q) = \mu_q = \|R_q\|, \tag{4.12}$$

where f_q is an eigenvector of norm 1 corresponding to the largest eigenvalue μ_q of R_q. We note in passing that the eigenvector f_q is not unique, even when the eigenspace is one-dimensional, but the value of $H_{dns}(q)$ is independent of which of the possible choices of the normalized f_q is used in the definition of w_{q,f_q} and $\tilde{w}_{q,f_q}$, provided that the same choice is made for both solutions.

We assume that $H'_{dns}(q) = 0$ for some $q \in \mathcal{L}^\infty(\Omega)$. One can show that for $h \in \mathcal{L}^\infty(\Omega)$,

$$H'_{dns}(q)[h] = \int_\Omega h.(w_{q,f_q}^2 - \tilde{w}_{q,f_q}^2). \tag{4.13}$$

Then

$$(w_{q,f_q} - \tilde{w}_{q,f_q})(w_{q,f_q} + \tilde{w}_{q,f_q}) = 0. \tag{4.14}$$

If $w_{q,f_q} - \tilde{w}_{q,f_q} = 0$ in Ω, then $H_{dns}(q) = 0$. Otherwise, for some $y_0 \in \Omega$, $w_{q,f_q}(y_0) - \tilde{w}_{q,f_q}(y_0) \neq 0$ and hence, by continuity, $w_{q,f_q}(x) - \tilde{w}_{q,f_q}(x) \neq 0$ in a neighbourhood of y_0. It follows from (4.14) that $w_{q,f_q} + \tilde{w}_{q,f_q} = 0$ in this neighbourhood. As $w_{q,f_q} + \tilde{w}_{q,f_q}$ is a solution of (4.5), by the unique continuation theorem (see, for example, [**3**, Lemma 8.5]) $w_{q,f_q} + \tilde{w}_{q,f_q} = 0$ in Ω. Thus

$$\begin{aligned} H_{dns}(q) &= 4 \int_{\Omega} |\nabla w_{q,f_q}|^2 + q w_{q,f_q}^2 \\ &= 4 \int_{\partial\Omega} w_{q,f_q} \partial_\nu w_{q,f_q} \\ &= -4 \int_{\partial\Omega} f_q \Lambda_Q^{-1} f_q. \end{aligned}$$

As Λ_Q and H_{dns} are non-negative, it follows that $H_{dns}(q) = 0$. □

This proposition has some interesting consequences. In particular, if one is given that Q is uniquely specified by the DN-map Λ_Q, it follows directly from essential convexity that H_{dns} has a unique critical point at Q. This is a very important fact if one is interested in computing Q by minimizing H_{dns}. It is also true that the critical set K_0 of H_{dns} consists precisely of absolute minima. This is a useful first step toward a uniqueness theorem. One can also show using critical point theory that the critical set here is path connected. The final step is to show that the critical set consists of just one point.

It is conjectured that both H_{dn} and its supremum analogue are also essentially convex. A proof of this conjecture would be an important step toward establishing uniqueness for a general $\mathcal{L}^\infty$ conductivity P.

In a different direction, the eigenfunctions f_q that ocurred in the above proof are of interest in their own right. The functions f represent current flux at the boundary, and there is evidence [**2, 7, 12, 13, 17**] that the quality of reconstruction can be greatly improved by an optimal choice of these functions. If we think of the value $G_{dns}(q, f)$ as representing the (squared) distance from q to Q using just the information in the pair $(f, \Lambda_Q(f))$, then $H_{dns}(q) = G_{dns}(q, f_q)$ represents the actual (squared) distance from q to Q, i.e. the distance that embodies all of the information in Λ_Q. In this sense, f_q is an optimal current flux function for this q; this may have numerical implications, with respect to the development of an efficient conductivity reconstruction algorithm.

References

[1] Thomas Budinger et al. *Mathematics and Physics of Emerging Biomedical Imaging.* National Research Council, Institute of Medicine, National Academic Press, Washington, D.C., 1996.

[2] Elena Cherkaeva and Alan C. Tripp. Inverse conductivity problem for inaccurate measurements. *Inverse Problems*, 12(6):869–883, 1996.

[3] David Colton and Rainer Kress. *Inverse Acoustic and Electromagnetic Theory.* Springer-Verlag, Berlin, 2nd edition, 1997.

[4] Klaus Deimling. *Nonlinear functional analysis.* Springer-Verlag, Berlin, 1985.

[5] R. S. Falk and P. B. Monk. Logarithmic convexity for discrete harmonic functions and the approximation of the Cauchy problem for Poisson's equation. *Math. Comp.*, 47(175):135–149, 1986.

[6] P. Franzone and E. Magenes. On the inverse problem of electrocardiology. *Calcolo*, 16:459–538, 1986.

[7] D. G. Gisser, D. Isaacson, and J. C. Newell. Electric current computed tomography and eigenvalues. *SIAM J. Appl. Math.*, 50:1623–1634, 1990.

[8] P. Grisvard. *Elliptic Problems in Nonsmooth Domains*. Pitman Publishing Inc., Massachusetts, 1985.

[9] J. Hadamard. *Bull. Univ. Princeton*, 13:49, 1902.

[10] J. Hadamard. *Lectures on the Cauchy Problem in Linear Partial Differential Equations*. Yale University Press, New Haven, 1923.

[11] Houde Han. The finite element method in a family of improperly posed problems. *Math. Comp.*, 38(157):55–65, 1982.

[12] Ping Hua, Eung Je Woo, John G. Webster, and Willis J. Tompkins. Iterative reconstruction methods using regularization and optimal current patterns in electrical impedance tomography. *IEEE Transactions on Medical Imaging*, 10(4):621–628, 1991.

[13] D. Isaacson. Distinguishability of conductivities by electric current conductivity. *IEEE Trans. on Medical Imaging*, MI-5:91–95, 1986.

[14] Victor Isakov. *Inverse problems for partial differential equations*. Springer-Verlag, New York, 1998.

[15] S. I. Kabanikhin and A. L. Karchevsky. Optimizational method for solving the Cauchy problem for an elliptic equation. *J. Inv. Ill-Posed Problems*, 3(1):21–46, 1995.

[16] Ian Knowles. A variational algorithm for electrical impedance tomography. *Inverse Problems*, 14(6):1513–1525, 1998.

[17] George K. Kyriacou, Christos S. Koukourlis, and John N. Sahalos. A reconstruction algorithm of electrical impedance tomography with optimal configuration of the driven electrodes. *IEEE Trans. on Medical Imaging*, 12:430–438, 1993.

[18] R. E. Langer. An inverse problem in differential equations. *Bull. Amer. Math. Soc.*, 39:814–820, 1933.

[19] R. Lattes and J. L. Lions. *Method of Quasi-Reversibility*. American Elsevier, New York, 1969.

[20] Adrian I. Nachman. Global uniqueness for a two-dimensional inverse boundary value problem. *Ann. of Math.*, 142:71–96, 1995.

[21] J. W. Neuberger. *Sobolev Gradients in Differential Equations*, volume 1670 of *Lecture Notes in Mathematics*. Springer-Verlag, New York, 1997.

[22] L. E. Payne. *Improperly Posed Problems in Partial Differential Equations*. SIAM, Philadelphia, 1975.

[23] William H. Press, Brian P. Flannery, Saul A. Teukolsky, and William T. Vetterling. *Numerical Recipes: The Art of Scientific Computing*. Cambridge University Press, Cambridge, 1989.

[24] Hans-Jürgen Reinhardt, Houde Han, and Dinh Nho Hào. Stability and regularization of a discrete approximation to the Cauchy problem for Laplace's equation. *SIAM J. Numer. Anal.*, 36(3):890–905 (electronic), 1999.

[25] John R. Rice and Ronald F. Boisvert. *Solving Elliptic Problems Using ELLPACK*. Springer-Verlag, Berlin, 1985.

[26] Gunther Uhlmann. Inverse boundary value problems for partial differential equations. In *Proceedings of the International Congress of Mathematicians, Vol. III (Berlin, 1998)*, number Extra Vol. III, pages 77–86 (electronic), 1998.

Department of Mathematics, University of Alabama at Birmingham, Birmingham, Alabama, 35294

E-mail address: iwk@math.uab.edu

Contemporary Mathematics
Volume **357**, 2004

GNGA: Recent Progress and Open Problems for Semilinear Elliptic PDE

John M. Neuberger

ABSTRACT. This article is a survey of successful numerical experiments which use variants of the Gradient Newton Galerkin Algorithm of Neuberger & Swift, together with a list of open problems and conjectures in semilinear elliptic PDE. The application is primarily to problems of the form $\Delta u + f(u) = 0$ on regions $\Omega \subset \mathbf{R}^2$ with zero Dirichlet boundary condition. By characterizing solutions to the PDE as critical points of an "action" functional, we consider Newton's method on the gradient of that functional. We use a Galerkin expansion, in eigenfunctions of the Laplacian, to find solutions of arbitrary Morse index. For general regions, generating this basis is one of the more computationally challenging portions of the algorithm. We often take $f'(0)$ to be a parameter, and analyze the resulting bifurcations. We also consider a one-parameter family of regions, where we then have two free bifurcation parameters. If the region or the nonlinearity has symmetry, it is often advantageous to use this information. We also report on several new numerical results and discuss related conjectures and on-going experiments.

1. Semilinear Elliptic BVP.

The Gradient Newton Galerkin Algorithm (GNGA, see [**36**]) is a general method for finding approximations of critical points of a functional. Given a Hilbert space H and a C^2 functional $J : H \to \mathbf{R}$, we work in a sufficiently large M-dimensional subspace $G \subset H$ and iteratively seek a function $u \in G$ so that the projection $P_G \nabla J(u) = 0$. For many of the semilinear elliptic Dirichlet problems that are our frequent objects of interest, it is reasonable to take $G = \text{span}\{\psi_i\}_{i=1,\dots,M}$, where $\{\psi_i\}_{i\in\mathbf{N}}$ are the eigenfunctions of $-\Delta$ with zero Dirichlet boundary condition in Ω, and

$$0 < \lambda_1 < \lambda_2 \le \lambda_3 \le \cdots \to \infty \tag{1.1}$$

1991 *Mathematics Subject Classification.* Primary 35A15; Secondary 65K10.
Key words and phrases. GNGA, Variational, Bifurcation, Symmetry, Sign-Changing Solutions, Elliptic PDE.
Partially supported by NSF DMS Grants 0074326 and 0124121.

are the corresponding eigenvalues. In this case we assume that the eigenfunctions are normal in $L_2 = L_2(\Omega)$ and of course orthogonal in both the Sobolev space $H = H_0^{1,2}(\Omega)$ and in L_2, with inner products

$$\langle u, v\rangle_H = \int_\Omega \nabla u \cdot \nabla v \quad \text{and} \quad \langle u, v\rangle_2 = \int_\Omega u\, v$$

respectively (see [**1**], [**24**], or [**42**]). It is a chief goal of ours to demonstrate the importance of these eigenfunctions to the theory and numerical solution of related nonlinear elliptic equations.

Our interest in variational methods for nonlinear elliptic boundary value problems (BVP) began with [**8**], where we proved the existence of a third nontrivial solution to a superlinear Dirichlet problem of the form

$$\begin{cases} \Delta u + f(u) = 0 & \text{in } \Omega \\ u = 0 & \text{on } \partial\Omega, \end{cases} \tag{1.2}$$

where among other hypotheses, we have $f(0) = 0$ and $f'(0) < \lambda_1$. This solution, if nondegenerate, is shown to be of Morse index (MI) 2 (see [**33**]). The techniques of proofs were natural extensions of the Mountain Pass Lemma (see [**2**]), and were similarly constructive. Recently, we extended this result in [**11**] to the case where $f'(0) < \lambda_2$. Similarly to the Mountain Pass Algorithm (MPA) of [**13**], we were able to develop the so-called Modified Mountain Pass Algorithm (MMPA) (see [**34**]) for approximating the new higher MI solution through a gradient descent method with two constraints. The proof in [**8**] and the MMPA were suitably modified in [**35**] to handle a slightly different non-linearity allowing $f(0) \neq 0$ where the variational structure is more complex than the frequently studied "Nehari Manifold". In [**15**], our steepest descent techniques are further modified to realize the Lyapunov-Schmidt reductions found in [**9**] and [**10**], where asymptotically linear problems are considered. In [**17**], we use both the MMPA and numerical high-linking methods (see [**16**]) together with gradient flow invariant sets to take advantage of symmetries of Ω to approximate solutions with MI greater than 2.

In seeking a general method to find solutions of MI greater than 2, we take advantage of the minimax nature inherent to Newton's Method in order to find saddle point type critical points. In [**36**], we developed the GNGA for exactly this purpose, and were able to describe in a fairly complete way all of the solutions of reasonable MI to a standard class of superlinear Dirichlet problems on the square. For the PDE (1.2) on a square region $\Omega = (0,1) \times (0,1)$, it is well-known that the (doubly indexed) eigenvalues and eigenfunctions of $-\Delta$ are

$$\lambda_{m,n} = (m^2 + n^2)\,\pi^2 \quad \text{and} \quad \psi_{m,n} = 2\sin(m\pi x)\sin(n\pi y), \tag{1.3}$$

where m and n range over all positive integers. We featured primarily the nonlinearity $f(u) = \lambda u + u^3$, and were thus able to analyze the possible types of symmetry on all of the branches bifurcating from the trivial branch at the first 6 eigenvalues of the Laplacian, as well as secondary and tertiary bifurcations off of those branches. In [**7**], we performed analogous experiments when Ω was a disk in $\mathbf{R}^2$, using the well known basis generated from Bessel functions in that case. In [**21**], we began an ambitious effort to understand the basins of attraction for Newton's Method in general, and the GNGA for our semilinear problems in particular. One could see this effort as an attempt to generalize the finite dimensional result in [**43**] to the significantly more complicated infinite dimensional case arising from the variational study of semilinear elliptic PDE. Efforts along these lines are still on-going. We are attempting to relate these experiments to the well known Lazer-Mckenna conjecture and subsequent Dancer counterexample (see [**29**] and [**18**]). In [**39**], we apply GNGA to systems of nonlinear elliptic BVP. In particular, we are able to approximate a triple junction solution to a vector equation of Ginzburg-Landau type (see also [**22**]). We are interested in and see the potential for applying our general methods to more complicated equations from physics, such as the full Ginzburg-Landau and Monge-Ampere equations (see [**42**], and [**4**]).

Our current efforts have focused on generating eigenfunctions of the Laplacian for general regions $\Omega \subset \mathbf{R}^2$ and then applying GNGA for (1.2) on that region. We present recent results for two interesting cases. In the first, we let $\Omega = \Omega_d$ be a family of Bunimovich stadia (see [**25**], [**45**], and [**6**]). In this case, the parameter d (radii of the end cap semicircles) is treated like a second bifurcation parameter. We observe the persistence of the well-known eigenvalue crossings and avoided crossings from the linear to the nonlinear case. Most recently, we considered the case where Ω is the Koch Snowflake. After first reproducing (and possibly improving upon) the approximations to the linear problem found in [**27**], we studied the nonlinear problem (1.2). Our focus here is on the symmetry of solutions, where we once again find guidance for the nonlinear problem by first understanding what happens in the linear case. These experiments rely heavily on the package ARPACK, an implementation of the Arnoldi process called the Implicitly Restarted Arnoldi Method, most useful in finding eigenvalues and eigenvectors to very large sparse matrices. At the time of this writing, we are porting both linear and nonlinear code to work in a parallel environment on a Linux cluster.

In Section 2 we will present the general GNGA algorithm, define precisely the nonlinearities for a class of semilinear elliptic BVP, and describe the variational framework for studying this class. Furthermore, we will provide some details for implementing the GNGA for these types of nonlinear PDE. In order to make the point of the generality of the numerical method, we also include a very simple ODE example where a basis other than eigenfunctions is used. In Section 3 we provide a brief historical overview of the constrained steepest descent methods (MPA, MMPA) used to find MI 1 and MI 2 solutions of problems of a similar type. In Section 4, we outline previous results where the GNGA was applied to BVP where the region Ω was a square or a disk. Section 5 contains an application to a system of PDE, with discussion of the applicability of the method to various scientific applications such as the full Ginzburg-Landau equation. Section 6 contains brief previews of several new results, where the basis of eigenfunctions for nonstandard regions is generated in order to further illuminate the relationship between linear and nonlinear problems. Section 7 contains several historical conjectures, some new conjectures, and describes ongoing experiments designed to further understanding of these conjectures. In Section 8, we list a number of open problems, as yet unperformed numerical experiments, and new ideas. Section 9 is the conclusion of this article.

2. GNGA and Variational Problems.

Suppose that one wants to find critical points of a C^2 functional $J : H \to \mathbf{R}$, where H is some Hilbert space. If one has good reason to believe that some of the critical points of interest lie almost entirely in an M-dimensional subspace $G \subset H$, then it should be fruitful to seek functions $u \in G$ so that the orthogonal projections $P_G \nabla J(u)$ are nearly zero. Let $G = \text{span}\{\psi_i\}_{i=1,\ldots,M}$ be such a subspace, where it is assumed that the ψ are orthogonal and suitably normalized. We use the notations $J(u) = J(\sum a_i \psi_i) = \hat{J}(a)$ interchangeably. In its most general form, we have damped Newton's Method with stepsize δ:

ALGORITHM 2.1. GNGA

(1) Choose initial coefficients $a = a^0 = \{a_k\}_{k=1}^M$, set $u = u^0 = \sum a_k \psi_k$, and set $n = 0$.
(2) Loop
  (a) Calculate $g = g^{n+1} = (J'(u)(\psi_k))_{k=1}^M \in \mathbf{R}^M$ (gradient vector).
  (b) Calculate $A = A^{n+1} = (J''(u)(\psi_j, \psi_k))_{j,k=1}^M$ (Hessian matrix).
  (c) Compute $\chi = \chi^{n+1} = A^{-1} g$ by computing inverse or pseudoinverse, solving system, or implementing least squares.
  (d) Set $a = a^{n+1} = a^n - \delta\chi$ and update $u = u^{n+1} = \sum a_k \psi_k$.
  (e) Increment loop counter n.
  (f) Calculate $\text{sig}(A(a))$ and $\hat{J}(a) = J(u)$ if desired.
  (g) Calculate approximation $\sqrt{g \cdot g}$ of $||\nabla J(u)||$; STOP if sufficiently small.

If the process converges to a nondegenerate saddle point u, then we expect that $\text{sig}(A(a^n))) := \{\text{number of negative eigenvalues of } A(a^n)\}$ to equal the MI of the critical point for n sufficiently large. The proper function spaces for semilinear elliptic BVP and other critical point problems are of course of paramount importance in both analytical and numerical investigations.

In the following simple ODE example, we take G to be spanned by the first M Legendre polynomials and $J(u) = \frac{1}{2}\int_{-1}^{1}((u+1)' - (u+1))^2\,dx = \frac{1}{2}\int_{-1}^{1}(u' - u - 1)^2\,dx$. Naturally, we are interested in global minima of J, satisfying the fundamental ODE $u' = u$. This simple type of problem is treated at length in [**42**], where the more obvious technique of steepest descent (in the proper Sobolev space) is used; we present the example here only to demonstrate the generality of GNGA. Let $\{\psi_k\}$ be the set of Legendre polynomials in $L_2(\Omega)$ with $\Omega = (-1, 1)$, normalized if desired. For our example, let $M = 4$ so that

$$G = \text{span}\{ \ 1.2247x, -0.7905 + 2.3717x^2, -2.8062x + 4.6770x^3, \\ 0.7954 - 7.9549x^2 + 9.2807x^4\}.$$

Since J is quadratic and hence $\nabla J(u)$ is linear, GNGA converges in one iteration to a nearby solution:

(1) Set $a = (0, 0, 0, 0)$ (for example) and form $u = a_1\psi_1 + \cdots + a_4\psi_4$.
(2) Loop (one iteration suffices)
   (a) Calculate $g_k = J'(u)(\psi_k) = \int_{-1}^{1}(u' - u - 1)(\psi_k' - \psi_k)\,dx$, $k = 1, \ldots, 4$.
   (b) Calculate $h_{j,k} = J'(u)(\psi_j, \psi_k) = \int_{-1}^{1}(\psi_j' - \psi_j)(\psi_k' - \psi_k)\,dx$, $j, k = 1, \ldots, 4$.
   (c) Compute χ satisfying $h\chi = g$.
   (d) Set $a = a - \chi$ and update $u = a_1\psi_1 + \cdots + a_4\psi_4$.
(3) Output a, $u + 1$, and $\sqrt{g \cdot g} = P_G\nabla J(u)$.

The above calculation yields a scalar multiple (solutions are not unique) of

$$u = 1 + 0.9962x + 0.4989x^2 + 0.1788x^3 + 0.04402x^4,$$

which is exactly $P_G y$ for an exact solution $y = e^x$, with $||y - u||_2 = 0.0007534$. Figure 1 shows plots for this example. If the above example is too obvious,

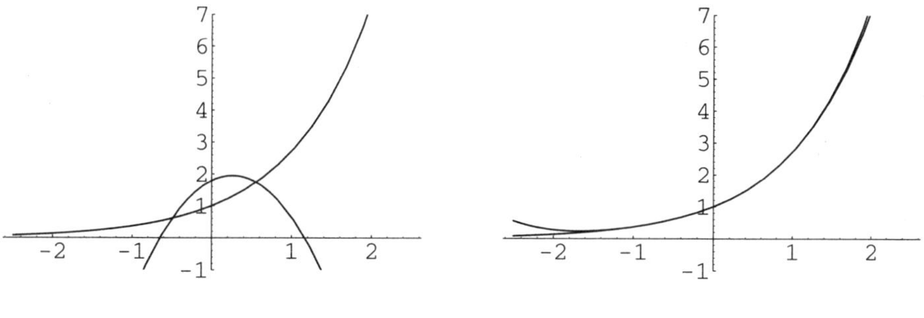

FIGURE 1. The graph on the left is of an initial guess and the 4th degree Taylor polynomial approximation of an exact solution $y = e^x$. The graph on the right compares the Taylor approximation to the function $u = P_G y$ obtained via the Newton iteration step.

we tested the method in a similar fashion on initial value problems such as $y' = 1 + y^2$, $y' = -xy$, and $(y')^2 + y^2 = 4$, where more than one iteration is required.

We now devote our attention to (1.2). Let $F(u) = \int_0^u f(s)\,ds$ for all $u \in \mathbf{R}$ define the primitive of f. We then define the action functional $J : H \to \mathbf{R}$ by

$$J(u) = \int_\Omega \left\{ \frac{1}{2}|\nabla u|^2 - F(u) \right\} dx. \tag{2.1}$$

We will make assumptions on f so that J is well defined and of class C^2 on H. Taking λ_i and ψ_i as in (1.1) and integrating by parts, a calculation shows that

$$J'(u)(\psi_i) = \int_\Omega \{\nabla u \cdot \nabla \psi_i - f(u)\,\psi_i\}\,dx = a_i\lambda_i - \int_\Omega f(u)\,v\,dx \tag{2.2}$$

and
(2.3)

$$J''(u)(\psi_i, \psi_j) = \int_\Omega \{\nabla \psi_i \cdot \nabla \psi_j - f'(u)\,\psi_i\,\psi_j\}\,dx = \lambda_i \delta_{ij} - \int_\Omega f(u)\,\psi_i\,\psi_j\}\,dx,$$

where δ_{ij} is the Kronecker delta function. In this case, note that there is no need for numerical integration when implementing Algorithm 2.1. This information is encoded in the solutions (eigenpairs) of the linear problem. It is well known that critical points of J are in fact solutions to (1.2) (see for example [**46**] and [**24**]). For the types of nonlinearities f we will consider, most if not all nontrivial solutions are saddle points of J. Roughly speaking, the Morse index is the number of linearly independent directions in function space in which J "curves down" at a (nondegenerate) solution u. This is the number of negative eigenvalues of the Hessian $D^2J(u)$, and for M and n in Algorithm 2.1 sufficiently large, also equal (one hopes) to $\mathrm{sig}(A)$. See [**33**] for more information about Morse index (MI).

We now give a precise statement of the hypothesis on the nonlinearity f that leads to existence and nodal structure theorems as well as the numerical approximation of solutions (see for example [**2**], [**8**], [**34**], [**9**], [**10**], [**11**], [**36**], [**47**], and [**48**]). In particular, we take $f \in C^1(\mathbf{R}, \mathbf{R})$ such that $f(0) = 0$. In general we assume that $f'(0) < \lambda_1$. This is necessary in order to find one-sign solutions. In [**11**] we assume that $f'(0) < \lambda_2$ and are able to prove the existence of a sign-changing exactly-once solution; we also demonstrated the existence of a solution which changes sign at most k times when $f'(0) < \lambda_k$.

We assume that there exist constants $A > 0$ and $p \in (1, \frac{N+2}{N-2})$ such that $|f'(u)| \leq A(|u|^{p-1} + 1)$ for all $u \in \mathbf{R}$. It follows that f is *subcritical*, i.e., there exists $B > 0$ such that $|f(u)| \leq B(|u|^p + 1)$. For $N = 1$ this condition is omitted, while for $N = 2$ it suffices to have $p \in (1, \infty)$ (see [**46**]). Also, we assume that there exists $m \in (0, 1)$ such that

$$\frac{m}{2} f(u)u \geq F(u), \tag{2.4}$$

(in fact this need only hold for $|u| > \rho$ for some $\rho > 0$), and that f is *superlinear*, i.e.,

$$\lim_{|u|\to\infty} \frac{f(u)}{u} = \infty. \tag{2.5}$$

Finally, we make the assumption that f satisfies

$$f'(u) > \frac{f(u)}{u} \quad \text{for } u \neq 0. \tag{2.6}$$

For convenience, we will call (1.2) with all of the above conditions on f "the superlinear problem". Recall that subcritical growth and the Sobolev Embedding Theorem (see [**1**]) imply that H is compactly embedded into L_{p+1}, which in turn shows that J is well defined on all of H. Under this hypothesis, J is in fact twice differentiable on H (see [**46**]).

In [**8**], the Mountain Pass argument used in [**2**] to find one-sign solutions to the superlinear problem was extended to find a sign-changing solution. Rather than one constraint, the theorem uses two constraints to find a MI 2 (if nondegenerate) solution to (1.2) using a subset of the well known Nehari manifold $S = \{u \in H \setminus \{0\} : J'(u)(u) = 0\}$:

THEOREM 2.1 (The CCN Theorem [**8**]). *If $f'(0) < \lambda_1$, then the superlinear problem (1.2) has at least three nontrivial solutions: $\omega_1 > 0$ in Ω, $\omega_2 < 0$ in Ω, and ω_3. The function ω_3 changes sign exactly once in Ω, i.e., $(\omega_3)^{-1}(\mathbf{R} \setminus \{0\})$ has exactly two connected components. If nondegenerate, the one-sign solutions are Morse index (MI) 1 critical points of J, and the sign-changing solution has MI 2. Furthermore,*

$$J(\omega_3) \geq J(\omega_1) + J(\omega_2).$$

For convenience, we call ω_3 the CCN solution. In this $f'(0) < \lambda_1$ superlinear case, the trivial solution $u = 0$ has MI 0 and is the only local minimum of J. All other critical points (solutions to (1.2)) are saddle points. With $u_+ = \max_\Omega\{u, 0\}$ and $u_- = \min_\Omega\{u, 0\}$, we can define the subset

$$S_1 = \{u \in S : u_+ \in S, u_- \in S\}, \tag{2.7}$$

which allows us to restate the CCN Theorem:

THEOREM 2.2 (CCN). *For the superlinear problem, given $f'(0) < \lambda_1$ there exist solutions $\omega_1 > 0$, $\omega_2 < 0$, and ω_3 which changes sign exactly once, where ω_1 and ω_2 are local (possibly global) minimizers of $J|_S$ and $J(\omega_3) = \min_{S_1} J$.*

In [**11**] we find that the MI 2 CCN solution persists when $f'(0) \in [\lambda_1, \lambda_2)$, which suggests that one let $\lambda = f'(0)$ vary as a bifurcation parameter. Using GNGA, we are able to accurately and thoroughly construct the portion of the bifurcation diagram corresponding to the first few eigenvalues (how few essentially depends only on the choice of M for the dimension of G). Ultimately, we seek to prove the existence and describe the nodal structure of all solutions to the superlinear problem (and indeed PDE (1.2) with other types of nonlinearities).

Although theory tells us that the so-called weak solutions in H are in fact in C^2, it is necessary to use the Hilbert space H as the domain of the functional J in order to find existence proofs. When considering numerical steepest descent methods, it is essential that one not use the poorly performing (and only densely defined) L_2 gradient (see [**42**]). We will be restricting our approximations to the finite dimensional subspace G of L_2, which consists of twice continuously differentiable functions. In this case, we may use the L_2 gradient $\nabla_2 J(u)$ and Hessian $\mathbf{D}_2^2 J(u)$, since the Newton search direction is seen to satisfy

$$\chi = (D_H^2 J(u))^{-1} \nabla_H J(u) = (D_2^2 J(u))^{-1} \nabla_2 J(u).$$

Thus, our implementation of Algorithm 2.1 approximates the infinite dimensional Newton iteration

$$u_{k+1} = u_k - (D_H^2 J(u_k))^{-1} \nabla_H J(u_k), \tag{2.8}$$

and the corresponding continuous infinite dimensional "flow" (ODE)

$$u'(t) = -(D_H^2 J(u(t)))^{-1} \nabla_H J(u(t)). \tag{2.9}$$

As a practical matter, since $D^2 J(u)$ may be noninvertible or ill conditioned, one should consider using singular value decomposition and pseudo inverses, least squares, or a system solver which handles our possible types of singularities when computing this search direction. At nondegenerate solutions, the Hessian is nonsingular in a neighborhood of the solution and the pseudoinverse is not needed. However, we are sometimes interested in degenerate solutions, for example at bifurcation points or non-radial solutions to the PDE on a disk where there is a continuum of solutions. We have found the algorithm to be surprisingly robust even when an actual inverse is computed and used very near a singularity, an interesting phenomena worthy of future study. In many cases the corresponding zero-eigenfunction directions are nearly orthogonal to nearby gradients, and hence had little effect on the "projected gradient" (Newton) search direction.

3. Mountain Pass and Modified Mountain Pass Algorithms

Historically, the MPA found in [**13**] represents the first numerical minimax technique applied to semilinear elliptic problems, although constrained optimization was not itself a new idea. The MPA is based on the Mountain Pass Lemma and the landmark existence theorem for one-sign solutions found in [**2**]. We briefly present a version of this algorithm as applied to the superlinear problem. We also describe the MMPA (see [**34**]) which likewise was developed from the constructive proof found in [**8**]. Both algorithms are applied here to the superlinear problem; different nonlinearities may significantly change the variational structure and hence necessitate modifications.

Since under the hypothesis of Theorem 2.1 the set S is a manifold, we can use the facts that given $u \neq 0$ there exists a unique $\hat{\alpha} > 0$ such that $\hat{\alpha} u \in S$ and that $J(\hat{\alpha} u) = \max_{\alpha > 0} J(\alpha u)$ to project nonzero elements of H on to S and sign-changing elements of H on to S_1. The Sobolev gradient $\nabla_H J(u) = u + z$ where $\Delta z = f(u)$ can be computed by solving a system via standard algorithms.

ALGORITHM 3.1. MPA

(1) Choose u of one sign (Typically $\pm c\psi_1$).
(2) Loop
   (a) Project u on to S by iterating $u + \sigma_1 P_u \nabla_H J(u) \to u$ (Ascent).
   (b) Solve a sparse system to compute $\nabla_H J(u) = u + (\Delta)^{-1} f(u)$.
   (c) Set $u - \sigma_2 \nabla_H J(u) \to u$ (Descent).
   (d) Stop when $||\nabla_H J(u)||_H$ is sufficiently small.

Similarly,

ALGORITHM 3.2. MMPA

(1) Choose u which changes sign exactly once (Typically $\pm c\psi_2$).
(2) Loop
   (a) Project u_+ on to S by iterating $u_+ + \sigma_1 P_{u_+} \nabla_H J(u_+) \to u_+$ (Ascent).
   (b) Project u_- on to S by iterating $u_- + \sigma_1 P_{u_-} \nabla_H J(u_-) \to u_-$ (Ascent).
   (c) Add $u = u_+ + u_- \in S_1$.
   (d) Solve a sparse system to compute $\nabla_H J(u) = u + (\Delta)^{-1} f(u)$.
   (e) Set $u - \sigma_2 \nabla_H J(u) \to u$ (Descent).
   (f) Stop when $||\nabla J(u)||_H$ is sufficiently small.

In both cases the step sizes σ_1 and σ_2 may be taken to be 1 if the initial guess is sufficiently close to a solution.

In general, the above methods will only find MI 1 and MI 2 solutions, respectively. It is easy to make guesses near higher MI solutions, and observe "instability" as iterates approach then fall away from such higher energy critical points. In **[17]** the MMPA is modified so that iterates u and corresponding gradients $\nabla_H J(u)$ are constrained to invariant subspaces corresponding to symmetry groups for a given region Ω. One simply inserts a step after the projection on to S_1 whereby u is replaced with Pu, the projection of u on to the invariant subspace. As an example, when $\Omega = (0,1) \times (0,1)$ for the superlinear problem with $f(u) = u^3$, we can find a MI 8 solution which is 90° rotationally invariant and changes sign exactly once. In practice, this is implemented by replacing an iterate with the average of all four 90° rotations.

So what does one do to find higher Morse index solutions? The existence of higher codimension sets analogous to S and S_1 is open. If we had such sets we would have the theory to prove the existence of many more solutions.

One scheme we have tried successfully (see [**42**]), if not efficiently, is to find global minimizers of the functional ϕ defined by

$$\phi(u) = \frac{1}{2}||\nabla_H J(u)||_2^2.$$

Obviously all critical points of J are now zeroes of the non negative functional ϕ. We suggest [**42**] as a reference for this and other steepest descent techniques. The potential exists for doing analysis on the behavior of ϕ (similar to that done on J in [**8**] and elsewhere) which might well lead to existence proofs. Since

$$\chi = (D^2 J(u))^{-1}\nabla J(u) = \sum \frac{1}{\beta_i} P_{e_i}\nabla J(u) \quad \text{with } D^2 J(u)e_i = \beta_i e_i,$$

Newton's method on the gradient can be viewed as a natural extension of the MPA and MMPA. In directions e_i where $\beta_i > 0$ (and hence J is concave up), one is doing steepest descent. Conversely, when $\beta_i < 0$, steepest ascent is performed. In this way, GNGA simulates constrained steepest descent when finding critical points of arbitrarily high MI.

4. Newton's Method and Morse Index: The Unit Square and Disk.

On regions other than those where the eigenfunctions are known in closed form there will be a substantial effort required to generate the basis, as described in Section 6. This effort will only have to be done once for each given region and eliminates the need for numerical differentiation when computing Newton search directions. However, when $\Omega = (0,1)^2$ or $B_1(0) \subset \mathbf{R}^2$, the exact basis is known in terms of sines, cosines, and/or Bessel functions. In this section we briefly describe the numerical results found in [**36**] and [**7**].

Newton's method converges very well given a good initial guess and is unpredictable given a poor initial guess. Understanding basins of attraction is difficult, interesting, and very important. Our rule of thumb for superlinear problems is to use an appropriate multiple of an eigenfunction having a prescribed nodal structure and signature. When f is odd, this works well provided the multiple is in the right ball park; we can use the numerical observation that $\lim_{\alpha\to\infty} \text{sig}(\alpha\psi_i) = \infty$ for all $i \in \mathbf{N}$ to obtain the appropriate multiplier.

When f is not so nice, solutions may have nodal structures that do not closely match that of eigenfunctions, whereby multiples of eigenfunctions may not be good starting points. Since our nonlinearity is of the form $f(u) = \lambda u + g(u)$ with $g'(0) = 0$, we can use λ as a bifurcation parameter and look for solutions which do closely resemble eigenfunctions by first starting near primary bifurcation points off of the trivial branch at λ just less than the corresponding eigenvalue. One then decrements (or increments) the bifurcation parameter to follow the branch towards the desired solution at (say) $\lambda = 0$. At each step we use the solution at the previous λ as the next initial guess. The matter is somewhat more delicate at secondary bifurcation points. In this case we add a small linear combination of zero-eigenfunctions of $D^2J(u)$ to a nearby element of the primary branch, where u is the singular bifurcation point. The same philosophy applies for finding tertiary branches off of secondary branches. In most of our experiments, we are able to use symmetry to make intelligent initial guesses. Our continuation techniques are effective but could no doubt be refined (see for example [**31**]).

In [**36**] we additionally provide an efficient method for implementing the GNGA scheme when f is polynomial. In that case, $\hat{J}$, g, and A are themselves polynomials in a, and once various integrals are calculated, no further integration needs to be performed. In that work we also investigated numerically polynomials which did not satisfy our exact CCN superlinear hypothesis. We present here a few excerpts of the superlinear results, in particular the bifurcation diagrams constructed therein when investigating the BVP (1.2) on the square $\Omega = (0,1) \times (0,1)$ with $f(u) = \lambda u + u^3$, where λ is treated as a bifurcation parameter. In [**36**], we spend considerable effort to use the symmetry of the square to explain and analyze the expected proliferation of solutions. Figure 2 depicts the first 9 primary branches that bifurcate from the origin at $\lambda \leq \lambda_{2,3} = 13\pi^2$.

At a bifurcation from a simple eigenvalue, there is a pitchfork bifurcation that creates two solutions with the same action J. Hence one branch is observed bifurcating from $2\pi^2$ and $8\pi^2$ in Figure 2. Double eigenvalues lead to two types of bifurcation, *pure modes* and *mixed modes*. Near the bifurcation, the pure modes are asymptotically a multiple of an eigenfunction $\pm\psi_{i,j}$ or $\pm\psi_{j,i}$. The mixed modes are asymptotic to a multiple of one of the four combinations of eigenfunctions $\pm\psi_{i,j} \pm \psi_{j,i}$. Naturally the behavior is more complicated at higher multiplicity eigenvalues. Of the branches seen in Figure 2, the solutions which bifurcate from $\lambda_{1,3}$ are the most interesting. Leaving the details to [**36**], we observe in Figures 3 and 4 secondary and tertiary bifurcations off of these primary branches. The tertiary branch we found contained solutions with no symmetry whatsoever. We are confident that we completely classified and numerically found all possible types of solutions found emmenating from this double eigenvalue, although we certainly do not have existence proofs. We do not include here our tests on the convergence of our results as $N_{\max}$ increases.

In a similar but technically more challenging application, we considered in [**7**] the case where $\Omega = B_1(0) \subset \mathbf{R}^2$. The eigenvalues of $-\Delta$ with zero Dirichlet boundary values on the disk are determined by the zeroes of the various Bessel functions of the first kind. The corresponding eigenfunctions, which form an orthonormal basis, are divided into three components. The radial component is defined by

$$\psi_i = a_i J_0(\sqrt{\lambda_i^0}\, r),$$

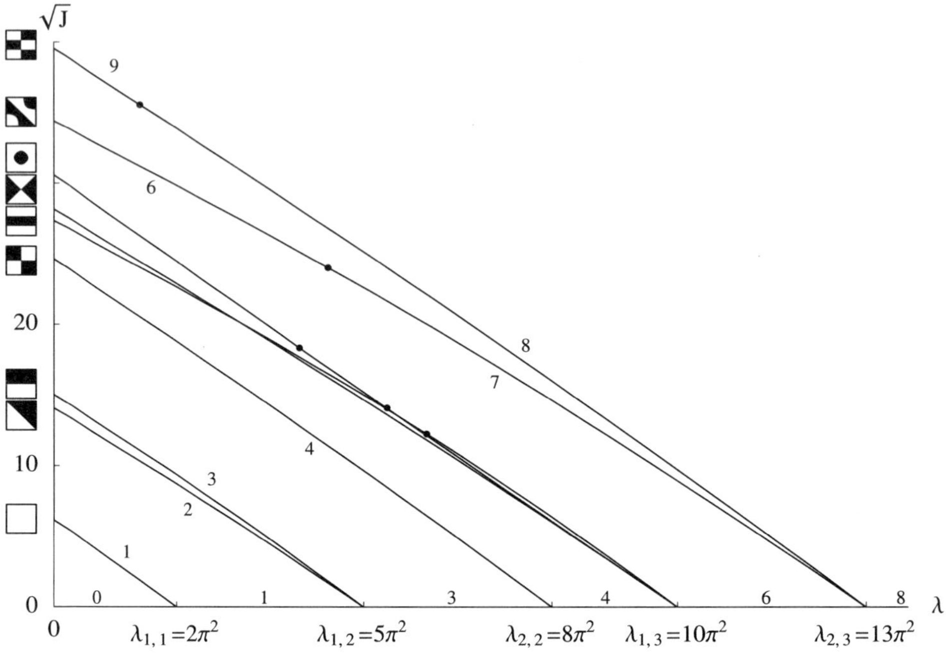

FIGURE 2. Bifurcation diagram showing $\sqrt{J}$ as a function of λ, for $f(u) = \lambda u + u^3$. Only the primary branches (the ones which bifurcate from the origin) are shown. The MI is indicated by the small numbers. The dots indicate where the MI changes at secondary bifurcations which create solutions that are not shown in this figure. More detail of the three branches that bifurcate from $\lambda_{1,3}$ is given in the next two figures. This figure is adapted from Figure 4 in [**36**], used with permission of the IJBC.

where J_0 is the Bessel Zero function, $\sqrt{\lambda_i^0}$ is the ith zero of $J_0(r)$, and $a_i = \frac{1}{\sqrt{\int_\Omega J_0(\sqrt{\lambda_i^0}\ r)^2}}$. That is to say, a_i, is calculated so that $\int \psi_i^2 = 1$. The remaining two components of the orthonormal basis are the nonradial components and are given by

$$\varphi_{i,j} = b_{i,j} J_i(\sqrt{\lambda_j^i}\ r)\cos(i\theta)$$

and

$$\chi_{i,j} = b_{i,j} J_i(\sqrt{\lambda_j^i}\ r)\sin(i\theta).$$

Here J_i denotes the various Bessel functions for $i \neq 0$, $\sqrt{\lambda_j^i}$ is the jth zero of the ith Bessel function, and $b_{i,j}$ is calculated so that $\int \varphi_{i,j}^2 = 1$, and $\int \chi_{i,j}^2 =$

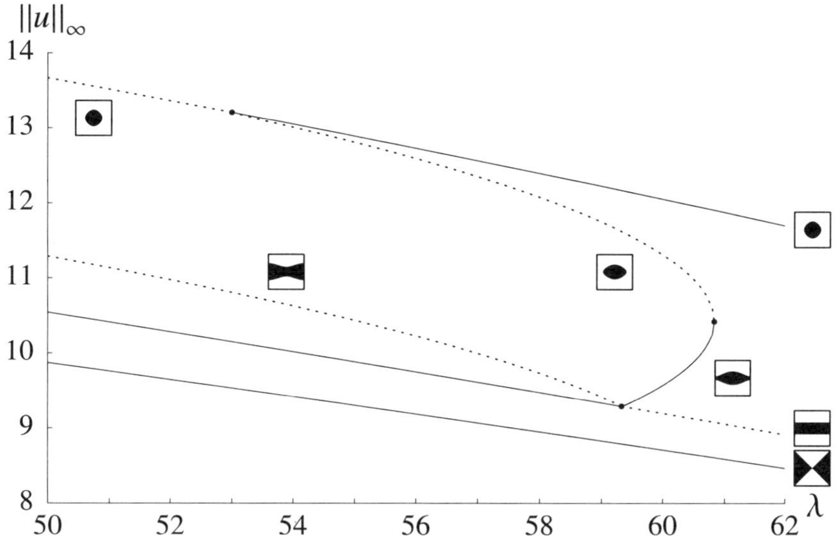

FIGURE 3. Detail of Figure 2 for the branches that bifurcate from $\lambda_{1,3}$. The supremum norm, $\|u\|_\infty$, is plotted against λ because it gives a good separation between the branches. The solutions with MI 5 are solid lines, and the solutions with MI 6 are dotted lines. This figure was computed with $N_{\max} = 9$ (81 Fourier modes). Figure adapted from Figure 5 in [**36**], with permission from the IJBC.

1. Thus, given $u \in L^2$ there exist constants $\{a_i\}$, $\{b_{i,j}\}$, and $\{c_{i,j}\}$ so that

$$u = \sum_i a_i \psi_i + \sum_i \sum_j b_{i,j} \varphi_{i,j} + \sum_i \sum_j c_{i,j} \chi_{i,j}.$$

Using this basis in exactly the same way as described above for the square, we were able to find many solutions to the nonlinear problem. In particular, we found the radial solutions guaranteed to exist by [**12**], as well as the expected nonradial ones. The symmetry can be analyzed as in [**36**], and bifurcations from the trivial branch predicted. At this time, we have not conclusively found any secondary bifurcations nor concluded that they do not exist. The matter remains open, and so we are not including a bifurcation diagram in this article. Figure 5 depicts a GNGA-approximated MI 2 CCN solution on the disk.

5. GNGA and Semilinear Systems.

In [**39**] we investigate the application of GNGA to systems of elliptic BVP. In particular, we find a triple junction solution conjectured to exist in

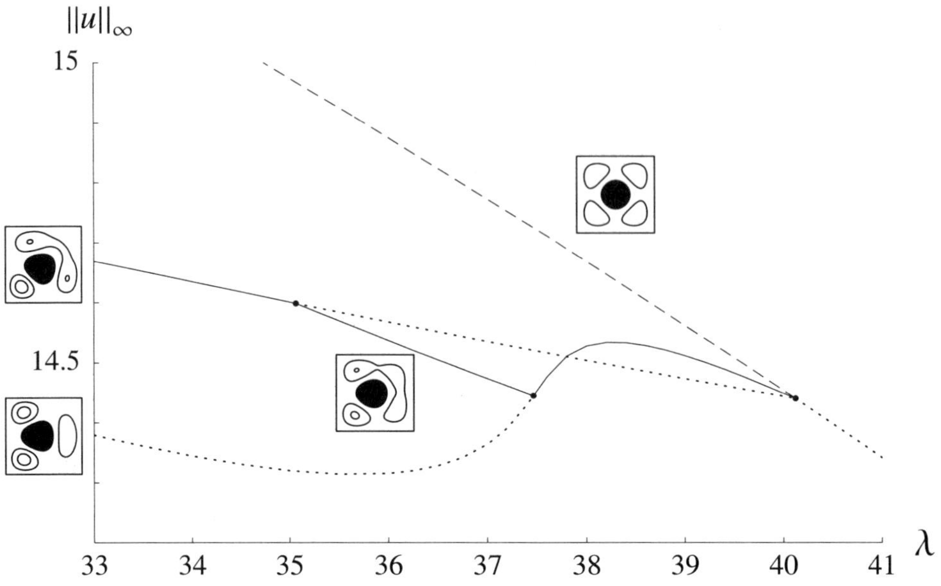

FIGURE 4. Further detail of Figure 2. The "radial" solution undergoes a bifurcation with $\mathbf{D}_4$ symmetry at $\lambda \approx 40$, creating the vertex and edge solutions. These secondary solutions in turn bifurcate, creating a branch of tertiary solutions that exist for λ in the approximate interval $[35.1, 37.5]$. These tertiary solutions have no symmetry. The Morse indices are indicated by the line type: MI 6 are dotted, MI 7 are solid, and MI 8 are dashed lines. Along with the black region, showing where $u < 0$, the contours at $u = 1/3\|u\|_\infty$ and $u = 2/3\|u\|_\infty$ are shown to better see the symmetry of the patterns. This figure was computed with $N_{\max} = 7$ (49 Fourier modes). Figure adapted from Figure 6 in [**36**], with permission from the IJBC.

[**22**] to the vector Ginzburg-Landau type problem:

$$(5.1)\qquad \begin{cases} -\epsilon^2 \Delta u + \nabla W(u) = 0 & \text{in } \Omega \\[2ex] \dfrac{\partial u}{\partial \eta} = 0 & \text{on } \partial\Omega, \end{cases}$$

where $\Omega = (0,1) \times (0,1)$, $u : \Omega \to \mathbf{R}^2$, $W : \mathbf{R}^2 \to [0, \infty)$ is a triple-well potential, η is the unit outer normal to the boundary, and $\epsilon > 0$ is a parameter. The gradient in $\nabla W(u)$ is with respect to u, hence $\nabla W = (\partial W / \partial u_1, \partial W / \partial u_2)$. The triple-well potential W achieves a global minimum of zero at 3 points, and has 7 critical points in total. Figure 6 shows the contour graph of the potential that we used in our experiments.

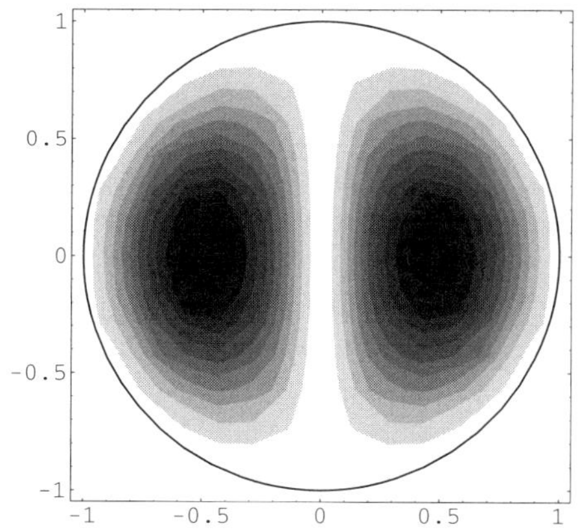

FIGURE 5. MI 2 CCN Solution on the unit disk.

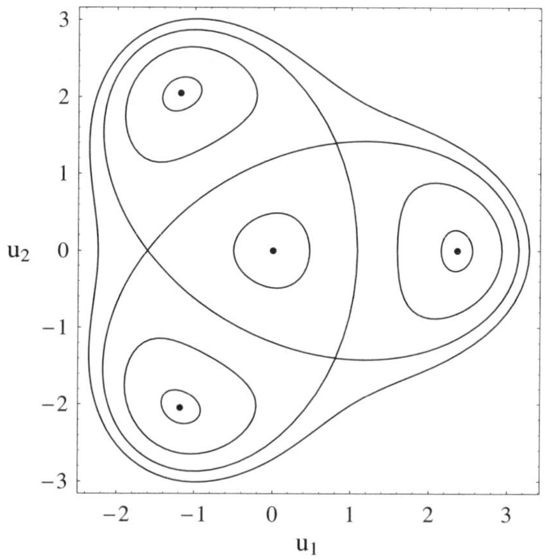

FIGURE 6. The triple-well potential W. The graph shows the minima and local maximum of W as dots, as well as the contours through the 3 saddle points and around the local extrema. Figure adapted from Figure 1 in [**39**], with permission from the IJBC.

We worked with sufficiently smooth functions $u \in L_2 \times L_2$. Since this problem is again on the square, we use known eigenfunctions similar to those found in (1.3) to form a basis for L_2, and then take a product to get a basis for $L_2 \times L_2$. Specifically, the (doubly-indexed) eigenvalues of $-\Delta$ are $\lambda_{m,n} = (m^2 + n^2)\pi^2$ and the eigenfunctions $\psi_{m,n}$ of $-\Delta$, normalized in L_2, are given by:

$$\psi_{m,n}(x,y) = \begin{cases} 1 & \text{for } m = n = 0 \\ \sqrt{2}\cos(m\pi x)\cos(n\pi y) & \text{for } m = 0, n \neq 0 \text{ or } n = 0, m \neq 0 \\ 2\cos(m\pi x)\cos(n\pi y) & \text{for } m \neq 0 \text{ and } n \neq 0, \end{cases}$$

where m and n range over all of the nonnegative integers. It is well known that $\{\psi_{m,n}\}$ forms a basis for L_2 and $H = H^{1,2}(\Omega)$ [Adams, 1975]. We obtain a finite sub-basis (basis for a subspace) of L_2 or H by choosing a positive integer k, and including all eigenfunctions whose eigenvalues are strictly less than $\pi^2 k^2$. These eigenvalues are then ordered and singly indexed as $\lambda_1 = 0 < \lambda_2 \leq \lambda_3 \leq ... \leq \lambda_{\widetilde{M}}$. This yields a singly indexed basis, $\{\psi_i\}$, of size $\widetilde{M}$. In our experiments, typically k is an even integer ranging from $k = 8$ ($\widetilde{M} = 56$) up to $k = 18$ ($\widetilde{M} = 269$). As a rule of thumb, we trust a numerical calculation if the results are not significantly changed when the cutoff parameter k is increased by 2. We use the following Galerkin subspace of $H \times H$ and $L_2 \times L_2$:

$$G = \text{Span}\{\Psi_i\}_{i=1}^{M} := \text{Span}\left\{\begin{pmatrix} \psi_i \\ 0 \end{pmatrix}\right\}_{i=1}^{\widetilde{M}} \bigcup \left\{\begin{pmatrix} 0 \\ \psi_i \end{pmatrix}\right\}_{i=1}^{\widetilde{M}},$$

which is of dimension $M = 2\widetilde{M}$. The energy functional whose critical points are solutions to (5.1) is defined by

$$J_\epsilon(u) = \frac{1}{\epsilon}\int_\Omega \left(\frac{\epsilon^2}{2}|\nabla u|^2 + W(u)\right) dx\,dy. \tag{5.2}$$

A calculation shows that the directional derivative of J at u in the Ψ_j direction is

$$\begin{aligned} g_j(u) := J'(u)(\Psi_j) &= \int_\Omega \left(\epsilon \nabla u \cdot \nabla \Psi_j + \frac{1}{\epsilon}\Psi_j \cdot \nabla W(u)\right) dx\,dy \\ &= \int_\Omega \left(\epsilon \sum_{i=1}^{M} a_i \left(\nabla \Psi_i \cdot \nabla \Psi_j\right) + \frac{1}{\epsilon}\Psi_j \cdot \nabla W(u)\right) dx\,dy \\ &= \epsilon a_j \lambda_j + \frac{1}{\epsilon}\int_\Omega \Psi_j \cdot \nabla W(u)\,dx\,dy. \end{aligned}$$

Second directional derivatives can be computed similarly, whence GNGA (Algorithm 2.1) can be implemented. When u is identically a constant critical number of W, we have a trivial solution. Figure 7 depicts a diagram of solutions bifurcating off of the trivial branches corresponding to the saddle and local maxima of W, as $\epsilon \to 0$. Figure 8 shows the various types of symmetries found in such solutions. Note that for small ϵ, the solutions tend towards step functions whose piecewise constants are critical numbers of the triple well potential W. It may well be that there are alternate choices for a basis with superior behavior in approximating the near step function solutions depicted in Figure 8, as there is no longer a dominant eigenmode in the expansion for such functions.

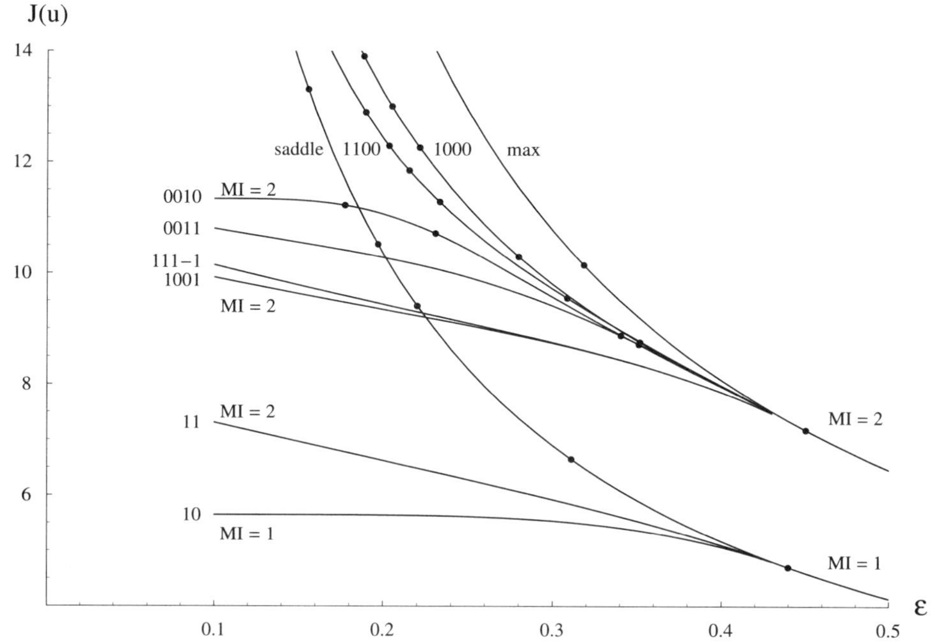

FIGURE 7. This bifurcation diagram shows the saddle solution, the maximum solution, and the solutions which bifurcate off these trivial solution branches at the first bifurcation as ϵ is decreased. The minimum solution, which is not shown, has $J = 0$ and MI 0 for all ϵ. The dots indicate a bifurcation point, where the MI changes. The lowest MI 2 secondary branch bifurcating off of the trivial maximum solution converges to the triple junction as $\epsilon \to 0$. The MI of the branches is indicated by "color"; subsequently we have preferred to identify branches by symmetry type rather than MI. Figure adapted from Figure 2 in [**39**], with permission from the IJBC.

We are also interested in the so-called full Ginzburg-Landau equation, other physical applications such as the Monge-Ampere equation, and finding a CCN-type solution for a system (see Section 8). In order to make this last problem statement precise, we present here a possible set of hypotheses to consider when looking for sign-changing solutions to superlinear systems.

We wish to generalize the hypotheses from [**8**] to the following system, in hopes that the techniques and results from that work will apply. In particular, we consider

$$\begin{cases} \Delta u + \nabla W(u) = 0 & \text{in } \Omega \\ u = 0 & \text{in } \partial\Omega, \end{cases} \tag{5.3}$$

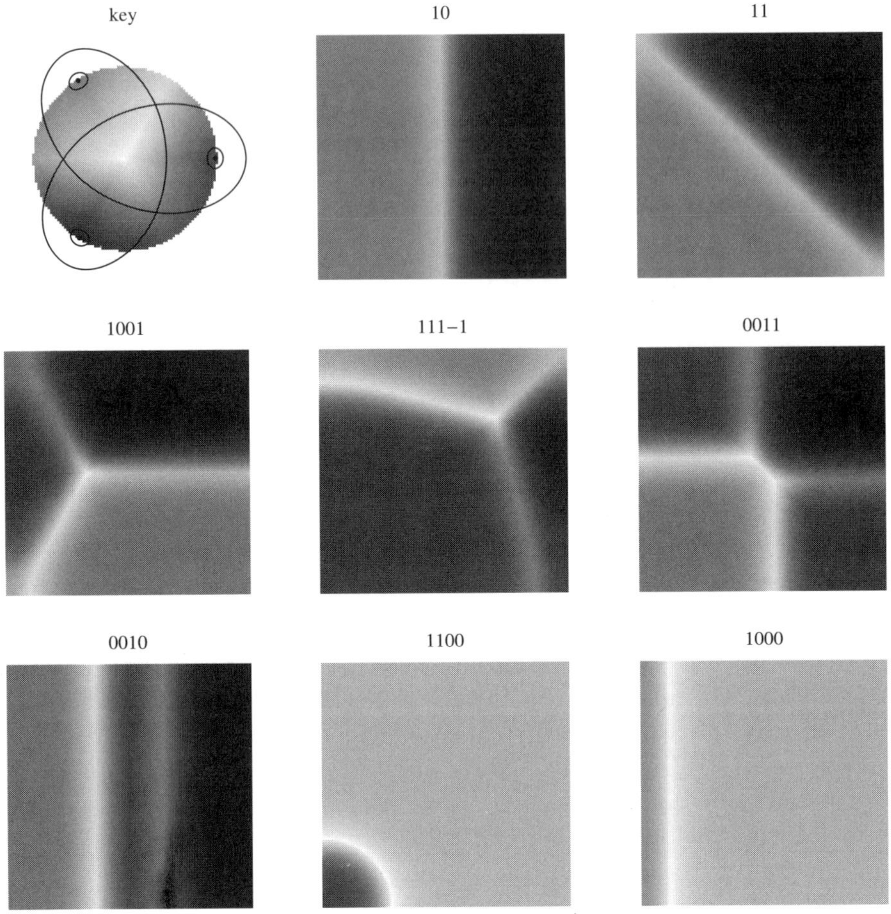

FIGURE 8. All of these solutions are computed at $\epsilon = 0.1$, where the nonlinear corrections to the critical eigenspace are significant. The key at the upper left shows a few contours of the potential W to indicate how the color codes the value of $u(x, y)$. The diagram with the $111 - 1$ symmetry is the MI 2 triple junction solution which bifurcates off of the maximum trivial branch, as conjectured in [**22**]. See [**39**] for the color version of these graphics which are somewhat more illuminating (used here with permission from the IJBC).

where $W : \mathbf{R}^2 \to \mathbf{R}$ is a C^2 function and $u : \Omega \to \mathbf{R}^2$. Let H be the Sobolev space $H_0^{1,2}(\Omega)$. We want a sufficient hypothesis on W so that the functional $J : H \times H \to \mathbf{R}$ defined by

$$J(u) = \int_\Omega \{\frac{1}{2}|\nabla u|^2 - W(u)\} \, dx$$

has the appropriate properties so that we may mimic the variational arguments found in [CCN]. Corresponding to the hypothesis made on f in [CCN], we may consider assuming that W satisfies the following conditions. Assume that

$$W \text{ has a local minimum } W(0) = 0.$$

Note that this implies that $\nabla W(0) = 0$, which corresponds to $f(0) = 0$. Let $\beta_1 \leq \beta_2$ be the eigenvalues of $D^2W(0)$, which are both positive since W has a local minimum at 0. Assume that

$$\beta_2 < \lambda_1,$$

a condition that parallels the CCN assumption that $f'(0) < \lambda_1$. Corresponding to the CCN convexity assumption $f'(v) > f(v)/v$ for all $|v| \neq 0$, we assume that

$$D^2W(u)u \cdot u > \nabla W(u) \cdot u.$$

The coercivity assumption $\frac{m}{2} v f(v) \geq F(v)$ for some $m \in (0, 1)$ where $F(v) = \int_0^v f(s)\, ds$ becomes

$$\frac{m}{2} u \cdot \nabla W(u) \geq W(u),$$

again for some $m \in (0, 1)$. In CCN, we required that f be superlinear, i.e., that $f(v)/v \to \infty$ as $|v| \to \infty$. Similarly, for the gradient system one would require that either

$$\frac{|\nabla W(u)|}{|u|} \to \infty \text{ or } \frac{\nabla W(u) \cdot u}{u \cdot u} \to \infty, \text{ as } |u| \to \infty.$$

We are not certain which is the most natural assumption. Both assumptions appear to imply that

$$\frac{W(u)}{u \cdot u} \to \infty \text{ as } |u| \to \infty,$$

where $|u|$ is the Euclidean norm of u in $\mathbf{R}^2$. Finally, we need to assume a subcritical growth of W. One possibility is to assume that there exists $A > 0$ so that

$$D^2W(u)u \cdot u \leq A(|u|^{p-1} + 1)|u|^2.$$

When $N = 2$ this assumption is unnecessary, but for higher N it is required in order for the application of the Sobolev Imbedding Theorem. The "critical hyperboloid" found in the literature suggests that perhaps the above subcritical condition might not be the best possible subcritical assumption.

The question now is whether a CCN-type solution exists to (5.3). If one takes this to mean the existence of a $w = (u, v) \in H \times H$ where both u and v change sign exactly once and w has the smallest J value among all such solutions, then numerical experiments suggest that the solution will be of MI 4 (and hence require 4 constraints for any minimax existence argument). In our research preliminary to the Ginzburg-Landau problem outlined above, we considered systems of scalar equations such as

$$\Delta u + u^3 + cuv = 0, \text{ and } \Delta v + v^3 + cuv = 0.$$

For various reasonable values of the coupling parameter c, we were able to numerically find solutions w where:

(1) w is MI 1, u is of one sign, and v is small.
(2) w is MI 1, v is of one sign, and u is small.
(3) w is MI 2, u and v are of one sign.
(4) w is MI 2, u is small, and v changes sign exactly once.
(5) w is MI 2, v is small, and u changes sign exactly once.
(6) w is MI 3, u is of one sign, and v changes sign exactly once.
(7) w is MI 3, v is of one sign, and u changes sign exactly once.
(8) w is MI 4, u and v change sign exactly once.

Clearly the results from [**8**] confirm the above results when the system is decoupled via setting $c = 0$. It seems reasonable to apply techniques similar to that found in [**8**] to prove the existence of the first three solutions. We are currently unsure as to how to proceed in looking for an existence proof for sign-changing solutions. Note that one could consider the third solution to be a MI 2 sign-changing solution if u and v were one sign solutions of opposite sign.

6. GNGA for General Regions.

We present here a preview of two new results. For general regions, one must first generate a suitable basis for the M dimensional subspace G. In the previous section, we saw an example where, although a basis of eigenfunctions of $-\Delta$ worked, it might not have been the best choice. Here, we are back in the territory of superlinear problems where the linear theory is clearly key in investigating the nonlinear phenomena.

Both experiments use the package ARPACK. At the time of this writing, we are porting both linear and nonlinear code to work in a parallel environment on a Linux cluster. Our scalar processing code uses LAPACK, BLAS, and ARPACK libraries. The parallel code uses MPI together with PARPACK, PBLAS, and SCALAPACK. Once our test platform is upgraded we will be able to repeat experiments with a much higher order of accuracy, as well as perform other experiments such as PDE on high dimensional regions Ω that currently are beyond our computational capabilities.

In [**25**], we consider the case where $\Omega = \Omega_d \subset \mathbf{R}^2$ is a Bunimovich stadion, or so-called billiards table or stadium (see [**6**] and Figure 9). The well-developed linear theory includes such interesting phenomena as crossing eigenvalues and the avoided crossing of eigenvalues. We are able to find many solutions to (1.2), construct bifurcation diagrams using either or both of the parameters $\lambda = f'(0)$ and the stadion dimension d describing the radii of the end caps of Ω_d. We do indeed observe the swapping of symmetry as d passes through a value where an eigenvalue crossing occurs and the corresponding eigenfunctions swap symmetry. The persistence of linear phenomena in nonlinear cases is a theme we return to time and time again. Figure 10 shows the well known symmetry swapping / avoided swapping

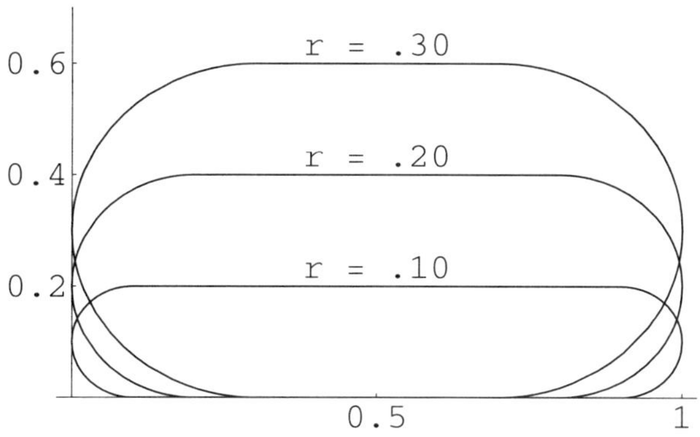

FIGURE 9. A family of Bunimovich stadions, often referred to as billiards or stadiums. Here the length is 1 for each stadium; the represented radii are labeled.

phenomena at a crossing and avoided crossing of eigenvalues. Figure 11 depicts solutions obtained by fixing $\lambda = f'(0)$ at the crossing value and varying the radius parameter d.

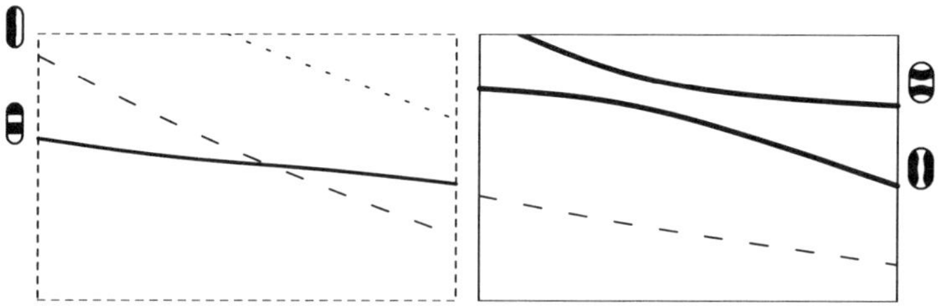

FIGURE 10. The figure on the left depicts the λ_4 and λ_5 crossing as the radius d ranges from $d = .19$ to $d = .21$, along with the accompanying symmetry swapping of ψ_4 and ψ_5. The symmetries are different, that is one branch starts as even in x and odd in y while the other starts as odd in x and even in y; symmetry swapping occurs at the multiple eigenvalue. The figure on the right depicts an avoided crossing of λ_8 and λ_9; the symmetries are the same (even in both variables) and no symmetry swapping occurs. In both graphs a third branch of a different symmetry type is nearby, but has no influence on the symmetry of linear and nonlinear solutions corresponding to the crossing / avoided crossing feature.

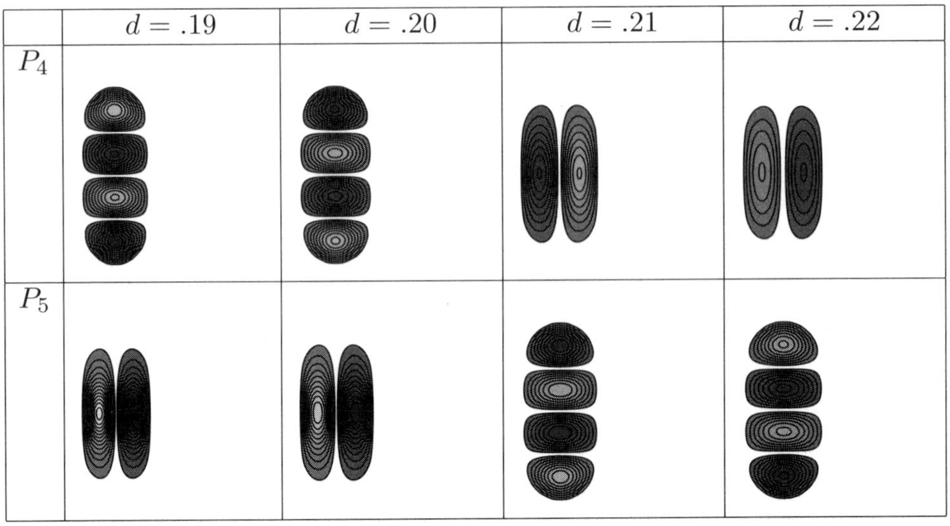

FIGURE 11. Contour plots of solutions on the two branches bifurcating from λ_4 and λ_5. Here, the bifurcation parameter $\lambda = f'(0)$ is held fixed near the multiple crossing eigenvalue and the second parameter d (radii of endcaps) is varied as in the left graphic in Figure 10. The swapping of symmetry for the nonlinear problem is thus evident. Solutions were selected at random from either the initial branch or the negative of that branch; either choice has the same symmetry and nodal structure, differing only by a multiple of -1. The article [**25**] contains more detailed bifurcation diagrams, including secondary bifurcations and deeper symmetry analysis.

We have investigated the case where Ω is the Koch Snowflake. The generation of eigenfunctions required us to essentially repeat the earlier work [**27**]. In our paper, we discuss possible refinements to or at least alternative methods for performing the linear computation. Once a basis was generated, we proceeded as in [**36**], [**7**], and [**25**]. Figure 13 shows a lower level grid approximating the Koch Snowflake. The key to invoking the ARPACK code to calculate eigenvalues of a linear map is in a user provided subroutine wherein the action of the map acting on a vector is coded. For the discretized $-\Delta$ operator and the equilateral grid, this can be easily done. As in Figure 12, let $e_1 = \frac{1}{2}\hat{\imath} + \frac{\sqrt{3}}{2}\hat{\jmath}$, $e_2 = \hat{\imath}$, and $e_3 = \frac{1}{2}\hat{\imath} - \frac{\sqrt{3}}{2}\hat{\jmath}$. Then $u''(x)(e_1, e_1) \approx \frac{a-2o+b}{h^2}$, $u''(x)(e_2, e_2) \approx \frac{c-2o+d}{h^2}$, and $u''(x)(e_3, e_3) \approx \frac{e-2o+f}{h^2}$. Using the bilinearity of $u''(x)$, a simple calculation shows that

$$-\Delta u(x) \approx \frac{2}{3h^2}(6o - (a + b + c + d + e + f)),$$

which is a generalization of the well-known 5-point rectangular second difference formula to this hexagonal situation. Figure 14 contains an example

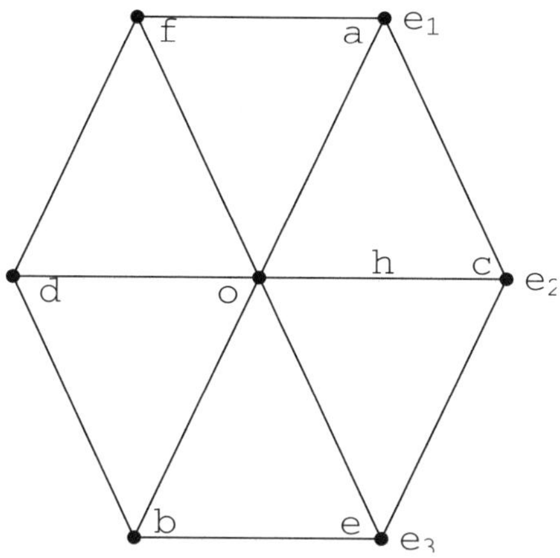

FIGURE 12. A center point x with value o and its 6 neighbors $x \pm he_1$, $x \pm he_2$, and $x \pm he_3$, with values a, b, c, d, e, and f. The gridsize h is proportional to 3^{-k}, where k is the level of the polygonal approximation to the fractal snowflake.

of automated bifurcation branch following output that can be obtained by careful application of the GNGA (see [**38**]). The provided diagram was obtained via repeated calls to an implementation of the GNGA which follows a single branch and saves off initial data for following secondary and tertiary branches when zero eigenvalues of the Hessian $D_2J(u)$ are encountered. Many refinements to our previous branch following methods are possible, for example by using knowledge of the lattice of possible symmetry types obtained by considering isotropy subgroups of the symmetry group of the snowflake $\mathbf{D}_6$ crossed with $\mathbb{Z}_2$ (given our odd choice of nonlinearity). For instance, we can seek only branches of a particular symmetry and/or compute only terms of gradients and Hessians corresponding to a branch's symmetry, hence gaining substantial computational efficiency and some added accuracy. Symmetry information can guide search direction choices at multiple bifurcation points, and be provided as output. Figure 14 contains solutions of 8 symmetry types, among the possible 23 proven to exist in [**38**]. The seven dots on the vertical axis indicate nontrivial solutions of distinct symmetry type; in Figure 15 we present these 7 solutions as well as a solution not found on this particular set of branches, namely a solution with no symmetry (type S_{23} in our notation). The article [**38**] describes in detail the symmetry analysis for this region, the techniques used to invoke the GNGA to perform automated branch following, and a complete graphical description of all solutions to (1.2) on the Koch Snowflake region up to solutions of a moderately high MI. The article [**37**] describes the method used for obtaining the basis of eigenfunctions used in that nonlinear effort via careful usage of ARPACK, as well as more details concerning our symmetry analysis of the region.

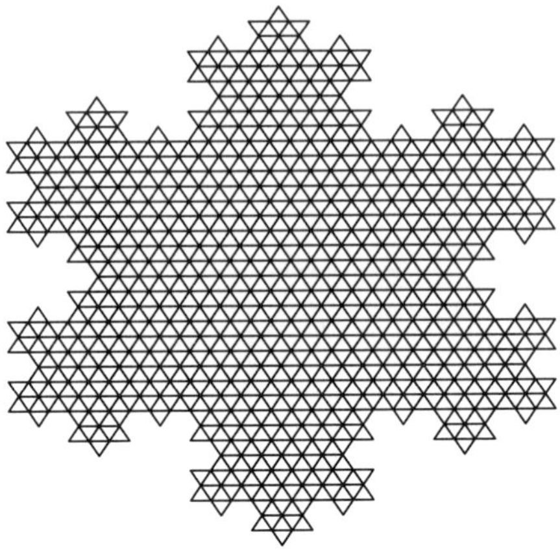

FIGURE 13. Level 3 grid of equilateral triangles used in the ARPACK eigenfunction generation code. We typically use Levels 4 and 5 in our nonlinear experiments and to compare eigenvalues with those previously calculated by [**27**]. Our current basis generation codes use a somewhat modified grid, placing the boundary of the region between grid points and using "ghost points" outside the region to enforce the zero Dirichlet (or Neumann) boundary conditions (see [**37**]).

7. The Lazer-McKenna Conjecture, Dancer Counterexample, Basins of attraction.

The interested reader should consult [**28**], [**29**], [**18**], and references both therein and subsequent by those authors. The issues discussed in these important papers have come to represent to this author the general question of existence and multiplicity of solutions to semilinear elliptic problems. In particular, complete resolutions of the explicit and implied questions within those articles would settle the matter of the conjectured existence of infinitely many solutions to our superlinear problem. Essentially, Lazer and McKenna conjectured that (1.2) should have $2k$ solutions when f had the asymptotically linear form

$$f(u) = \lambda u + g(u) - h_1(x) - s\phi_1,$$

where ϕ_1 is a (normalized) eigenfunction corresponding to the first eigenvalue λ_1, h_1 is orthogonal to ϕ_1 in L_2, s is sufficiently large, and $f \in C^1$ is such that $\limsup_{u\to-\infty}(\lambda u+g(u))/u < \lambda_1$ and $\lambda_k < \lim_{u\to\infty}(\lambda u+g(u))/u < \lambda_{k+1}$. This would prove the existence of arbitrarily many solutions to a subclass of problems of type (1.2), as the number of enclosed eigenvalues increased. However, Dancer proved via a now famous counterexample that

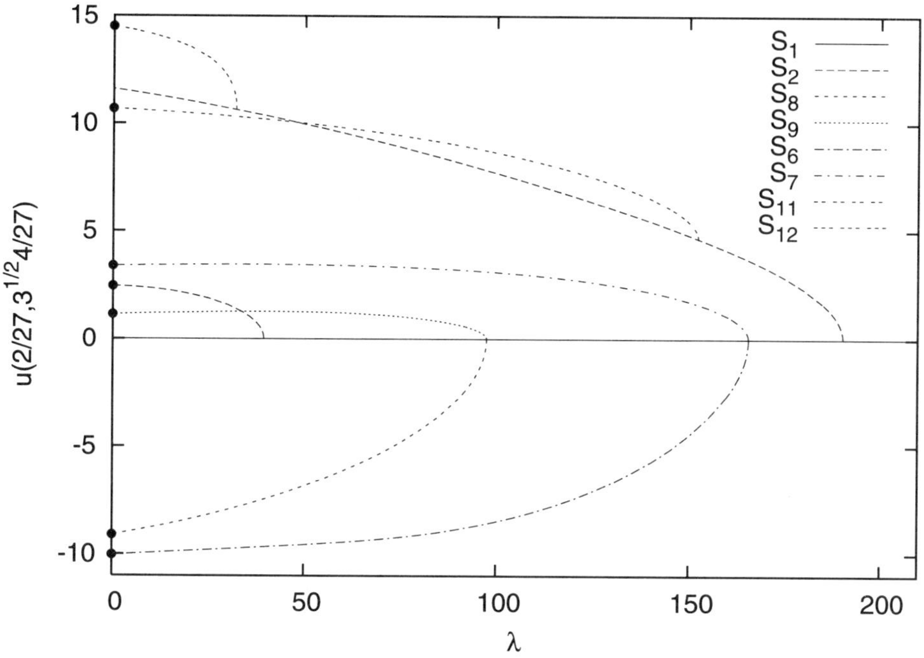

FIGURE 14. Bifurcation diagram for the first four primary branches of (1.2) on the Koch Snowflake. The x-axis is the bifurcation parameter $\lambda = f'(0)$; the y-axis is the value of the solution evaluated at a generic point of the fractal domain, a technique which yields good visual separation of branches. Two secondary bifurcations and 8 of the 23 possible symmetry types are depicted in this graphic. See Figure 15 for contour plots of solutions of the 7 nontrivial symmetry types, that is, not including the zero solution with symmetry type S_1. The fourth primary branch is the second occurrence of symmetry type S_2; its contour plot is also not included. See [**38**] for a complete list of all symmetry types and examples of bifurcation diagrams where each type of solution can be found.

this conjecture is in general false. A key ingredient in the counterexample's construction was the usage of multiple eigenvalues. Assuming that Ω was a ball, $f \in C^\infty$, and that $k+1$ eigenvalues were crossed, Dancer provided an example where only 4 solutions existed for all large s.

This author wishes to state a slightly different conjecture, and to provide a high-level description of some ideas that might be useful in investigating the conjecture's veracity.

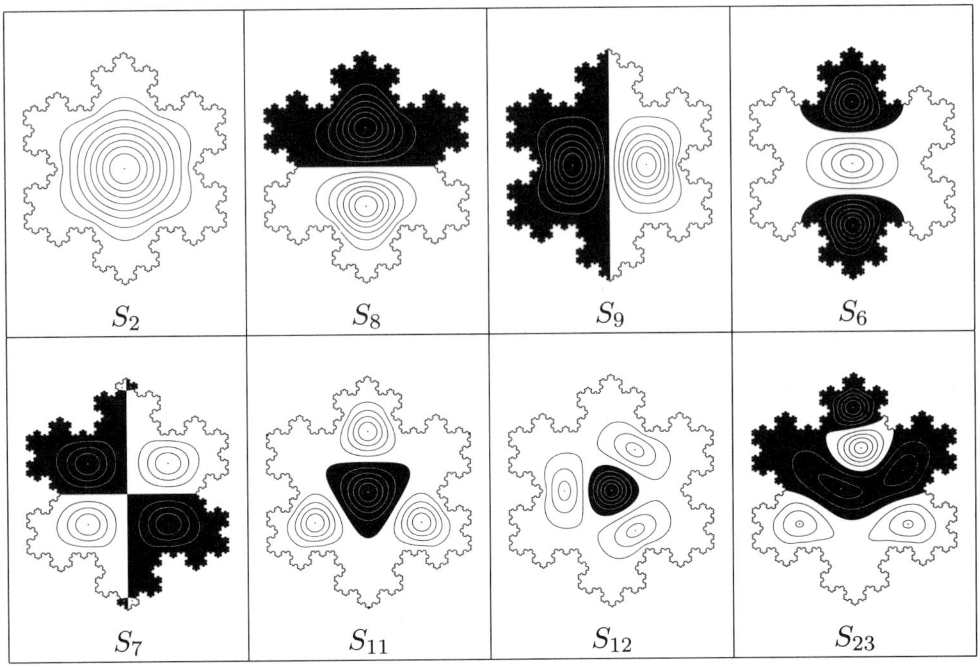

FIGURE 15. Contour plots of solutions corresponding to the 7 nontrivial symmetries found in Figure 14. The third plot is the CCN solution (symmetry type S_9). The eighth plot depicts a non-symmetric solution (symmetry type S_{23}) found on a higher branch.

CONJECTURE 7.1. Let $f \in C^1$ satisfy $f(0) = 0$ and assume that there exists $m \in (0, 1)$ such that $\frac{m}{2} f(u)u \geq F(u)$, for $|u| > \rho$, some $\rho > 0$, and that $f'(u) > \frac{f(u)}{u}$ for $u \neq 0$. If the interval (α, β) defined by $\alpha = f'(0) < \lambda_1$ and $\lambda_j < \beta = \lim_{u \to \pm\infty} f(u)/u < \lambda_{j+1}$ contains k *distinct* eigenvalues, then there exist $2k + 1$ solutions to (1.2).

Under this hypothesis (see Figure 16), one of these solutions is the trivial solution $u \equiv 0$. It is not difficult to see that the superlinear arguments can be applied to get the 3 nontrivial solutions found in [**2**] and [**8**] even to this asymptotically linear problem, provided that $\beta > \lambda_2$.

One cannot help but gain confidence in the existence of many solutions to (1.2) when observing them numerically using the GNGA. It appears that given a proof of existence of a solution, it should be possible to prove that the GNGA converges to that solution for a suitable initial guess. Conversely, if one could prove that GNGA converged for a given guess, one might be able to conclude analytically that the solution exists. This line of reasoning was the starting point for the Research Experience for Undergraduates (REU) work of Joel Fish (see [**21**]), where we set out to better understand basins

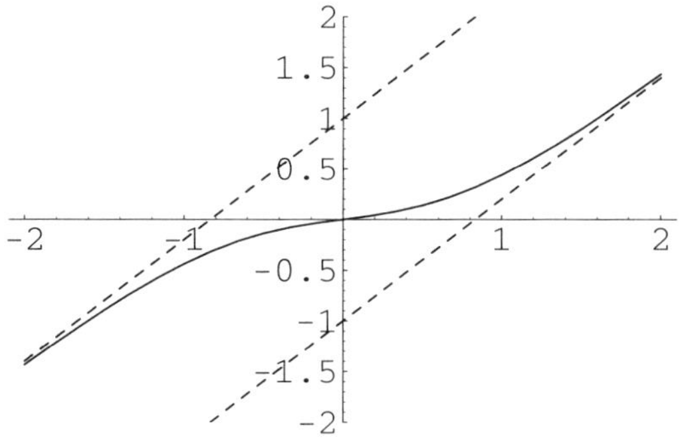

FIGURE 16. Asymptotically linear $f \in C^1$ with $f'(0) = c$, $f'(\pm\infty) = 1 + c \geq 0$ defined by $f(x) = (1 + c)x - \tanh x$.

of attraction of continuous Newton's method. To see what is possible, refer to [**43**] for a very simple finite-dimensional example where the basins of attraction are completely understood. A potentially far-reaching analytical result using continuous Newton's method as a tool can be found in [**44**]; therein an elegant result nearly analogous to the celebrated Nash-Mosher theory is proven in a vastly simpler fashion. We must admit up front that the remainder of this section is almost completely conjecture, only marginally supported by numerical evidence. Even if true, the proofs may well be intractable. Never the less, we find that the material suggests intriguing research directions that may well be quite fruitful.

We have observed that the low MI solutions are of simple nodal structure, and thus should have eigenfunction (Fourier) expansions with the first few coefficients dominating. It thus seems reasonable that for sufficiently large $M \in \mathbf{N}$, the $M \times M$ matrix $A = A(u) = (J''(u)(\psi_i, \psi_j))_{i,j=1}^M$ should approximate $D^2J(u)$ well. One can view the approximation as good in the following sense. Firstly, one should have $\text{sig}(D^2J(u)) = \text{sig}(A)$ and that the first few eigenvalues of $D^2J(u)$ and A closely agree. Secondly, the first few eigenvectors of A in $\mathbf{R}^M$, thought of as coordinates in the Galerkin space $G = \text{span}\{\psi_i\}_{i=1}^M$, appear to closely represent the first few eigenfunctions of $D^2J(u)$ in H. If $D^2J(u)$ is invertible this seems provable. It is not clear how to handle the case where the Hessian has one or more zero eigenvalues. Given a nondegenerate solution $v \in H$, we propose that given $\epsilon > 0$ there exist $M \in \mathbf{N}$, $v^* = \sum_{i=1}^M a_i\psi_i \in G = G_M$, and $\delta > 0$ such that $\nabla\hat{J}(a) = 0$, $||v^* - v|| < \epsilon$, and Newton's method (continuous or discrete) converges to $v^* \in G$ given any initial approximation v_0 with $||v^* - v_0|| < \delta$. Experimentally, one uses a small step size to approximate the continuous Newton's flow

$$u'(s) = -(D^2J(u(s)))^{-1}\nabla J(u), \quad u(0) = u_0.$$

Multiplying by $D^2J(u(s))$ and undoing the chain rule results in the initial value problem

$$((\nabla J)(u))'(s) = -\nabla J(u(s)), \quad \nabla J(u(0)) = \nabla J(u_0).$$

The solution is $\nabla J(u(s)) = \nabla J(u_0)e^{-s}$, and so the gradient goes to zero and

$$u'(s) = -(D^2J(u(s)))^{-1}\nabla J(u_0)e^{-s}.$$

If the inverse (pseudoinverse) of $D^2J(u(s))$ could be controlled, then convergence of the flow to a critical point might be proven. The proof that J satisfies the Palais-Smale condition (see [**46**] or [**10**]) or a similar argument might be useful in showing that $\lim_{s\to\infty} u(s)$ exists. The coercivity of J (see [**8**]) might also be appealed to. It seems reasonable that the signature should be constant along these flows, at least when the limit (solution) is nondegenerate. Perhaps the most tractable of our conjectures is that there exist initial values of arbitrarily large signature, whereby convergence along signature-invariant flows would provide the existence of infinitely many solutions.

It is known that the basins of attraction for continuous Newton's method are generally more straightforward than those of discrete Newton's method, typically lacking the fractal boundaries and accompanying dynamical complexities. As in simple cases (where it can be easily proven), continuous Newton's method when applied to the variational formulation of (1.2) appears to have connected basins with measure zero boundary. Obviously if one could describe the basins analytically and if there were infinitely many of them, then one would have the highly desirable result that infinitely many solutions exist.

Together with our Summer 1999 REU student Joel Fish, we have made some progress towards understanding these basins and their boundaries. There are simple examples where continuous Newton flows terminate in finite time to points that are not roots, e.g., where a zero derivative is encountered. The inflection sets (where $D^2J(u(s))$ is not invertible) almost certainly contain such points. The experimental and novel work of our REU student suggests that the collection of initial points that converge to such "bad points" themselves belong to the set

$$\Gamma = \{u \in H : J'(u)(e_i) = 0 \text{ for some eigenfunction } e_i \text{ of } D^2J(u(s))\},$$

which may itself be composed of infinitely many orthogonally intersecting manifolds. It appears to be the case that the inflection sets form part of the boundaries of the basins of attraction for our Newton flows, and that part of Γ forms the rest of the boundaries. Certainly it is clear that all solutions must lie in Γ, possibly at the points where infinitely many (all but one) manifolds intersect orthogonally. Degenerate solutions which belong to both an inflection set and Γ exist and add to the difficulty of finding a proof. Finally, we have experimentally observed a subset of Γ that very closely resembles the manifold S used in CCN in that it appears to be a codimension

1 submanifold diffeomorphic to the unit sphere which contains all nontrivial solutions to (1.2). This very interesting set can be characterized as

$$\Gamma^* = \{u \in H : J'(u)(e_1) = 0, \beta_1 < 0\},$$

where β_1 is the smallest eigenvalue of $D^2J(u)$ and e_1 is the corresponding eigenfunction.

The components of Γ corresponding to a particular eigenfunction e_i (at least locally where unambiguously defined) apparently are both gradient and Newton flow invariant sets. We have a numerical experiment where the one-sign solutions for both the $\Omega = (0,1)$ and $\Omega = (0,1) \times (0,1)$ cases have been found by following such "flows". Specifically, starting with a small guess $c\psi_1$, we iteratively project on to the set $\{u \in H : D^2J(u)\nabla J(u) = \beta_1 \nabla J(u)\}$ and take small gradient steps. This procedure terminates at a one-sign solution approximation where the 1-dimensional set intersects Γ^*. Since $\nabla J(u)$ is parallel to $e_1(u)$ along this set, we sometimes refer to this process as "following the eigenflow". We have had limited success in finding higher MI solutions via this method. For the ODE, the only stumbling block is in the smoothness of the map $e_i = e_i(u)$. It is easy to see that $e_i(0) = \psi_i$, but it does not work to continue to define e_i as the ith eigenfunction of $D^2J(u)$. When a multiplicity is encountered, care is needed to continue in the correct direction, often requiring a reordering of the eigenvalues of $D^2J(u)$. For the PDE, the matter is even more complicated. Our experiments on the unit square have so far failed to work for all but the MI 1 solutions, likely due to the multiplicity of the 2nd eigenvalue.

We believe that eigenflows corresponding to distinct simple eigenvalues λ_i should converge to one of two solutions, corresponding to the directions $\pm\psi_i = \pm e_i(0)$. This contributes $2p$ solutions towards the conjectured $2k+1$, for the p simple eigenvalues found among the k distinct eigenvalues in $\lambda_1 < \lambda_2 \leq \cdots \leq \lambda_j < \lambda_{j+1}$. The other $2(k-p)$ solutions are much more complicated to understand. Ironically, in most of the situations that we have encountered multiplicity leads to a proliferation of solutions of all possible symmetries, rather than leading to nonexistence. For an eigenvalue λ_i of multiplicity $r > 1$, we envision an r-dimensional eigenflow-invariant manifold emanating from the origin, intersecting the S-like set Γ^* in a $r-1$ dimensional submanifold. The 2 solutions contributed by this distinct eigenvalue would be a minimizer and maximizer of J restricted to this submanifold of Γ^*. We have some numerical evidence that a solution corresponding to a eigenfunction ψ_i will have a MI in the range $\{i, \ldots, i+r+s\}$, where r and s are the multiplicities of λ_i and the next distinct eigenvalue, respectively.

8. Open Problems and Future Directions.

We now present open problems requiring analytical investigation, feasible new numerical experiments, and new ideas that merit further study. We proceed with the former:

A1 **Zero-Set Conjecture.** All of our numerical experiments support the conjecture that the CCN solution to the superlinear problem has an internal zero set which intersects the boundary $\partial\Omega$. In particular, if proven this would imply that the MI 2 CCN solution on the ball is non-radial. When $\Omega = (0,1) \times (0,1)$, the solution possessing 90° rotationally invariant symmetry (where the zero-set is a closed loop in Ω) is the one corresponding to $\psi_{1,3} + \psi_{3,1}$ (see [**17**] and [**36**]). Our numerical experiment provides a MI greater than 2, supporting that this is not the CCN solution. A starting point for proving this conjecture is the theorem by Melas (see [**32**]) where it is proven that if Ω is a bounded and convex region in $\mathbf{R}^2$, then the second eigenfunction ψ_2 has this property, i.e., its internal zero set intersects the boundary in exactly two points. A related question is whether or not the zero-set is connected. The author feels that geometric arguments similar to the "Moving Planes" of [**23**] may be the key (albeit some sign-changing variant).

A2 **The Second CCN Solution.** As a first step towards finding more solutions to our superlinear problem, one could try to find a second CCN solution. This solution roughly corresponds to the negative of the solution already proven to exist. As in the "8 solutions on the square" problem below, one would generally expect more solutions depending on the symmetry of the region Ω. Finding this solution might be very important, as a linking or mountain pass argument restricted to S_1 (see (2.7)) might then lead to a higher energy MI 3 solution. Potentially, if that argument in turn was useful in obtaining a second MI 3 solution, there might be a snowball effect whereby infinitely many solutions could be found.

A3 **Infinitely Many Solutions.** It is widely (but not universally) believed that the superlinear problem has infinitely many solutions. This is the logical extension to the facts that there are indeed infinitely many solutions when $\Omega \subset \mathbf{R}$, Ω is a ball in $\mathbf{R}^N$, or when f is odd. We find that a series of papers by Lazer-McKenna and Dancer is essential reading when considering this open question (see for example [**18**] and [**28**], and Section 7 above.) This problem's difficulty is confirmed by the more than half a century of sustained effort mathematicians have exerted since Ljusternik and Schnirrleman's celebrated "f odd" result. Ambrosetti, Castro, Costa, Dancer, and Rabinowitz, to name just a few involved in this effort, have all given the problem considerable thought. Item C1 below suggests a possible research direction to follow in considering this problem.

A4 **Sublinear Dirichlet Problem.** When the nonlinearity f is taken to be sublinear, the hypothesis of Theorem 2.1 can be restated in a natural way. For example, $f(0) = 0$, $f'(0) > \lambda_k$, $\frac{f(u)}{u} \to 0$ as

$|u| \to \infty$, $\frac{m}{2}uf(u) \leq F(u)$, and $f'(u) < \frac{f(u)}{u}$ for $u \neq 0$. Our numerical experiments suggest that a sign-changing exactly-once solution analogous to the CCN solution persists. Like the situation in [**11**], the Nehari manifold is degenerate at 0. Different techniques must be used. It may be that the Lyapunov-Schmidt reduction method will again prove useful.

A5 **Eight Solutions on the Unit Square.** In many of our numerical experiments it has become obvious how to completely classify certain low energy, low MI solutions. For example, when $\Omega = (0,1) \times (0,1)$, one might prove the existence of the 8 sign-changing exactly-once solutions (4 MI 2 CCN solutions and 4 MI 3 solutions) depicted on the 2nd and 3rd branches bifurcating at $\lambda_{12} = \lambda_{21} = 5\pi^2$ in Figure 2. Note that if f is odd, then one can "stitch together" one-sign solutions on subregions and tile patterns to get these solutions and many more.

A6 **Is S_1 a Manifold?** If not, how and where does it fail? If S_1 (see (2.7)) were a manifold, the existence proof found in [**8**] could be considerably simplified. It appears that the lack of differentiability of the map $u \to u_+$ plays a role here. A related but easier question is whether or not S_1 is connected; we conjecture that this is so.

A7 **Convergence of GNGA.** It should be possible (for at least some special cases) to prove that given a solution ω to (1.2) and $\epsilon > 0$ that there exists a natural number M and a real number $\delta > 0$ so that if $||\omega - u_0|| < \delta$, $u_0 \in G = \text{span}\{\psi_1, \ldots, \psi_M\}$, then the GNGA with the initial guess u_0 will converge to a function $\hat{\omega} \in G$ with $||\omega - \hat{\omega}|| < \epsilon$. A starting point for proving this might be found in [**19**] and references both subsequent and therein. Using their theory, it seems reasonable that one could prove that given $u \in G$, a natural number k, and a real number $\epsilon > 0$, there exists M sufficiently large so that the first k eigenvalues of the $M \times M$ Hessian approximation matrix A from (2.1) will be within ϵ of the first k eigenvalues of the true Hessian $D_2^2 J(u)$. This problem is likely very difficult, although partial results may be more tractable.

The author hopes that the above problems have been stated precisely enough so that the interested researcher could begin or resume thinking about them immediately. The following list contains suggested numerical experiments for the future. All of them could be investigated using the GNGA; indeed some of them have already been partially attempted.

B1 **Higher Dimensional Regions for GNGA.** There is no reason that one cannot implement GNGA for $\Omega \subset \mathbf{R}^N$, $N > 2$. The difficulties are largely related to efficiency and accuracy. ARPACK will handle the eigenbasis generation on arbitrary regions, given proper computational power. It may be that Monte-Carlo techniques will be useful when performing numerical integration in calculating the gradient and Hessian. Some thought has been given to perhaps generating a random grid, where the vertices provide the Monte-Carlo points and the graph (irregular grid) could be used by ARPACK to approximate eigenfunctions.

B2 **More General Region Experiments.** Using generic eigenfunction generating code (such as ARPACK) for arbitrary regions in the plane (or indeed in higher dimensions), we can continue to repeat our nonlinear experiments for other interesting shapes, such as dumb-bells, where the eigenfunctions may not be known in closed form. One hopes that the effects of the choice of Ω on the bifurcation diagrams will continue to be enlightening. A region that interests the author and may present implementational challenges is the Mandlebrot set. In general, one wonders how much effect fractal boundaries have on solutions to (1.2). For example, in the Koch Snowflake experiments, would the most interesting phenomena be captured by studying the far simpler hexagon region with apparently the same symmetries?

B3 **Fucik Spectrum and GNGA.** There is every reason to believe that numerical experiments for nonlinearities of the form $f(u) = au_+ - bu_-$ will be effective in shedding light on many open problems concerning Fucik spectrum.

B4 **Secondary Bifurcations for Superlinear Problems on the Disk.** Our initial experiments are inconclusive. It should be fairly straightforward to determine numerically if there is or is not secondary bifurcation analogous to that found in [**36**] when Ω is a disk in $\mathbf{R}^2$.

B5 **Alternate Basis for GNGA.** For our superlinear problem, it seems clear that eigenfunctions of the Laplacian are the natural choice. For other problems, such as the Ginzburg-Landau equation in [**39**], this may not be the case. Whenever solutions tend towards step functions, have "vortices", or other more pathological features, Fourier expansions leave something to be desired. We presented in Section 2 a trivial example where Legendre polynomials were used. Among other possibilities, wavelet and frame theory may suggest suitable alternatives.

B6 **Full Ginzburg-Landau Application of GNGA.** In [**42**] the functional for this substantially difficult problem is given:

(8.1)
$$E(u, A) = \int_\Omega \left(\frac{1}{2}|(\nabla - iA)u|^2 + \frac{1}{2}|\nabla \times A - H_0|^2 + \frac{\kappa^2}{4}((1 - |u|^2)^2 \right) dx,$$

The region Ω is a subset of $\mathbf{R}^p$, for $p = 2$ or $p = 3$. Here the unknowns are the order parameter $u \in H^{1,2}(\Omega, \mathbf{C})$ and the vector potential $A \in H^{1,2}(\Omega, \mathbf{R}^p)$. If the applied magnetic field H_0 is identically zero, then we can assume that $A(x) = 0$ and the

Euler-Lagrange equation of (8.1) reduces to (5.1), where $\kappa = 1/\epsilon$. One challenge that presents itself in investigating this problem is in dealing with the natural boundary conditions

$$(\nabla \times A) \times \eta = H_0 \times \eta \text{ and } ((\nabla - iA)u) \cdot \eta = 0,$$

where η is as before the outward unit normal. There is a great body of literature considering various facets of the analysis of Ginzburg-Landau problems, of which we have referenced only a few select works. References to numerical investigations of these problems are somewhat sparse. Very little can be found concerning analysis or numerics in the "full Ginzburg-Landau" case. We find that GNGA using some suitable basis should be an effective tool for investigating this important equation, as a similar but competitor technique to those found in [**42**]. The same comment applies to the following problem.

B7 **Monge-Ampere Application of GNGA.** We can find almost no literature concerning the numerical investigation of the Monge-Ampere equation $|D^2u| = f(x, u)$ on $\Omega \subset \mathbf{R}^N$ (with suitable boundary conditions). Indeed, there appears to be almost no analytical work in cases where the choice of f leads to sign-changing solutions and the principal part becomes in places non-elliptic. We have successfully performed preliminary GNGA experiments on this equation, for simple examples such as the case $\Omega = (0, 1) \times (0, 1)$, with zero-Dirichlet boundary conditions, $f \equiv 1$, while seeking only negative solutions.

B8 **Nonlinear Partial Difference Equations (PdE) on Graphs.** The Laplacian operator L can be defined and has been well studied on graphs $G = (V, E)$ with $m = |V|$ and $n = |E|$ finite (see for example [**5**]). When the graph is a regular grid, it can be shown that this linear operator converges to the usual Laplacian (with zero Neumann boundary conditions) as the mesh goes to zero. In [**40**], we seek solutions $u : V \to \mathbf{R}$, identified with vectors $u \in \mathbf{R}^m$, to the semilinear elliptic partial difference equation (PdE)

$$-Lu + f(u) = 0$$

on the graph G. Using a variational approach entirely analogous to that used when studying the PDE (1.2), we successfully investigated the PdE by finding critical points of the functional $J : \mathbf{R}^m \to \mathbf{R}$ defined by

$$J(u) = \frac{1}{2} Du \cdot Du - \sum_{i=1}^{m} F(u_i),$$

where $F' = f$. In [**40**] and following [**8**] and [**17**], we are able to prove the existence of one-sign and sign-changing solutions and to prove the existence of solutions with symmetry. Additionally and following [**34**] and [**15**], we apply variants of the MPA and MMPA to approximate low MI solutions numerically. Following [**36**], we also apply a version of the GNGA suitably adapted for PdE to plot bifurcation curves and provide MI, symmetry, and nodal structure information.

In the spirit of our most recent works [**37**] and [**38**], it remains to modify the ARPACK basis generating code to compute bases for large graphs, analyze the symmetry of the resulting eigenfunctions, apply our automated branch following techniques, and further analyze and exploit the lattice of symmetries for the nonlinear PdE. This project appears to be as open-ended as the subject of graph theory itself.

B9 **Parallel implementation of GNGA.** Larger problems and more details may be handled by first using PARPACK, the parallel version of ARPACK, to generate bases of eigenfunctions. One could use SCALAPACK or other parallel libraries of matrix routines to port the GNGA to this environment, but we are suggesting a simpler scheme. Namely, it seems efficient to have scalar routines implement the following of a single branch on each node, concurrently, and then rely on a master script on the main node to farm out new branch following jobs for secondary and tertiary branches as the need to do so occurs. In this way, our methodology and implementation from [**38**] hardly changes; we just follow one branch per node at the same time rather than proceed through the growing list of branches sequentially.

The above problems can be attempted immediately with GNGA, although additional analytical insight might be required in picking appropriate bases and initial guesses. The following problems are less well defined, representing only suggested future research directions.

C1 **Following Eigenflows.** Since our interest is primarily in proving existence theorems in elliptic PDE, we are hopeful that GNGA will be useful in providing insight (such as the MI of solutions) and may be of direct use in proving such theorems by analyzing continuous Newton flows. The algorithm suggested in Section 7 for following "eigenflows" needs to be refined and more carefully studied, particularly for the PDE and multiple eigenvalues.

C2 **CCN Solutions for Systems.** It seems reasonable to prove the existence of sign-changing solutions to elliptic semilinear problems for systems. The variational structure is clearly more complicated than in the scalar case. In particular, the pseudo-linearity of the maps $u \to u_\pm$ used in [**8**] in conjunction with the separating property of S_1 does not appear to persist. Experiments show that a solution $u = (u_1, u_2)$ where both u_1 and u_2 change sign exactly once should be of MI 4, necessitating more constraints. There is a good, fairly well defined problem to solve here. See the ending note in Section 5 for a precise statement of one possible hypothesis to consider when looking for a CCN solution to a superlinear system, as well as a brief report of an unpublished numerical experiment.

C3 **Semilinear Parabolic Problems.** It should be possible to apply what we now know and what we conjecture for semilinear elliptic problems to the non-steady state case. For example, can we find and understand an unstable flow where the time slices are CCN solutions to elliptic problems? We have extended the idea of PdE (see item B8 above) to heat equations. Will it be fruitful to further develop this to nonlinear parabolic PdE? Can pattern formation and symmetry be explored in a new way in this setting?

9. Conclusion.

GNGA can be an effective tool for investigating nonlinear phenomena. One needs a natural variational formulation, a suitable basis, and some intuition guiding the choosing of initial guesses. The technique seems particularly well suited to semilinear elliptic problems, where one expects a high degree of interaction with the linear problem whose solutions constitute the basis. Thus, for these types of problems we have all three essential ingredients in order to apply GNGA. Tongue in cheek, we like to think of the method as "counter constructive", i.e., it may at times suggest existence proofs just as a constructive proof suggests an algorithm. We believe that eventually, and with possibly great effort, consideration of the basins of attraction for continuous Newton's method as applied to finding critical points of variational functionals for semilinear elliptic PDE will lead to progress towards solving many of the subject's open questions.

References

[1] Adams, R., *Sobolev Spaces*, Academic Press: New York (1975).

[2] Ambrosetti, A.; Rabinowitz, P. *Dual variational methods in critical point theory and applications.*, J. Functional Analysis **14** (1973), p349-381.

[3] Argyros, I., *The Asymptotic Mesh Independence Principal for Inexact Newton-Galerkin-Like Methods*, Preprint (1999).

[4] Aubin, T., *Nonlinear analysis on manifolds. Monge-Ampre equations*, Grundlehren der Mathematischen Wissenschaften [Fundamental Principles of Mathematical Sciences], **252**, Springer-Verlag, New York, (1982).

[5] Bapat, R. B., *The Laplacian matrix of a graph*, Math. Student **65** (1996), no. 1-4, p214-223.

[6] Bunimovich, L. A., *On the ergodic properties of nowhere dispersing billiards*, Comm. Math. Phys. **65** (1979), no. 3, p295-312.

[7] Michael Butros, *Newton's method for semilinear BVP on the disk*, M.S. Thesis (2000), Northern Arizona University.

[8] Castro, A.; Cossio, J.; Neuberger, John M., *Sign-Changing Solutions for a Superlinear Dirichlet Problem*, **27** (1997), no. 4, p1041-1053.

[9] Castro, A.; Cossio, J.; Neuberger, John M., *On Multiple Solutions of a Nonlinear Dirichlet Problem*, Proceedings of the Second World Congress of Nonlinear Analysts, Part 6 (Athens, 1996). Nonlinear Anal. **30**, no. 6 (1997), p3657–3662.

[10] Castro, A.; Cossio, J.; Neuberger, John M., *A Minmax Principle, Index of the Critical Point, and Existence of Sign-Changing Solutions to Elliptic Boundary Value Problems*, Electron. J. Differential Equations, No. 02 (1998), 18 pp.

[11] Castro, A.; Drabek, P.; Neuberger, John M., *Sign-Changing Solutions for a Superlinear Dirichlet Problem, II*, Proceedings of the Fifth Mississippi State Conference on Differential Equations and Computational Simulations, EJDE **10** (2003).

[12] Castro, A.; Kurepa, A., *Infinitely many radially symmetric solutions to a superlinear Dirichlet problem in a Ball*, Proc. of the AMS **101** (1987), p57-64.

[13] Choi, Y. S.; McKenna, P. J., *A Mountain Pass Method for the Numerical Solutions of Semilinear Elliptic Problems*, Nonlinear Analysis, **20** (1993), p417-437.

[14] Chow, S. N.; Hale, J. K., *Methods of Bifurcation Theory*, Springer-Verlag - Berlin, New York (1982).

[15] *A Numerical Reduction Method for Investigating Semilinear Elliptic PDE*, Proceedings of the Second World Congress of Nonlinear Analysts, (Catania, 2000). Nonlinear Anal. **47**, p3379-3390.

[16] Ding, Z.; Costa, D.; Chen, G., *A high-linking algorithm for sign-changing solutions of semilinear elliptic equations.*, Nonlinear Anal. **38** (1999), no. 2, Ser. A: Theory Methods, p151-172.

[17] Costa, D.; Ding, Z.; Neuberger, John M., *A Numerical Investigation of Sign-Changing Solutions to Superlinear Elliptic Equations on Symmetric Domains*, J. Comput. Appl. Math. **131** (2001), no. 1-2, p299-319.

[18] Dancer, E. N., *A counterexample to the Lazer-McKenna conjecture*, Nonlinear Anal. **13** (1989), no. 1, p19-21.

[19] Descloux, J.; Nassif, N.; Rappaz, J., *On spectral approximation. I. The problem of convergence*, RAIRO Anal. Numr. **12** (1978), no. 2, p97-112, iii.

[20] Kinderlehrer, D.; Stampacchia, G., *Introduction to Variational Inequalities and Their Applications*, Academic Press : New York (1979).

[21] Fish, J.; Neuberger, John M., *Eigenflows and basins of attraction for Newton's method*, unpublished REU report (2000).

[22] Flores, G.; Padilla, P.; Tonegawa, Y., *Higher energy solutions in the theory of phase transitions: a variational approach*, Special issue in celebration of Jack K. Hale's 70th birthday, Part 3 (Atlanta, GA/Lisbon, 1998), J. Differential Equations **169** (2001), no. 1, p190-207.

[23] Gidas, B.; Ni, W. M.; Nirenberg, L. *Symmetry of positive solutions of nonlinear elliptic equations in R^n*, Mathematical analysis and applications, Part A, Adv. in Math. Suppl. Stud., 7a, Academic Press, New York-London, (1981), p369-402.

[24] Gilbarg, D; Trudinger, N., *Elliptic Partial Differential Equations of Second Order*, Springer-Verlag: Berlin, New York (1983).

[25] Hineman, J.; Neuberger, John M.; Swift, James W., *Numerical Solutions to Semilinear Elliptic BVP on Bunimovich Stadia*, Work in Progress.

[26] Johnson, L.; Riess, R., *Numerical Analysis*, Addison-Wesley : Reading, Mass. (1982).

[27] Lapidus, M.; Neuberger, J. W.; Renka, R.; Griffith, C., *Snowflake harmonics and computer graphics: numerical computation of spectra on fractal drums*, Internat. J. Bifur. Chaos Appl. Sci. Engrg. **6** (1996), no. 7, p1185-1210.

[28] Lazer, A. C.; McKenna, P. J. *On the number of solutions of a nonlinear Dirichlet problem.*, J. Math. Anal. Appl. **84** (1981), no. 1, p282-294.

[29] Lazer, A. C.; McKenna, P. J., *On a conjecture related to the number of solutions of a nonlinear Dirichlet problem*, Proc. Roy. Soc. Edinburgh Sect. A **95** (1983), no. 3-4, p275-283.

[30] Ljusternik , L.; Schnirrelmann, L., *Methodes Topologique dans les Problems Variational*, Hermann and Cie, Paris (1934).
[31] Z. Mei, *Numerical bifurcation analysis for reaction-diffusion equations*, Springer Series in Computational Mathematics, **28**. Springer-Verlag, Berlin (2000), xiv+414 pp.
[32] Melas, A., *On the nodal line of the second eigenfunction of the Laplacian in R^2*, J. Differential Geom. **35** (1992), no. 1, p255-263.
[33] Milnor, J., *Morse Theory*, Princeton University Press : Princeton (1963).
[34] Neuberger, John M., *A Numerical Method for Finding Sign-Changing Solutions of Superlinear Dirichlet Problems*, Nonlinear World **4**, no. 1 (1997), p73-83.
[35] Neuberger, John M., *A Sign-Changing Solution for a Superlinear Dirichlet Problem with a Reaction Term Nonzero at Zero*, Nonlinear Anal. **33**, no. 5 (1998), p427-441.
[36] Neuberger, John M.; Swift, James W. , *Newton's method and Morse index for semilinear elliptic PDEs*, Internat. J. Bifur. Chaos Appl. Sci. Engrg. **11** (2001), no. 3, p801-820.
[37] Neuberger, John M.; Sieben, N.; Swift, James W., *Computing Eigenfunctions on the Koch Snowflake: A New Grid and Symmetry*, submitted for publication (2004).
[38] Neuberger, John M.; Sieben, N.; Swift, James W., *A semilinear elliptic PDE on a Fractal Domain*, Work in Progress (2004).
[39] Neuberger, John M.; Rice, Dennis; Swift, James W. , *Numerical Solutions to a Vector Ginzburg-Landau Equation with Triple-Well Potential*, Int. J. Biff. Chaos. **13**, No. 11 (2003) p3295-3306.
[40] Neuberger, John M., *Nonlinear Elliptic Partial Difference Equations on Graphs*, preprint, submitted (2003).
[41] Neuberger, J. W., *Constructive Variational Methods for Differential Equations*, Numerical Analysis, Theory, Methods, and Applications, **13**, no. 4 (1988), p413-428.
[42] Neuberger, J. W., *Sobolev Gradients and Differential Equations*, Springer Lecture Notes, (1997).
[43] Neuberger, J. W., *Continuous Newton's method for polynomials*, Math. Intelligencer **21** (1999), no. 3, p18-23.
[44] Neuberger, J. W., *A Nash-Moser theorem with near-minimal hypothesis*, Int. J. Pure Appl. Math. **4** (2003), no. 3, p269-280.
[45] Neuberger, B.; Neuberger, J. W.; Noid, D. W., *Eigenfunctions on a Stadium Associated with Avoided Crossings of Energy Levels*, http://arxiv.org/abs/math.NA/0105217 (2001).
[46] Rabinowitz, P., *Minimax Methods in Critical Point Theory with Applications to Differential Equations*, Regional Conference Series in Mathematics, **65**, AMS : Providence, R.I. (1986).
[47] Tehrani, H., *H^1 versus C^1 local minimizers on manifolds*, Nonlinear Anal. **26**, no. 9 (1996), p1491-1509.
[48] Wang, Z. Q., *On a Superlinear Elliptic Equation*, Ann. Inst. H. Poincare Analyse Non Lineaire **8** (1991), p43-57.

Department of Mathematics and Statistics, Northern Arizona University PO Box 5717, Flagstaff, AZ 86011-5717, USA
E-mail address: John.Neuberger@nau.edu

Contemporary Mathematics
Volume **357**, 2004

Critical Nonlinearities and Symmetric Solutions

Florin Catrina

ABSTRACT. In this article we discuss certain nonlinear elliptic PDE's. The problems are variational and we inquire about the existence of least energy solutions in various Hilbert spaces. We present a delicate phenomenon related to critical exponents in presence of symmetries, and list a number of open questions.

1. Introduction

In this paper we look at problems

$$\begin{cases} -\,div(|x|^{-2a}\nabla u) = |x|^{-2^*a}K(x)u^{2^*-1} \\ u > 0, \quad u \in \mathcal{D}_a(\mathbb{R}^N) \end{cases} \tag{1.1}$$

where

$$N \geq 3, \quad -\infty < a < \frac{N-2}{2}, \quad \text{and } 2^* = \frac{2N}{N-2}. \tag{1.2}$$

Here $\mathcal{D}_a(\mathbb{R}^N)$ denotes the Hilbert space obtained as the completion of smooth functions with compact support in $\mathbb{R}^N$ under the norm induced by the inner product

$$(u, v) = \int_{\mathbb{R}^N} |x|^{-2a}\nabla u \cdot \nabla v \, dx.$$

We shall assume that K is a positive continuous function satisfying

$$K(0) = \lim_{|x|\to\infty} K(x) = 1. \tag{1.3}$$

1991 *Mathematics Subject Classification.* Primary: 35J20, 35J70; Secondary: 35J60.
Key words and phrases. critical exponents, invariant solutions, least energy.
The author gratefully acknowledges partial support from the organizers of this conference.

Nontrivial solutions of (1.1) are critical points of the energy functional

$$E_a(u) = \frac{\displaystyle\int_{\mathbb{R}^N} |x|^{-2a}|\nabla u|^2\,dx}{\left(\displaystyle\int_{\mathbb{R}^N} |x|^{-2^*a}K(x)u^{2^*}\,dx\right)^{\frac{2}{2^*}}} \tag{1.4}$$

which is defined on $\mathcal{D}_a(\mathbb{R}^N)\setminus\{0\}$. Let

$$\overline{K} = \sup_{x\in\mathbb{R}^N} K(x) \geq 1.$$

Using a weighted version of the Sobolev inequality (e.g. the interpolation inequalities of Caffarelli, Kohn, and Nirenberg in [**CKN**]), one obtains

$$E_a(u) \geq \frac{S_a}{\overline{K}^{\frac{2}{2^*}}},$$

where S_a is a positive constant described below in Theorem 1.1. We shall label the critical points of E_a as bound states. Also, if the infimum of E_a is achieved by some function u, we call this a ground state.

We remark that in the case $a = 0$ the problem is

$$\begin{cases} -\Delta u = K(x)u^{2^*-1} \\ u > 0, \quad u \in \mathcal{D}_0(\mathbb{R}^N). \end{cases} \tag{1.5}$$

Equation (1.5) is known as the prescribed scalar curvature problem in $\mathbb{R}^N$. Existence of a solution, amounts to existence of a metric conformal to the standard metric in $\mathbb{R}^N$, with scalar curvature K. Problem (1.5) has been intensively studied (see Chapter 6 in [**A1**], and the references therein). One asks: What are the possible functions K such that (1.5) admits a solution? To the author's knowledge, there is no complete answer for this question.

In the even more restrictive case $K \equiv 1$, the equation $-\Delta u = u^{2^*-1}$ is called the Yamabe problem in $\mathbb{R}^N$. It was obtained independently by Aubin [**A**] and Talenti [**T**] that the only positive solutions in $\mathcal{D}_0(\mathbb{R}^N)$ are the functions

$$U_{0,\varepsilon,y} = (N(N-2))^{\frac{N-2}{4}}\left(\frac{\varepsilon}{\varepsilon^2+|x-y|^2}\right)^{\frac{N-2}{2}}. \tag{1.6}$$

One can calculate

$$S_0 = E_0(U_{0,\varepsilon,y}) = \frac{\displaystyle\int_{\mathbb{R}^N} |\nabla U_{0,\varepsilon,y}|^2\,dx}{\left(\displaystyle\int_{\mathbb{R}^N} U_{0,\varepsilon,y}^{2^*}\,dx\right)^{\frac{2}{2^*}}} = \pi N(N-2)\left(\frac{\Gamma\left(\frac{N}{2}\right)}{\Gamma(N)}\right)^{\frac{2}{N}}. \tag{1.7}$$

The problem (1.1) in the case $K \equiv 1$, under conditions (1.2), has been studied by several authors. The theorem below is known (see [**CW1**] and the references therein).

THEOREM 1.1. *In (1.1), assume $K \equiv 1$. Then the following are true.*
(i) For any $a < \frac{N-2}{2}$ problem (1.1) has radial solutions

$$U_{a,\varepsilon}(x) = \left(\frac{N(N-2-2a)^2}{N-2}\right)^{\frac{N-2}{4}} \left(\frac{\varepsilon}{\varepsilon^2 + |x|^{\frac{2(N-2-2a)}{N-2}}}\right)^{\frac{N-2}{2}}.$$

The energy is

$$R_a = E_a(U_{a,\varepsilon}) = \frac{\int_{\mathbb{R}^N} |x|^{-2a}|\nabla U_{\varepsilon,y}|^2 \, dx}{\left(\int_{\mathbb{R}^N} |x|^{-2^*a} U_{\varepsilon,y}^{2^*} \, dx\right)^{\frac{2}{2^*}}}$$

$$= \frac{\pi N(N-2-2a)^2}{N-2} \left(\frac{(N-2)\Gamma\left(\frac{N}{2}\right)}{(N-2-2a)\Gamma(N)}\right)^{\frac{2}{N}}.$$

(ii) For $a \in (0, \frac{N-2}{2})$ the solutions in (i) are the only solutions of (1.1), and they are ground states (achieve the infimum $S_a = R_a$ of the energy E_a).
(iii) For $a = 0$ the only nontrivial solutions in $\mathcal{D}_0(\mathbb{R}^N)$, are the ground state solutions given by (1.6) with energy S_0 given by (1.7).
(iv) For $a < 0$ the infimum of the energy is $S_a \equiv S_0$ independent of a, and it is not achieved (there is no ground state).

Regarding the problem (1.1) with K satisfying (1.3) non-constant, we shall mention two recent studies. Felli and Schneider (in [**FS**]) consider $K(x) = 1 + \varepsilon k(x)$. For all $a < \frac{N-2}{2}$ satisfying

$$a \neq \frac{N-2}{2}\left(1 - \sqrt{\frac{j(N-2+j)}{N-1}}\right), \quad \text{for some } j = 1, 2, \ldots$$

they prove that if $\lim_{|x|\to\infty} k(x) = k(0) = 0$ (K satisfies (1.3)) then for $|\varepsilon|$ small enough (1.1) has solution. For the case $0 < a < \frac{N-2}{2}$, problem (1.1) has been also studied by Smets in [**S**]. The main theorem in [**S**] is in dimension $N = 4$. Denote by σ the stereographic projection from $\mathbb{S}^4$ to $\mathbb{R}^4$. The result states that provided that $K \circ \sigma \in C^2(\mathbb{S}^4)$, then for $0 < a < \frac{N-2}{2}$ the problem (1.1) has solution. The solutions obtained by either method in [**FS**] or [**S**] cannot always be a ground state.

Our results are about the infimum of the energy E_a restricted to subspaces $\mathcal{D}_{a,G}(\mathbb{R}^N)$ of $\mathcal{D}_a(\mathbb{R}^N)$. Let G be a closed subgroup of $\mathbf{Z}_2 \times \mathbf{O}(N)$. We fix an action $\rho : G \times \mathbb{R}^N \setminus \{0\} \to \mathbb{R}^N \setminus \{0\}$ such that the nontrivial element of $\mathbf{Z}_2$ acts by inversion $x \to \frac{x}{|x|^2}$, and the orthogonal part of the group moves points in $\mathbb{R}^N$ by standard linear action. This action of G on $\mathbb{R}^N \setminus \{0\}$ induces an action

$$\tilde{\rho} : G \times \mathcal{D}_a(\mathbb{R}^N) \to \mathcal{D}_a(\mathbb{R}^N) \quad (gu)(x) := |x|^{(\epsilon-1)\frac{N-2-2a}{2}} u(g^{-1}x),$$

where $\epsilon \in \{-1, 1\}$ is the $\mathbf{Z}_2$ component of g. Denote by $\mathcal{D}_{a,G}(\mathbb{R}^N)$ the space of all functions in $\mathcal{D}_a(\mathbb{R}^N)$ which are fixed points of the action $\tilde{\rho}$. Assuming that K satisfies (1.3) and it is ρ-invariant (i.e. $K(x) = K(gx)$ for all $g \in G$ and $x \in \mathbb{R}^N$), we ask whether the infimum of E_a restricted to subspaces $\mathcal{D}_{a,G}(\mathbb{R}^N)$ is achieved, i.e. whether there exists a ground state in $\mathcal{D}_{a,G}(\mathbb{R}^N)$. Note that if u achieves the infimum of E_a in $\mathcal{D}_{a,G}(\mathbb{R}^N)$, then so does $|u|$. By the principle of symmetric criticality, $|u|$ is a critical point of E_a in the full space $\mathcal{D}_a(\mathbb{R}^N)$. By rescaling, $|u|$ is a weak solution of (1.1) and by regularity theory u is a positive classical solution in $\mathbb{R}^N \setminus \{0\}$.

The paper is organized as follows. In Section 2 we present a class of problems equivalent to (1.1). This gives more insight on the problem, and makes certain aspects easier to understand. In this section we also state a compactness result for minimizing sequences. Section 3 contains the proof of the following:

THEOREM 1.2 (Fixed Points). *Assume the ρ-action of G has a fixed point $y \in \mathbb{R}^N \setminus \{0\}$. If K is ρ-invariant, satisfies (1.3), and $K(y) = \overline{K}$, then for $a \leq 0$ we have*

$$\inf_{u \in \mathcal{D}_{a,G}(\mathbb{R}^N) \setminus \{0\}} E_a(u) = \frac{S_0}{\overline{K}^{\frac{N-2}{N}}} \tag{1.8}$$

and a ground state solution exists in $\mathcal{D}_{a,G}(\mathbb{R}^N)$ if and only if $a = 0$ and $K \equiv 1$.

In Section 4 we prove a theorem in the spirit of [**BCS**] (see also [**CW2**]). We give sufficient conditions on K so that the least energy in $\mathcal{D}_{a,G}(\mathbb{R}^N)$ is achieved. It is essential that the solution we obtain is multi-bump. We shall assume that the action of G has no fixed points in $\mathbb{R}^N \setminus \{0\}$ and that there is a finite smallest orbit $\{y_1, \dots, y_k\}$ (with $k \geq 2$ being the minimum number of points in an orbit). Let

$$S_{a,G} = \inf_{u \in \mathcal{D}_{a,G}(\mathbb{R}^N) \setminus \{0\}} \frac{\displaystyle\int_{\mathbb{R}^N} |x|^{-2a} |\nabla u|^2 \, dx}{\left(\displaystyle\int_{\mathbb{R}^N} |x|^{-2^* a} u^{2^*} \, dx \right)^{\frac{2}{2^*}}}. \tag{1.9}$$

Assume K is ρ-invariant and for some minimal orbit $\{y_1, \dots, y_k\}$ we have

$$K(y_i) = \overline{K} \geq k^{\frac{2}{N-2}} \left(\frac{S_0}{S_{a,G}} \right)^{\frac{N}{N-2}}. \tag{1.10}$$

Moreover, in a neighborhood of $\{y_1, \dots, y_k\}$ we need to assume the following flatness condition

(1.11)

There are $\alpha \in (N-2, N)$, and $\delta > 0$, such that $K(x) \geq \overline{K} - \delta |x - y_i|^\alpha$ for any x within a distance $c > 0$ (independent of i) of y_i.

THEOREM 1.3 (Free Action). *Assume that the action of* G *has no fixed points in* $\mathbb{R}^N \setminus \{0\}$, K *has property (1.3), is* ρ*-invariant, and relative to a minimal orbit* $\{y_1, \ldots, y_k\}$ *it satisfies conditions (1.10), and (1.11). Then the following are true.*

(i) For any $0 \leq a < \dfrac{N-2}{2}$ *there is a ground state in* $\mathcal{D}_{a,G}(\mathbb{R}^N)$.

(ii) For $N = 3$, *there exists* $a_0 < 0$ *such that for any* $a_0 < a < \dfrac{1}{2}$ *there is a ground state in* $\mathcal{D}_{a,G}(\mathbb{R}^3)$.

We remark that the case $a = 0$ is the case discussed in [**BCS**]. While essentially the same method applies for $a > 0$, the situation in the case $a < 0$ is much more delicate.

We conclude in Section 5 with a series of remarks and open questions.

2. Equivalent Problem

Even if all results can be stated and proven in $\mathbb{R}^N$, we find it insightful to have a different perspective on problem (1.1). As in [**CW1**], for certain calculations it may be advantageous to work with an equivalent equation, set on the cylinder $\mathcal{C} = \mathbb{R} \times \mathbb{S}^{N-1} \subset \mathbb{R}^{N+1}$. We consider the conformal diffeomorphism

$$\vartheta : \mathbb{R}^N \setminus \{0\} \to \mathcal{C} \quad \text{given by} \quad \vartheta(x) = \left(-\ln|x|, \frac{x}{|x|} \right) = (t, \theta).$$

The transformation

$$\Upsilon_a : H^1_a(\mathcal{C}) \to \mathcal{D}_a(\mathbb{R}^N) \qquad \Upsilon_a v = u \quad \text{with} \quad u(x) = |x|^{-\frac{N-2-2a}{2}} v(\vartheta(x))$$

is a Hilbert space isomorphism. Here $H^1_a(\mathcal{C})$ is obtained by completion of smooth functions with compact support in $\mathcal{C}$ under the norm

$$||v||^2 = \int_{\mathcal{C}} |\nabla v|^2 + \left(\frac{N-2-2a}{2} \right)^2 v^2 \, d\mu.$$

We remark that as a function space, $H^1_a(\mathcal{C})$ is independent of a. We keep the index a however, to indicate which inner product is used. If u is solution of (1.1) then $v = \Upsilon_a^{-1} u$ satisfies the equation

$$\text{(2.1)} \qquad \begin{cases} -\Delta v + \left(\dfrac{N-2-2a}{2} \right)^2 v = K(t, \theta) v^{p-1}, \\ v > 0 \text{ in } \mathcal{C}, \; v \in H^1(\mathcal{C}) \end{cases}$$

where

$$K(t, \theta) = K(\vartheta(x)) := K(x).$$

On $\mathcal{C}$, the differential operators ∇, div, Δ, and the volume form $d\mu$, are the ones associated to the canonical metric induced from $\mathbb{R}^{N+1}$. The energy is

$$E_a(\Upsilon_a v) = F_a(v) = \frac{\displaystyle\int_{\mathcal{C}} |\nabla v|^2 + \left(\frac{N-2-2a}{2}\right)^2 v^2 \, d\mu}{\left(\displaystyle\int_{\mathcal{C}} K(t,\theta)|v|^{2^*} \, d\mu\right)^{2/2^*}}. \tag{2.2}$$

We remark that by comparing the coefficients of the $L^2(\mathcal{C})$ part in the expression of the energy (2.2), it follows that if $a_1 < a_2 < \frac{N-2}{2}$ then

$$F_{a_1}(v) > F_{a_2}(v) \text{ for any } v \in H^1_{a_1}(\mathcal{C}) = H^1_{a_2}(\mathcal{C}).$$

This implies

$$E_{a_1}(\Upsilon_{a_1}\Upsilon_{a_2}^{-1}u) > E_{a_2}(u) \text{ for any } u \in \mathcal{D}_{a_2}(\mathbb{R}^N).$$

The action ρ of a closed subgroup G of $\mathbf{Z}_2 \times \mathbf{O}(N)$ on $\mathbb{R}^N \setminus \{0\}$ induces an action $\rho : G \times \mathcal{C} \to \mathcal{C}$ which makes ϑ equivariant (i.e. $\vartheta(gx) = g\vartheta(x)$).

We denote by $H_{a,G}$ the space of all functions in $H^1_a(\mathcal{C})$ which are ρ-invariant, i.e. $v(y) = (gv)(y) := v(g^{-1}y)$ for any $g \in G$ and any $y = (t,\theta) \in \mathcal{C}$. Note that in the case G has only the null element we have $H_{a,G} = H^1_a(\mathcal{C})$. Also, note that Υ_a restricted to $H_{a,G}$ is a Hilbert space isomorphism between $H_{a,G}$ and $\mathcal{D}_{a,G}(\mathbb{R}^N)$. From (1.9) we get

$$S_{a,G} = \inf_{v \in H_{a,G}} \frac{\displaystyle\int_{\mathcal{C}} |\nabla v|^2 + \left(\frac{N-2-2a}{2}\right)^2 v^2 \, d\mu}{\left(\displaystyle\int_{\mathcal{C}} |v|^{2^*} \, d\mu\right)^{2/2^*}}.$$

Denote by k (≥ 1) the smallest number of points in an orbit ($k = \infty$ if there is no finite orbit). The following compactness result can be proved by repeating the arguments in [**BCS**] on $\mathcal{C}$.

THEOREM 2.1. *If $K(\vartheta(x))$ satisfies (1.3), it is ρ-invariant, and*

$$\inf_{H_{a,G}} F_a < \min\left\{\frac{k^{\frac{2}{N}} S_0}{\overline{K}^{\frac{N-2}{N}}}, S_{a,G}\right\}$$

then there exists $v \in H_{a,G}$, $v > 0$, such that $F_a(v) = \inf_{H_{a,G}} F_a$, i.e. there exists a ground state in $H_{a,G}$.

Theorem 2.1 converts into a theorem on $\mathbb{R}^N$ just by replacing $H_{a,G}$ by $\mathcal{D}_{a,G}(\mathbb{R}^N)$ and F_a by E_a.

3. Fixed Points

PROOF OF THEOREM 1.2. Without loss of generality we can assume that the fixed point y of the action of G is on the unit sphere in $\mathbb{R}^N$. For

$a = 0$ and $K \equiv 1$, the function $U_{0,1,0}(x)$ in (1.6) is in any $\mathcal{D}_{0,G}(\mathbb{R}^N)$ and it has energy S_0. Since

$$\inf_{\mathcal{D}_{a,G}(\mathbb{R}^N)} E_a \geq S_0,$$

we get that the least energy is S_0 and it is achieved in $\mathcal{D}_{0,G}(\mathbb{R}^N)$.

If K is not constant, for any $u \in \mathcal{D}_{0,G}(\mathbb{R}^N)$ we have

$$E_0(u) > \frac{\int_{\mathbb{R}^N} |\nabla u|^2\, dx}{\left(\int_{\mathbb{R}^N} \overline{K}|u|^{2^*}\, dx\right)^{\frac{2}{2^*}}} \geq \frac{S(0)}{\overline{K}^{\frac{N-2}{N}}}. \tag{3.1}$$

Consider the functions

$$u_\varepsilon(x) = \left(\frac{\varepsilon}{\varepsilon^2 + |(\varepsilon^2+1)^{\frac{1}{2}}x - y|^2}\right)^{\frac{N-2}{2}}. \tag{3.2}$$

We have that $u_\varepsilon \in \mathcal{D}_{0,G}(\mathbb{R}^N)$ for any G leaving y fixed. Moreover,

$$\lim_{\varepsilon\to 0} E_0(u_\varepsilon) = \frac{S_0}{\overline{K}^{\frac{N-2}{N}}}.$$

This together with (3.1) implies that $\inf_{\mathcal{D}_{0,G}(\mathbb{R}^N)} E_0 = \frac{S_0}{\overline{K}^{\frac{N-2}{N}}}$. The fact that the infimum is not achieved follows from the strict inequality in (3.1).

In the case $a < 0$ we first show that for any $u \in \mathcal{D}_{a,G}(\mathbb{R}^N) \setminus \{0\}$, we have

$$E_a(u) \geq \frac{S_0}{\overline{K}^{\frac{N-2}{N}}}. \tag{3.3}$$

Assuming that there exists $u \in \mathcal{D}_{a,G}(\mathbb{R}^N) \setminus \{0\}$ such that the strict reversed inequality (3.3) holds, we get a contradiction as follows.

$$\frac{S_0}{\overline{K}^{\frac{N-2}{N}}} \leq E_0(\Upsilon_0\Upsilon_a^{-1}u) < E_a(u) < \frac{S_0}{\overline{K}^{\frac{N-2}{N}}}. \tag{3.4}$$

Therefore

$$\inf_{\mathcal{D}_{a,G}(\mathbb{R}^N)\setminus\{0\}} E_a \geq \frac{S_0}{\overline{K}^{\frac{N-2}{N}}}. \tag{3.5}$$

To show that in (3.5) we actually have equality, we construct a minimizing sequence using the functions (3.2) above. Since u_ε may not belong to $\mathcal{D}_{a,G}(\mathbb{R}^N)$, we consider $v_\varepsilon = \Upsilon_a\Upsilon_0^{-1}u_\varepsilon \in \mathcal{D}_{a,G}(\mathbb{R}^N)$, i.e. $v_\varepsilon(x) = |x|^a u_\varepsilon(x)$. A short computation shows

$$\lim_{\varepsilon\to 0} E_a(v_\varepsilon) = \frac{S_0}{\overline{K}^{\frac{N-2}{N}}}.$$

Again, the strict inequality in

$$\frac{S_0}{\overline{K}^{\frac{N-2}{N}}} \leq E_0(\Upsilon_0\Upsilon_a^{-1}u) < E_a(u)$$

for any $u \in \mathcal{D}_{a,G}(\mathbb{R}^N) \setminus \{0\}$ implies that no ground state exists. □

4. Finite Orbits

PROOF OF THEOREM 1.3. We shall use Theorem 2.1. Since from (1.10) we have that

$$\min\left\{\frac{k^{\frac{2}{N}}S_0}{\overline{K}^{\frac{N-2}{N}}}, S_{a,G}\right\} = \frac{k^{\frac{2}{N}}S_0}{\overline{K}^{\frac{N-2}{N}}},$$

we only need to prove that there exist functions $u \in \mathcal{D}_{a,G}(\mathbb{R}^N)$ such that

$$E_a(u) < \frac{k^{\frac{2}{N}}S_0}{\overline{K}^{\frac{N-2}{N}}}.$$

For $\varepsilon > 0$ fixed, let

$$U_i(x) = C\left(\frac{\varepsilon}{\varepsilon^2 + |(\varepsilon^2+1)^{\frac{1}{2}}x - y_i|^2}\right)^{\frac{N-2}{2}},$$

where y_i belongs to the minimal orbit in (1.10) and (1.11). The constant C is such that $||U_i||_{L^{2^*}(\mathbb{R}^N)} = 1$ (and therefore $||U_i||_{\mathcal{D}_a(\mathbb{R}^N)} = S_0$). Note that up to C, the functions U_i are the functions u_ε in (3.2), centered at y_i. Define

$$U_\varepsilon(x) = \max\{U_1(x), \dots, U_k(x)\} \in \mathcal{D}_{0,G}(\mathbb{R}^N).$$

Again, let $V_\varepsilon = \Upsilon_a\Upsilon_0^{-1}U_\varepsilon$. Then

$$E_a(V_\varepsilon) = E_a(\Upsilon_a\Upsilon_0^{-1}U_\varepsilon) = \frac{\int_{\mathbb{R}^N} |\nabla U_\varepsilon|^2 + \gamma|x|^{-2}U_\varepsilon^2\, dx}{\left(\int_{\mathbb{R}^N} K(x)U_\varepsilon^{2^*}\, dx\right)^{2/2^*}}, \tag{4.1}$$

where

$$\gamma = \left(\frac{N-2-2a}{2}\right)^2 - \left(\frac{N-2}{2}\right)^2 = -a(N-2-a) > -\left(\frac{N-2}{2}\right)^2.$$

Denote $\Omega_i = \{x \in \mathbb{R}^N \mid \operatorname{dist}(x, y_i) < \operatorname{dist}(x, y_j) \text{ for all } j = 1, \dots, k,\ j \neq i\}$. We have

$$\int_{\mathbb{R}^N} |\nabla U_\varepsilon|^2 + \gamma|x|^{-2}U_\varepsilon^2\, dx = \sum_{i=1}^k \int_{\Omega_i} |\nabla U_i|^2 + \gamma|x|^{-2}U_i^2\, dx = kS_0(1 - A(\varepsilon)),$$

where

$$A(\varepsilon) := \frac{1}{kS_0}\sum_{i=1}^k \left(\int_{\mathbb{R}^N\setminus\Omega_i} |\nabla U_i|^2\, dx - \gamma\int_{\Omega_i} |x|^{-2}U_i^2\, dx\right).$$

Also

$$\int_{\mathbb{R}^N} K(x)U_\varepsilon^{2^*}\, dx = \sum_{i=1}^k \int_{\Omega_i} K(x)U_i^{2^*}\, dx = k\overline{K}(1 - B(\varepsilon)),$$

where

$$B(\varepsilon) := \frac{1}{k}\sum_{i=1}^{k}\left(\int_{\Omega_i}\frac{\overline{K}-K(x)}{\overline{K}}U_i^{2^*}\,dx + \int_{\mathbb{R}^N\setminus\Omega_i}U_i^{2^*}\,dx\right).$$

Direct calculation shows that there are positive constants c_1, c_2, c_3, c_4 such that for all $i = 1, \ldots, k$ and ε small we have

$$\int_{\mathbb{R}^N\setminus\Omega_i}|\nabla U_i|^2\,dx \geq c_1\varepsilon^{N-2}, \quad \int_{\Omega_i}|x|^{-2}U_i^2\,dx \leq c_2\varepsilon,$$

$$\int_{\Omega_i}\frac{\overline{K}-K(x)}{\overline{K}}U_i^{2^*}\,dx \leq c_3\varepsilon^{\alpha}, \quad \int_{\mathbb{R}^N\setminus\Omega_i}U_i^{2^*}\,dx \leq c_4\varepsilon^N.$$

From (4.1) and the estimates above we conclude that for $\gamma \leq 0$ (i.e. $a \geq 0$) we have

$$E_a(V_\varepsilon) \leq \frac{k^{\frac{2}{N}}S_0}{\overline{K}^{\frac{N-2}{N}}}\frac{1-\varepsilon^{N-2}c_1/S_0}{(1-\varepsilon^{\alpha}c_3-\varepsilon^N c_4)^{\frac{N-2}{N}}} < \frac{k^{\frac{2}{N}}S_0}{\overline{K}^{\frac{N-2}{N}}}$$

for small ε. The same argument works in dimension $N = 3$ *even for negative* a provided that $-a$ is not too large. Indeed, if a_0 is the negative root of $x^2 - x - \frac{c_1}{c_2} = 0$, then for any $a_0 < a < \frac{1}{2}$ we have $c_1 - \gamma c_2 > 0$ and the estimates above give

$$E_a(V_\varepsilon) \leq \frac{k^{\frac{2}{3}}S_0}{\overline{K}^{\frac{1}{3}}}\frac{1-\varepsilon(c_1-\gamma c_2)/S_0}{(1-\varepsilon^{\alpha}c_3-\varepsilon^3 c_4)^{\frac{1}{3}}} < \frac{k^{\frac{2}{3}}S_0}{\overline{K}^{\frac{1}{3}}}$$

for small ε. □

5. Conclusions and Open Questions

Our first comment is that the test functions V_ε in the proof of Theorem 1.3 have to be multi-peaked. For this reason we do not allow fixed points of the action. At the expense of imposing stronger conditions on K, one can relax the requirements on the action. This was done in [**CW2**] for the prescribed scalar curvature problem (the case $a = 0$). We can show that if $\{y_1, \ldots, y_k\}$ ($k \geq 2$) is *any* finite orbit of the action, and if y_i are *local* maxima of K, and if K satisfies the flatness condition (1.11) *plus an extra* peakness condition, then for a as in Theorem 1.3 one can find solutions of (1.1) which are *local minima* of E_a in $\mathcal{D}_{a,G}(\mathbb{R}^N)$. This is a local result. It is independent of the value of $S_{a,G}$ and of whether there are smaller orbits (with fewer points) or not.

It is worth making the following

REMARK 5.1. If $G = \mathbf{Z}_2\times$ a subgroup of $\mathbf{O}(n)$, i.e. the first component of G is the whole $\mathbf{Z}_2$, then $S_{a,G}$ in Theorem 2.1 may be replaced by $2^{\frac{2}{N}}S_{a,G}$. Therefore condition (1.10) required by Theorem 1.3 can be relaxed to

$$\overline{K} \geq \left(\frac{k}{2}\right)^{\frac{2}{N-2}}\left(\frac{S_0}{S_{a,G}}\right)^{\frac{N}{N-2}}.$$

A very important question seems to be the following:

PROBLEM 5.2. Can $S_{a,G}$ be calculated?

The case when the $\mathbf{O}(N)$ component of G is the identity element only, i.e. G is the trivial group or $\mathbf{Z}_2 \times \{I\}$, is the only situation known to the author in which $S_{a,G}$ is explicitly known (we have $S_{a,G} = S_a$, where S_a is discussed in Theorem 1.1). So far, we can make the following observations.

(i) For a fixed action, $S_{a,G}$ is nondecreasing in the argument $-a$.

(ii) If H is a subgroup of G, then $S_{a,H} \leq S_{a,G}$.

(iii) For $0 \leq a < \frac{N-2}{2}$ and any group G, we have $S_{a,G} = S_a = R_a$.

Let $G = \mathbf{Z}_2 \times \{I, -I\} \subset \mathbf{Z}_2 \times \mathbf{O}(N)$. In view of Remark 5.1, and because $S_{a,G} \geq S_0$ for $a < 0$, in dimension $N = 3$ we can apply Theorem 1.3 for a negative, and $K \equiv 1$. Therefore, for $a < 0$ and $|a|$ sufficiently small $S_{a,G}$ is achieved by some $U_{a,G}$.

PROBLEM 5.3. Is $U_{a,G}$ radial (and therefore $S_{a,G} = R_a$)?

PROBLEM 5.4. What is the exact range for a so that the ground state $U_{a,G}$ exists?

The radial solution can be ground state in $\mathcal{D}_{a,G}(\mathbb{R}^3)$ only if $R_a \leq 2^{\frac{2}{N}} S_0$, therefore if the answer to Problem 5.3 is yes, then the answer to Problem 5.4 is a finite number. More generally, one can ask the following.

PROBLEM 5.5. In the range $a < 0$, is $S_{a,G}$ achieved for other group actions and/or in higher dimensions?

Another interesting question concerns the range $a \ll -1$. The existence results in [**S**] do not apply in this range. The results in [**FS**] assume that K is close to $K \equiv 1$ and it is likely that the solutions they obtain are close to the radial solutions $U_{a,\varepsilon}$ in Theorem 1.1 (i). We inquire about solutions with energy much lower than that of the radial solutions.

PROBLEM 5.6. Is there a sequence $0 < \underline{K} \leq K_n \leq \overline{K}$ (with symmetry, or without) satisfying (1.3) such that for a sequence $a_n \to -\infty$, the problems (1.1) have solutions u_n with $E_{a_n}(u_n) \leq M$ for some $M < \infty$ (M independent of n)?

References

[A] T. Aubin, *Problèmes isopérimétriques et espaces de Sobolev,* J. Diff. Géo. **11** (1976), 573–598.

[A1] T. Aubin, *Some Nonlinear Problems in Riemannian Geometry,* Springer-Verlag, New York, NY, 1998.

[BCS] G. Bianchi, J. Chabrowski and A. Szulkin, *On* symmetric solutions of an elliptic equation with a nonlinearity involving critical Sobolev exponent, Nonl. Anal., T.M.A. **25** (1995), 41–59.

[CKN] L. A. Caffarelli, R. Kohn and L. Nirenberg, *First order interpolation inequalities with weights,* Compositio Mathematica **53** (1984), 259–275.

[CW1] F. Catrina and Z.-Q. Wang, *On the Caffarelli-Kohn-Nirenberg inequalities: sharp constants, existence (and nonexistence) and symmetry of extremal functions*, Comm. Pure and Appl. Math. **54** (2001), 229–258.

[CW2] F. Catrina and Z.-Q. Wang, *Symmetric solutions for the prescribed scalar curvature problem*, Indiana Univ. Math. Journal **49** (2000), 779–813.

[FS] V. Felli and M. Schneider, *Perturbation results of critical elliptic equations of Caffarelli-Kohn-Nirenberg type*, preprint.

[S] D. Smets, *Nonlinear Schrödinger equations with Hardy potential and critical nonlinearities*, preprint.

[T] G. Talenti, *Best constant in Sobolev inequality*, Ann. Mat. Pura Appl. **110** (1976), 353–372.

Department of Mathematics, University of Rochester, Rochester, New York 14627

E-mail address: fcatrina@math.rochester.edu

Contemporary Mathematics
Volume **357**, 2004

Non-convergent Radial Solutions of a Semilinear Elliptic Equation in $\mathbb{R}^N$

Joseph A. Iaia

Abstract. We examine the bounded radial solutions of $\Delta u + f(u) = 0$ in $\mathbb{R}^N$ with $N > 1$. We show for many nonlinearities f that if $\lim_{r\to\infty} u(r) =$ *does not exist* then $f \equiv 0$ on some nonempty open interval and $N = 2$.

1. Introduction

In this paper, we consider radial solutions of:

$$\Delta u + f(u) = 0 \quad \text{in} \quad \mathbb{R}^N \tag{1.1}$$

where $f\colon \mathbb{R} \to \mathbb{R}$ is locally Lipschitz. That is, $u = u(r)$ where $r = |x|$, $x \in \mathbb{R}^N$ and $u \in \mathbf{C}^2[0, \infty)$ satisfies:

$$u'' + \frac{N-1}{r}u' + f(u) = 0 \text{ for } 0 < r < \infty, \tag{1.2}$$

$$u'(0) = 0, \ u(0) = d \ \text{ for some } \ d \in \mathbb{R}. \tag{1.3}$$

Throughout the rest of this paper, we will consider the more general situation where the N in (1.2) represents a *real* parameter with $N > 1$.

It is well-known and easy to show that if $\lim_{r\to\infty} u(r) = u_0 \in \mathbb{R}$ then $f(u_0) = 0$ (see Lemma 2.1 below). However, it need *not* always be the case that $\lim_{r\to\infty} u(r) =$ *exists.* In fact, in 1994 S. Maier ([1]) gave an example with $N = 2$ of a nonlinearity f and a bounded solution u of (1.2)-(1.3) for which $\lim_{r\to\infty} u(r) =$ *does not exist.* Maier also proved necessary conditions on f in order that there exist a bounded solution u of (1.2)-(1.3) for which $\lim_{r\to\infty} u(r) =$ *does not exist.* Denoting $c_1 \equiv \liminf_{r\to\infty} u(r)$ and $c_2 \equiv \limsup_{r\to\infty} u(r)$, Maier ([1]) showed that if $\lim_{r\to\infty} u =$ *does not exist* then $c_1 < c_2$, $F(c_1) = F(c_2) \equiv \alpha$, $F(u) \le \alpha$ for $u \in [c_1, c_2]$ (where $F(u) = \int_0^u f(t)\, dt$), and $f(c_1) \le 0 \le f(c_2)$. In addition, f must attain positive values on $(c_2, c_2 + \varepsilon)$ for every $\varepsilon > 0$ and f must attain negative values on $(c_1 - \varepsilon, c_1)$ for every $\varepsilon > 0$.

2000 *Mathematics Subject Classification.* Primary 34B15: Secondary 35J65.
Key words and phrases. Nonconvergent, radial.

In this paper, we show for many nonlinearities that if $N > 2$ then $\lim_{r\to\infty} u(r) =$ exists. We in fact prove following:

MAIN THEOREM. *Let u be a bounded solution of (1.2)-(1.3) with $N > 1$ and suppose $\lim_{r\to\infty} u(r) =$ does not exist. Denoting $c_1 \equiv \liminf_{r\to\infty} u(r)$ and $c_2 \equiv \limsup_{r\to\infty} u(r)$, then we must have either $f(c_1) = 0$ or $f(c_2) = 0$.*

In addition, if $f(c_1) < 0$ and $f(c_2) = 0$ with f nondecreasing on $(c_2, c_2+\varepsilon_2)$ for some $\varepsilon_2 > 0$, OR if $f(c_2) > 0$ and $f(c_1) = 0$ with f nondecreasing on $(c_1 - \varepsilon_1, c_1)$ for some $\varepsilon_1 > 0$, OR if $f(c_1) = f(c_2) = 0$ with f nondecreasing on $(c_1 - \varepsilon_1, c_1)$ and on $(c_2, c_2 + \varepsilon_2)$ for some $\varepsilon_1 > 0$ and $\varepsilon_2 > 0$, then $f \equiv 0$ on $[c_1, c_2]$ and $1 < N \leq 2$.

REMARK: So we see that if there is a bounded nonconvergent solution of (1.2)-(1.3) and $N > 2$ then f must oscillate infinitely often in either *every* neighborhood to the right of c_2 or *every* neighborhood to the left of c_1. This contrasts remarkably with $1 < N \leq 2$ where there are many examples of bounded nonconvergent solutions of (1.2)-(1.3) where the corresponding f is monotonic. In fact in Maier's original example in [1] with $N = 2$, the nonlinearity is monotonic. Further, this example extends easily to $1 < N \leq 2$. And more generally, it is not difficult to show that if $f(u) = 0$ for $-1 \leq u \leq 1$ with $f(u) > 0$ for $u > 1$ and $f(u) < 0$ for $u < -1$, then the solution of (1.2)-(1.3) with $|d| > 1$ satisfies: $\lim_{r\to\infty} u(r) =$ *does not exist.* (The key to this argument is that the nonconstant solutions to (1.2) on the set where $f(u) \equiv 0$ are unbounded functions for large r if and only if $N \leq 2$).

This paper is organized as follows: in section 2, we establish some preliminary facts about solutions of (1.2)-(1.3). In section 3, we establish some facts about solutions of (1.2)-(1.3) for which $\lim_{r\to\infty} u(r) =$ *does not exist.* In section 4 we prove the main theorem. Finally, in section 5, we give examples of nonlinearities f for which *all* solutions of (1.2)-(1.3) converge.

2. Preliminaries

Let u be a solution of (1.2)-(1.3) with $N > 1$ and let $F(u) = \int_0^u f(s)\, ds$. Define:

$$E = \frac{1}{2}(u')^2 + F(u) \quad \text{for} \quad r \geq 0. \tag{2.1}$$

Differentiating (2.1) and using (1.2) gives:

$$E' = -\frac{N-1}{r}(u')^2 \leq 0 \quad \text{for} \quad r > 0. \tag{2.2}$$

Thus, we see that E is a nonincreasing function.

REMARK: If u is *not* a constant function then E is *strictly decreasing* for if $E(r_1) = E(r_2)$ for some $r_1 < r_2$ then since $E' \leq 0$, we see that E is constant on (r_1, r_2) and therefore $-\frac{N-1}{r}(u')^2 = E' \equiv 0$ on (r_1, r_2). Hence u is constant on (r_1, r_2). But then by uniqueness of solutions of initial value problems (see [2]), u is a constant function on $[0, \infty)$ - a contradiction.

LEMMA 2.1. *Let u be a solution of (1.2)-(1.3) with $N > 1$ and suppose there exists $u_0 \in \mathbb{R}$ such that $\lim_{r \to \infty} u(r) = u_0$. Then $f(u_0) = 0$.*

PROOF. Since $\lim_{r \to \infty} u(r) = u_0$ and F is continuous, it follows that $\lim_{r \to \infty} F(u(r)) = F(u_0)$. Further, since $E(r) \geq F(u(r))$ and E is nonincreasing, after taking limits as $r \to \infty$, we obtain $\alpha \equiv \lim_{r \to \infty} E(r) \geq F(u_0)$. Thus $\lim_{r \to \infty} |u'(r)| = \sqrt{2(\alpha - F(u_0))}$. Now using this and taking limits in (1.2) yields $\lim_{r \to \infty} u''(r) = -f(u_0)$. If $f(u_0) \neq 0$, say $f(u_0) < 0$, then for $r > r_0$ we would have $u''(r) \geq \frac{|f(u_0)|}{2}$. After two integrations this yields $u(r) \geq u(r_0) + u'(r_0)(r - r_0) + \frac{|f(u_0)|}{4}(r - r_0)^2 \to \infty$ as $r \to \infty$ contradicting $\lim_{r \to \infty} u(r) = u_0$. Similarly, we obtain a contradiction if $f(u_0) > 0$. Thus, we must have $f(u_0) = 0$. $\square$

3. Nonconvergent Solutions

We now assume that u is a solution of (1.2)-(1.3) with $N > 1$ such that:

$$u \text{ is } \textit{bounded} \text{ and } \lim_{r \to \infty} u(r) = \textit{does not exist}. \tag{3.1}$$

Taking limits in (1.2) as $r \to 0^+$ and using (1.3) gives $u''(0) = -\frac{f(d)}{N}$. If $f(d) = 0$, then by uniqueness of solutions of initial value problems (see [2]), $u(r) \equiv d$ for all $r > 0$ violating (3.1) and so we must have $f(d) \neq 0$. Without loss of generality, we assume $f(d) > 0$. (If $f(d) < 0$ then $v = -u$ satisfies the equation $\Delta v + g(v) = 0$ where $g(u) = -f(-u)$ and $g(-d) > 0$). Therefore, $u''(0) = -\frac{f(d)}{N} < 0$ and so u is initially decreasing. If $u' < 0$ for all $r > 0$ then (3.1) is violated, hence u must have a first local minimum, $m_1 > 0$, with $u' < 0$ on $(0, m_1)$, $u'(m_1) = 0$, and $u''(m_1) \geq 0$. (In fact $u''(m_1) > 0$ hence $f(u(m_1)) < 0$ otherwise u is constant by uniqueness of solutions of initial value problems - see [2]).

Next, we note that $u(r) < d$ for all $r > 0$ for if there exists an $r_0 > 0$ such that $u(r_0) = d$ then since u is nonconstant we know by the remark after inequality (2.2) that E is strictly decreasing so that $E(r_0) < E(0)$. That is, $\frac{1}{2}(u'(r_0))^2 + F(d) < F(d)$ which implies $(u'(r_0))^2 < 0$ which is impossible. A similar argument shows that $u(r) > u(m_1)$ for all $r \geq 0$ with $r \neq m_1$.

Using a similar argument as for the existence of m_1, it follows that there exists a first positive local maximum $M_1 > m_1$ such that $u' > 0$ on (m_1, M_1), $u''(M_1) < 0$ hence $f(u(M_1)) > 0$ and also $u(m_1) < u(M_1) < d$.

Continuing in this fashion and using (3.1), we see that we obtain:

$$0 = M_0 < m_1 < M_1 < m_2 < M_2 < \cdots \to \infty \tag{3.2}$$

where the $\{m_k\}$ are local minima of u, the $\{M_k\}$ are local maxima of u, with $u' > 0$ on (m_k, M_k) and $u' < 0$ on (M_k, m_{k+1}). (These must tend to ∞ for if these converged to say $M > 0$ then we would have $u'(M) = u''(M) = 0$ and again by uniqueness of solutions of initial value problems (see [2]), u would then reduce to a constant function - contradicting (3.1)). In addition, using an argument similar to the proof that $u(m_1) < u(M_1) < d$, it follows that $u(m_1) < u(m_2) < u(M_2) < u(M_1)$ and in general that the $\{u(M_k)\}$ form a decreasing sequence which is bounded below, the $\{u(m_k)\}$ form an increasing sequence which is bounded above, and that $u(m_k) < u(M_k)$. Hence there exist real numbers c_1, c_2 with $c_1 \leq c_2$ such that:

$$\lim_{k\to\infty} u(M_k) = c_2 \text{ and } \lim_{k\to\infty} u(m_k) = c_1. \tag{3.3}$$

And in order for (3.1) to hold we must have

$$c_1 < c_2. \tag{3.4}$$

Also, since

$$f(u(m_k)) < 0 < f(u(M_k)), \tag{3.5}$$

after taking limits in these inequalities and using (3.3) we see that:

$$f(c_1) \leq 0 \leq f(c_2). \tag{3.6}$$

Further, since $u(m_1) \leq u(r) \leq d$ for all $r \geq 0$ we see that $E(r) \geq \min\limits_{[u(m_1),d]} F(u)$ (and the right hand side is finite since F is continuous and $[u(m_1), d]$ is compact). Hence we have:

$$\alpha \equiv \lim_{r\to\infty} E(r) \geq \min_{[u(m_1),d]} F(u). \tag{3.7}$$

In addition, since E is strictly decreasing, we have:

$$E(0) > E(m_1) > E(M_1) > E(m_2) > \cdots \to \alpha.$$

That is,

$$F(d) > F(u(m_1)) > F(u(M_1)) > F(u(m_2)) > \cdots \to \alpha,$$

and so this and (3.3) imply:

$$F(c_2) = F(c_1) = \alpha. \tag{3.8}$$

Thus, our solution u oscillates infinitely often with the local maximum values of u monotonicly decreasing to c_2 and the local minimum values of u monotonicly increasing to c_1.

LEMMA 3.1. *Let u satisfy (3.1). Then $F(u) \leq \alpha$ for all $u \in [c_1, c_2]$.*

REMARK: The proof we present here is taken from [1].

PROOF. Suppose there is a $c_0 \in (c_1, c_2)$ such that $F(c_0) > \alpha$. Hence, there exists an $\varepsilon_0 > 0$ such that $F(c_0) = \alpha + \varepsilon_0$. Since u attains the value of c_1 and c_2 infinitely many times and E converges to α as $r \to \infty$, there exists an $r_0 > 0$ such that $u(r_0) = c_0$ and $E(r_0) < \alpha + \frac{\varepsilon_0}{2}$. On the other hand, $E(r_0) = \frac{1}{2}(u'(r_0))^2 + F(c_0) \geq \alpha + \varepsilon_0$ and this is a contradiction. Hence the lemma is proved. □

So we see that there exist numbers $\{r_k\}$ with $r_{4k-2} < m_k < r_{4k-1}$ and $r_{4k} < M_k < r_{4k+1}$ such that $u(r_{4k-2}) = u(r_{4k-1}) = c_1$ with $u(r) < c_1$ on (r_{4k-2}, r_{4k-1}), $u(r_{4k}) = u(r_{4k+1}) = c_2$ with $u(r) > c_2$ on (r_{4k}, r_{4k+1}), and $c_1 < u(r) < c_2$ with u strictly monotonic on (r_{2k-1}, r_{2k}).

LEMMA 3.2. *Let u satisfy (3.1). Then on (r_{4k}, r_{4k+1}) we have $F(u(r)) \leq F(u(M_k))$ and on (r_{4k-2}, r_{4k-1}) we have $F(u(r)) \leq F(u(m_k))$.*

PROOF. For $r \in [M_k, r_{4k+1})$ we have $F(u(r)) \leq \frac{1}{2}(u'(r))^2 + F(u(r)) = E(r) \leq E(M_k) = F(u(M_k))$ since E is decreasing (by (2.2)). Thus the first part of the lemma holds on $[M_k, r_{4k+1})$. On the other hand, if there exists a $t_0 \in (r_{4k}, M_k)$ with $F(u(t_0)) > F(u(M_k))$, then there exists a $t_1 \in (M_k, r_{4k+1})$ with $u(t_0) = u(t_1)$ (because $u(r_{4k}) = u(r_{4k+1}) = c_2$ with u increasing on (r_{4k}, M_k) and u decreasing on (M_k, r_{4k+1})) and hence $F(u(t_0)) = F(u(t_1))$. But we have ruled out the existence of such a t_1. Thus $F(u(r)) \leq F(u(M_k))$ on (r_{4k}, r_{4k+1}) and the first half of the lemma is proved. The second half of the lemma proceeds similarly. □

We now define

$$F_\alpha(u) \equiv F(u) - \alpha \quad \text{and} \quad E_\alpha(r) \equiv E(r) - \alpha \tag{3.9}$$

and we observe from lemma 3.1 that $F_\alpha \leq 0$ on (r_{2k-1}, r_{2k}) and from (2.2) and (3.7) that $E_\alpha > 0$ with E_α decreasing and $\lim_{r \to \infty} E_\alpha = 0$.

LEMMA 3.3. *Let u satisfy (3.1). Suppose $f(u) \not\equiv 0$ on $[c_1, c_2]$. Then there exists a $c_0 > 0$ such that for large $k \in \mathbb{N}$:*

$$r_{2k}E_\alpha(r_{2k}) - r_{2k-1}E_\alpha(r_{2k-1}) \leq -c_0 < 0.$$

PROOF.
Differentiating rE_α we see that:

$$(rE_\alpha)' = E_\alpha + rE'_\alpha = (\frac{3}{2} - N)(u')^2 + F_\alpha$$

and thus:

$$r_{2k}E_\alpha(r_{2k}) - r_{2k-1}E_\alpha(r_{2k-1}) = (\frac{3}{2} - N)\int_{r_{2k-1}}^{r_{2k}} (u')^2\,dr + \int_{r_{2k-1}}^{r_{2k}} F_\alpha(u(r))\,dr. \tag{3.10}$$

Since E_α is decreasing (by (2.2)) we have that $E_\alpha(r) \leq E_\alpha(r_{2k-1})$ for $r > r_{2k-1}$ and therefore:

$$\frac{|u'|}{\sqrt{2}\sqrt{E_\alpha(r_{2k-1}) - F_\alpha(u)}} \leq 1 \text{ for } r > r_{2k-1}. \tag{3.11}$$

Since $F_\alpha(u(r)) \leq 0$ for $r \in (r_{2k-1}, r_{2k})$ (by lemma 3.1) and u is monotonic for $r \in (r_{2k-1}, r_{2k})$ we obtain using (3.11):

$$\int_{r_{2k-1}}^{r_{2k}} F_\alpha(u)\,dr \leq \int_{r_{2k-1}}^{r_{2k}} \frac{F_\alpha(u)|u'|}{\sqrt{2}\sqrt{E_\alpha(r_{2k-1}) - F_\alpha(u)}}\,dr$$
$$= \int_{c_1}^{c_2} \frac{F_\alpha(t)}{\sqrt{2}\sqrt{E_\alpha(r_{2k-1}) - F_\alpha(t)}}\,dt. \tag{3.12}$$

As $k \to \infty$, $\frac{F_\alpha(t)}{\sqrt{E_\alpha(r_{2k-1}) - F_\alpha(t)}} \to -\sqrt{|F_\alpha(t)|}$ uniformly on $[c_1, c_2]$ hence the right hand side of (3.12) converges to:

$$-\frac{1}{\sqrt{2}}\int_{c_1}^{c_2} \sqrt{|F_\alpha(t)|}\,dt < 0. \tag{3.13}$$

(The inequality in (3.13) is strict because by assumption $F'_\alpha(u) = f(u) \not\equiv 0$ on $[c_1, c_2]$).

Thus for $N \geq 3/2$ we see from (3.10), (3.12), and (3.13) that for large k:

$$r_{2k}E_\alpha(r_{2k}) - r_{2k-1}E_\alpha(r_{2k-1}) \leq -c_0 \tag{3.14}$$

where

$$c_0 = \frac{1}{2\sqrt{2}} \int_{c_1}^{c_2} \sqrt{|F_\alpha(t)|}\, dt.$$

This completes the proof of lemma 3.3 for $N \geq 3/2$.

For $1 < N < 3/2$ we also need to estimate the $\int_{r_{2k-1}}^{r_{2k}} (u')^2\, dr$ term in (3.10). Here we see using (3.11) that:

$$\int_{r_{2k-1}}^{r_{2k}} (u')^2\, dr \leq \sqrt{2} \int_{r_{2k-1}}^{r_{2k}} \sqrt{E_\alpha(r_{2k-1}) - F_\alpha(u)}|u'|\, dr$$

$$= \sqrt{2} \int_{c_1}^{c_2} \sqrt{E_\alpha(r_{2k-1}) - F_\alpha(t)}\, dt. \tag{3.15}$$

Thus we see from (3.10), (3.12), and (3.15) that:

$$r_{2k}E_\alpha(r_{2k}) - r_{2k-1}E_\alpha(r_{2k-1}) = (\frac{3}{2} - N) \int_{r_{2k-1}}^{r_{2k}} (u')^2\, dr + \int_{r_{2k-1}}^{r_{2k}} F_\alpha(u)\, dr$$

$$\leq (\frac{3}{2} - N)\sqrt{2} \int_{c_1}^{c_2} \sqrt{E_\alpha(r_{2k-1}) - F_\alpha(t)}\, dt + \int_{c_1}^{c_2} \frac{F_\alpha(t)}{\sqrt{2}\sqrt{E_\alpha(r_{2k-1}) - F_\alpha(t)}}\, dt \tag{3.16}$$

As in (3.12)-(3.13), we see that the right hand side of (3.16) converges as $k \to \infty$ to:

$$(\frac{3}{2} - N)\sqrt{2} \int_{c_1}^{c_2} \sqrt{|F_\alpha(t)|}\, dt - \frac{1}{\sqrt{2}} \int_{c_1}^{c_2} \sqrt{|F_\alpha(t)|}\, dt$$

$$= -\sqrt{2}(N-1) \int_{c_1}^{c_2} \sqrt{|F_\alpha(t)|}\, dt < 0 \tag{3.17}$$

Thus for $1 < N < 3/2$ we see from (3.16)-(3.17) that for large k:

$$r_{2k}E_\alpha(r_{2k}) - r_{2k-1}E_\alpha(r_{2k-1}) \leq -c_0$$

where

$$c_0 = \frac{\sqrt{2}(N-1)}{2} \int_{c_1}^{c_2} \sqrt{|F_\alpha(t)|}\, dt.$$

This completes the proof of lemma 3.3 for $1 < N < 3/2$ and so combined with the earlier proof for $N \geq 3/2$ we see that this concludes the proof of lemma 3.3. □

For the rest of section 3 we will estimate rE_α on the set where $F_\alpha(u) \geq 0$. We will find it useful to have the following estimate:

$$M_k - r_{4k} \leq r_{4k+1} - M_k.$$

To obtain this, observe that for $r \in (M_k, r_{4k+1})$ we have (by (2.2)) $E_\alpha(r) \leq E_\alpha(M_k)$, i.e. $\frac{1}{2}(u')^2 + F_\alpha(u) \leq F_\alpha(u(M_k))$. Thus, for $r \in (M_k, r_{4k+1})$ we have

$\frac{|u'|}{\sqrt{2}\sqrt{F_\alpha(u(M_k))-F_\alpha(u(r))}} \leq 1$. Integrating this on (M_k, r_{4k+1}) where u is monotonic gives:

$$\frac{1}{\sqrt{2}}\int_{c_2}^{u(M_k)} \frac{dt}{\sqrt{F_\alpha(u(M_k)) - F_\alpha(t)}} \leq r_{4k+1} - M_k. \tag{3.18}$$

On the other hand, for $r \in (r_{4k}, M_k)$ we have $E_\alpha(r) \geq E_\alpha(M_k)$ and thus for $r \in (r_{4k}, M_k)$ we have $\frac{|u'|}{\sqrt{2}\sqrt{F_\alpha(u(M_k))-F_\alpha(u(r))}} \geq 1$. Integrating this on (r_{4k}, M_k) where u is monotonic gives:

$$\frac{1}{\sqrt{2}}\int_{c_2}^{u(M_k)} \frac{dt}{\sqrt{F_\alpha(u(M_k)) - F_\alpha(t)}} \geq M_k - r_{4k}. \tag{3.19}$$

Comparing (3.18) and (3.19) we see that:

$$M_k - r_{4k} \leq r_{4k+1} - M_k. \tag{3.20}$$

A very similar argument shows:

$$m_k - r_{4k+2} \leq r_{4k+3} - m_k. \tag{3.21}$$

LEMMA 3.4. *If there exists a constant $c_\delta > 0$ and $\varepsilon > 0$ such that $f(t) \geq c_\delta(t - c_2)$ on $(c_2, c_2 + \varepsilon)$ (in particular if $f(c_2) > 0$ or if $f(c_2) = 0$, f diff'ble, and $f'(c_2) > 0$) then*

$$\limsup_{k\to\infty}\, [r_{4k+1}E_\alpha(r_{4k+1}) - r_{4k}E_\alpha(r_{4k})] \leq 0.$$

PROOF. Let v be the solution of:

$$v'' + \frac{N-1}{r}v' + c_\delta(v - c_2) = 0 \quad \text{for} \quad r > M_k,$$

$$v'(M_k) = 0, v(M_k) = u(M_k).$$

It then follows from the *Sturm comparison test* (see for example [2]) that:

$$r_{4k+1} - M_k \leq s_{4k+1} - M_k \tag{3.22}$$

where s_{4k+1} is such that $v(r) > c_2$ on $[M_k, s_{4k+1})$ and $v(s_{4k+1}) = c_2$.

The reason we know that such a s_{4k+1} exists is that letting $w(r) = r^{\frac{N-2}{2}}[v(r/\sqrt{c_\delta}) - c_2]$, it follows that w satisfies:

$$w'' + \frac{1}{r}w' + [1 - \frac{((N-2)/2)^2}{r^2}]w = 0 \text{ for } r > \sqrt{c_\delta}M_k.$$

This is *Bessel's equation* of order $\frac{N-2}{2}$ and it is well-known (see [2]) that w has an infinite number of zeros on $[\sqrt{c_\delta}M_k, \infty)$ and that the distance between consecutive zeros of w approaches π as $r \to \infty$.

Since w has an infinite number of zeros on $[\sqrt{c_\delta}M_k, \infty)$ it follows that $v - c_2$ has an infinite number of zeros on $[M_k, \infty)$ and in particular a first zero greater than M_k which we have denoted s_{4k+1}.

Now since $v > c_2$ on $[M_k, s_{4k+1})$ it follows that $w > 0$ on $[\sqrt{c_\delta}M_k, \sqrt{c_\delta}s_{4k+1})$. And since $\sqrt{c_\delta}M_k \to \infty$ as $k \to \infty$ we see from (3.22) and the above remarks about *Bessel's equation* that for large k:

$$\sqrt{c_\delta}[r_{4k+1} - M_k] \leq \sqrt{c_\delta}[s_{4k+1} - M_k] \leq 2\pi. \tag{3.23}$$

Now combining (3.20) and (3.23) we see that:

$$r_{4k+1} - r_{4k} \le 2(r_{4k+1} - M_k) \le \frac{4\pi}{\sqrt{c_\delta}}. \tag{3.24}$$

For $N \ge 3/2$, integrating $(rE_\alpha)'$ on (r_{4k}, r_{4k+1}), using lemma 3.2, (3.10), and (3.24) gives:

$$r_{4k+1}E_\alpha(r_{4k+1}) - r_{4k}E_\alpha(r_{4k}) = (\frac{3}{2} - N)\int_{r_{4k}}^{r_{4k+1}} (u')^2\,dr + \int_{r_{4k}}^{r_{4k+1}} F_\alpha(u)\,dr \tag{3.25}$$

$$\le \int_{r_{4k}}^{r_{4k+1}} F_\alpha(u)\,dr \le F_\alpha(u(M_k))(r_{4k+1} - r_{4k}) \le \frac{4\pi}{\sqrt{c_\delta}} F_\alpha(u(M_k))$$

$$\equiv \frac{4\pi}{\sqrt{c_\delta}}[F(u(M_k)) - \alpha].$$

Therefore as $k \to \infty$ we see from (3.3) and (3.8) that

$$\limsup_{k\to\infty}\, [r_{4k+1}E_\alpha(r_{4k+1}) - r_{4k}E_\alpha(r_{4k})] \le 0.$$

This completes the proof of lemma 3.4 for $N \ge 3/2$.

For $1 < N < 3/2$ we again need to estimate the $\int_{r_{2k-1}}^{r_{2k}} (u')^2\,dr$ term in (3.10). Using that $F_\alpha(u(r)) \ge 0$ for $r \in (r_{4k}, r_{4k+1})$ and that (by (2.2)) $E_\alpha(r) \le E_\alpha(0)$ (i.e. $\frac{1}{2}(u')^2 + F_\alpha(u) \le F_\alpha(d)$) we obtain:

$$\int_{r_{4k}}^{r_{4k+1}} (u')^2\,dr \le \sqrt{2}\int_{r_{4k}}^{r_{4k+1}} \sqrt{F_\alpha(d) - F_\alpha(u)}\,|u'|\,dr$$

$$= \sqrt{2}\int_{r_{4k}}^{M_k} \sqrt{F_\alpha(d) - F_\alpha(u)}\,|u'|\,dr + \sqrt{2}\int_{M_k}^{r_{4k+1}} \sqrt{F_\alpha(d) - F_\alpha(u)}\,|u'|\,dr \tag{3.26}$$

$$= 2\sqrt{2}\int_{c_2}^{u(M_k)} \sqrt{F_\alpha(d) - F_\alpha(t)}\,dt \le 2\sqrt{2}[u(M_k) - c_2]\sqrt{F(d) - \alpha}.$$

Combining (3.25)-(3.26) we see that:

$$r_{4k+1}E_\alpha(r_{4k+1}) - r_{4k}E_\alpha(r_{4k})$$
$$\le 2\sqrt{2}(\frac{3}{2} - N)[u(M_k) - c_2]\sqrt{F(d) - \alpha} + \frac{4\pi}{\sqrt{c_\delta}}[F(u(M_k)) - \alpha].$$

Thus from (3.3) and (3.8) we see that:

$$\limsup_{k\to\infty}\, [r_{4k+1}E_\alpha(r_{4k+1}) - r_{4k}E_\alpha(r_{4k})] \le 0.$$

This completes the proof of lemma 3.4 for $1 < N < 3/2$ and along with the earlier proof for $N \ge 3/2$ we see that this completes the proof of lemma 3.4. □

Similarly we have:

LEMMA 3.5. *If there exists a constant $c_\delta > 0$ and $\varepsilon > 0$ such that $f(t) \le c_\delta(t - c_1)$ on $(c_1 - \varepsilon, c_1)$ (in particular if $f(c_1) < 0$ or if $f(c_1) = 0$, f diff'ble, $f'(c_1) > 0$) then*

$$\limsup_{k\to\infty} \left[r_{4k+3}E_\alpha(r_{4k+3}) - r_{4k+2}E_\alpha(r_{4k+2})\right] \le 0.$$

LEMMA 3.6. *Suppose $f(c_2) = 0$ and for some $\varepsilon_2 > 0$ that $f(t)$ is nondecreasing on $(c_2, c_2 + \varepsilon_2)$. Then*

$$\limsup_{k\to\infty} \left[r_{4k+1}E_\alpha(r_{4k+1}) - r_{4k}E_\alpha(r_{4k})\right] \le 0.$$

PROOF. We begin by first defining:

$$L_k = \sup_{(r_{4k}, r_{4k+1})} \frac{r^2 F_\alpha(u(r))}{u(r) - c_2}. \tag{3.27}$$

Since $f(c_2) = 0$, it follows that $\lim\limits_{u\to c_2} \frac{F_\alpha(u)}{u - c_2} = 0$. Also, $F_\alpha(u) \ge 0$ on $(c_2, c_2 + \varepsilon_2)$ thus the supremum defined in (3.27) is achieved at some $t_k \in (r_{4k}, r_{4k+1})$. Further, $t_k \in [M_k, r_{4k+1})$ for if $t_k \in (r_{4k}, M_k)$ then there would exist an $s_k \in (M_k, r_{4k+1})$ such that $u(t_k) = u(s_k)$ because u is increasing on (r_{4k}, M_k), u is decreasing on (M_k, r_{4k+1}), and $u(r_{4k}) = u(r_{4k+1}) = c_2$. Therefore $\frac{F_\alpha(u(t_k))}{u(t_k)-c_2} = \frac{F_\alpha(u(s_k))}{u(s_k)-c_2}$ and thus $\frac{s_k^2 F_\alpha(u(s_k))}{u(s_k)-c_2} > \frac{t_k^2 F_\alpha(u(t_k))}{u(t_k)-c_2}$ contradicting the definition of t_k.

We now define the following two sets:

$$\mathbf{A} = \{k \in \mathbb{N} \mid L_k \le 2N^2\} \tag{3.28}$$

and

$$\mathbf{B} = \{k \in \mathbb{N} \mid L_k > 2N^2\}. \tag{3.29}$$

For $k \in \mathbf{A}$ we then have that $F_\alpha(u(r)) \le \frac{2N^2[u(r)-c_2]}{r^2}$ for $r \in (r_{4k}, r_{4k+1})$. Also for $r \in (r_{4k}, r_{4k+1})$ we have $0 \le u(r) - c_2 \le u(M_k) - c_2 \to 0$ as $k \to \infty$ (by (3.3) and the remarks after (3.2)) so that $F_\alpha(u(r)) \le \frac{2N^2}{r^2}$ for $r \in (r_{4k}, r_{4k+1})$ for large k and $k \in \mathbf{A}$. Thus, from (3.10) we have for $N \ge 3/2$:

$$r_{4k+1}E_\alpha(r_{4k+1}) - r_{4k}E_\alpha(r_{4k}) = (\frac{3}{2} - N)\int_{r_{4k}}^{r_{4k+1}} (u')^2\,dr + \int_{r_{4k}}^{r_{4k+1}} F_\alpha(u)\,dr$$

$$\le \int_{r_{4k}}^{r_{4k+1}} F_\alpha(u(r))\,dr \le \int_{r_{4k}}^{r_{4k+1}} \frac{2N^2}{r^2}\,dr \le \frac{2N^2}{r_{4k}} \to 0 \text{ as } k \to \infty. \tag{3.30}$$

If $k \in \mathbf{B}$ we begin by estimating E_α on $(t_k, r) \subset (t_k, r_{4k+1})$:

$$\begin{aligned} E_\alpha(t_k) - E_\alpha(r) &= \int_{t_k}^{r} \frac{N-1}{t}(u'(t))^2\,dt \\ &\le \frac{N-1}{t_k}\int_{t_k}^{r} \sqrt{2}\sqrt{E_\alpha(t_k) - F_\alpha(u(t))}|u'(t)|\,dt \\ &= \frac{\sqrt{2}(N-1)}{t_k}\int_{u(r)}^{u(t_k)} \sqrt{E_\alpha(t_k) - F_\alpha(s)}\,ds. \end{aligned} \tag{3.31}$$

The inequality in (3.31) comes from the fact that E_α is decreasing so that $\frac{1}{2}(u'(t))^2 + F_\alpha(u(t)) = E_\alpha(t) \le E_\alpha(t_k)$ for $t > t_k$.

Since by assumption $f(c_2) = 0$ and f is nondecreasing on $(c_2, c_2 + \varepsilon_2)$ it follows that

$$F_\alpha \text{ is nondecreasing and convex on } (c_2, c_2 + \varepsilon_2). \tag{3.32}$$

We now use the fact that $2ab \le a^2 + b^2$ for any two real numbers a, b and (3.32) in (3.31) to obtain:

$$E_\alpha(t_k) - F_\alpha(u(r)) \le \frac{1}{2}(u'(r))^2 + \frac{\sqrt{2}(N-1)}{t_k}\sqrt{E_\alpha(t_k) - F_\alpha(u(r))}[u(t_k) - u(r)]$$

$$\le \frac{1}{2}(u'(r))^2 + \frac{1}{2}[E_\alpha(t_k) - F_\alpha(u(r))] + \frac{(N-1)^2}{t_k^2}[u(t_k) - u(r)]^2.$$

Thus,

$$E_\alpha(t_k) - F_\alpha(u(r)) \le (u'(r))^2 + \frac{2(N-1)^2}{t_k^2}[u(t_k) - u(r)]^2$$

and since $E_\alpha(t_k) \ge F_\alpha(u(t_k))$ we obtain:

$$F_\alpha(u(t_k)) - F_\alpha(u(r)) \le (u'(r))^2 + \frac{2(N-1)^2}{t_k^2}[u(t_k) - u(r)]^2.$$

Rewriting this gives:

$$\frac{[u(t_k) - u(r)]}{t_k^2}\left[\frac{t_k^2[F_\alpha(u(t_k)) - F_\alpha(u(r)]}{u(t_k) - u(r)} - 2(N-1)^2[u(t_k) - u(r)]\right] \le (u'(r))^2. \tag{3.33}$$

Since $F_\alpha(t)$ is convex on $(c_2, c_2 + \varepsilon_2)$ (by (3.32)), it follows that:

$$\frac{[F_\alpha(u(t_k)) - F_\alpha(u(r))]}{u(t_k) - u(r)} \ge \frac{F_\alpha(u(t_k))}{u(t_k) - c_2} \text{ for } r \in (t_k, r_{4k+1}).$$

Inserting this into (3.33) gives:

$$\frac{[u(t_k) - u(r)]}{t_k^2}\left[\frac{t_k^2 F_\alpha(u(t_k))}{[u(t_k) - c_2]} - 2(N-1)^2[u(t_k) - u(r)]\right] \le (u'(r))^2,$$

i.e.

$$\frac{[u(t_k) - u(r)]}{t_k^2}\left[L_k - 2(N-1)^2[u(t_k) - u(r)]\right] \le (u'(r))^2.$$

Now from (3.3) and the remarks preceding (3.3) we see that $0 \le u(t_k) - u(r) \le u(M_k) - c_2 \to 0$ as $k \to \infty$ for $r \in (t_k, r_{4k+1})$ and since $k \in \mathbf{B}$ we have $L_k \ge 2N^2$. Thus, on (t_k, r_{4k+1}) with $k \in \mathbf{B}$ and k large we have:

$$L_k - 2(N-1)^2[u(t_k) - u(r)] \ge 2N^2 - 2(N-1)^2[u(M_k) - c_2] \ge N^2$$

hence

$$\frac{N}{t_k} \le \frac{|u'|}{\sqrt{u(t_k) - u(r)}}.$$

Integrating this on (t_k, r_{4k+1}) gives:

$$\frac{r_{4k+1}}{t_k} \leq 1 + \frac{2}{N}\sqrt{u(t_k) - c_2} \leq 1 + \frac{2}{N}\sqrt{u(M_k) - c_2}$$

and so by (3.3) we have for $k \in \mathbf{B}$:

$$\lim_{k \to \infty} \frac{r_{4k+1}}{t_k} = 1. \tag{3.34}$$

We now estimate $\frac{t_k}{M_k}$.

We begin by rewriting equation (1.2) as:

$$u'' + \frac{N-1}{r}u' + \frac{f(u)}{u - c_2}(u - c_2) = 0$$

and we define v to be the solution of:

$$v'' + \frac{N-1}{r}v' + \frac{f(u(t_k))}{u(M_k) - c_2}(v - c_2) = 0 \quad \text{for} \quad r > M_k,$$

$$v'(M_k) = 0, \; v(M_k) = u(M_k).$$

Since $\frac{f(u)}{u-c_2} \geq \frac{f(u(t_k))}{u(M_k)-c_2}$ for $u \in [u(t_k), u(M_k)]$, it then follows from the *Sturm comparison test* (see [2]) that $u(r) < v(r)$ for $r \in (M_k, t_k)$. Denoting x_k as the first value of $r > t_k$ such that $v(x_k) = u(t_k)$, we then have that: $t_k - M_k \leq x_k - M_k$. The reason we know the number x_k exists is that letting:

$$w(r) = r^{\frac{N-2}{2}}[v(r/\sqrt{\frac{f(u(t_k))}{u(M_k) - c_2}}) - c_2]$$

it follows that w satisfies *Bessel's equation* of order $\frac{N-2}{2}$, i.e.

$$w'' + \frac{1}{r}w' + [1 - \frac{((N-2)/2)^2}{r^2}]w = 0$$

and as is well known, solutions of *Bessel's equation* have an infinite number of zeros. Also the distance between any two consecutive zeros of w is bounded by $c(N)$ where $c(N)$ is a constant depending only on N (see [2]). Since $v > c_2$ on $[M_k, x_k]$ it follows that $w > 0$ on $[M_k\sqrt{\frac{f(u(t_k))}{u(M_k)-c_2}}, x_k\sqrt{\frac{f(u(t_k))}{u(M_k)-c_2}}]$ and by the remarks about *Bessel's equation* we have that:

$$\sqrt{\frac{f(u(t_k))}{u(M_k) - c_2}}[x_k - M_k] \leq c(N).$$

Thus we see that:

$$\sqrt{\frac{f(u(t_k))}{u(M_k) - c_2}}[t_k - M_k] \leq \sqrt{\frac{f(u(t_k))}{u(M_k) - c_2}}[x_k - M_k] \leq c(N).$$

Therefore,

$$\frac{t_k^2 f(u(t_k))}{u(M_k) - c_2}(1 - \frac{M_k}{t_k})^2 \leq [c(N)]^2. \tag{3.35}$$

Now since f is nondecreasing on $(c_2, c_2+\varepsilon_2)$ it follows that $F_\alpha(u) \leq f(u)(u-c_2)$. Therefore for $k \in \mathbf{B}$ we obtain:

$$\frac{t_k^2 f(u(t_k))}{u(M_k)-c_2} \geq \frac{t_k^2 F_\alpha(u(t_k))}{[u(t_k)-c_2][u(M_k)-c_2]} = \frac{L_k}{u(M_k)-c_2} \geq \frac{2N^2}{u(M_k)-c_2} \tag{3.36}$$

hence by (3.3) we see that the right hand side of (3.36) goes to infinity as k goes to infinity and therefore $\frac{t_k^2 f(u(t_k))}{u(M_k)-c_2} \to \infty$ as $k \to \infty$.

Thus from (3.35) and (3.36) we obtain for $k \in \mathbf{B}$:

$$\lim_{k\to\infty} \frac{t_k}{M_k} = 1. \tag{3.37}$$

Multiplying (3.34) and (3.37) we finally obtain for $k \in \mathbf{B}$:

$$\lim_{k\to\infty} \frac{r_{4k+1}}{M_k} = 1. \tag{3.38}$$

Next we observe that

$$(r^{2N-2}E_\alpha(r))' = (2N-2)r^{2N-3}F_\alpha(u(r)).$$

Integrating on $(M_k, r) \subset (M_k, r_{4k+1})$ and using that F_α is increasing on $(c_2, c_2+\varepsilon_2)$ (by (3.32)) gives:

$$r^{2N-2}E_\alpha(r) - M_k^{2N-2}E_\alpha(M_k) = \int_{M_k}^{r} (2N-2)t^{2N-3}F_\alpha(u(t))\,dt$$

$$\geq F_\alpha(u(r))[r^{2N-2} - M_k^{2N-2}].$$

Rewriting this gives:

$$\frac{1}{2}r^{2N-2}(u')^2 \geq M_k^{2N-2}[F_\alpha(u(M_k)) - F_\alpha(u(r))].$$

Thus:

$$\frac{|u'(r)|}{\sqrt{2}\sqrt{F_\alpha(u(M_k)) - F_\alpha(u(r))}} \geq \frac{M_k^{N-1}}{r^{N-1}}.$$

Integrating on (M_k, r_{4k+1}) gives:

$$(\frac{M_k}{r_{4k+1}})^{N-1}(r_{4k+1}-M_k) \leq \int_{M_k}^{r_{4k+1}} \frac{M_k^{N-1}}{r^{N-1}} \leq \frac{1}{\sqrt{2}}\int_{c_2}^{u(M_k)} \frac{1}{\sqrt{F_\alpha(u(M_k)) - F_\alpha(t)}}\,dt.$$

Since F_α is convex (by (3.22)), it follows that:

$$(\frac{M_k}{r_{4k+1}})^{N-1}(r_{4k+1}-M_k) \leq \frac{1}{\sqrt{2}}\frac{\sqrt{u(M_k)-c_2}}{\sqrt{F_\alpha(u(M_k))}}\int_{c_2}^{u(M_k)} \frac{1}{\sqrt{u(M_k)-t}}\,dt.$$

Rewriting gives:

$$(r_{4k+1}-M_k)F_\alpha(u(M_k)) \leq \sqrt{2}(\frac{r_{4k+1}}{M_k})^{N-1}(u(M_k)-c_2)\sqrt{F_\alpha(u(M_k))}. \tag{3.39}$$

Finally, using (3.10), (3.20), and (3.39) we have for $N \geq 3/2$:

$$r_{4k+1}E_\alpha(r_{4k+1}) - r_{4k}E_\alpha(r_{4k}) = \int_{r_{4k}}^{r_{4k+1}} (rE_\alpha)'\,dr$$

$$\leq \int_{r_{4k}}^{r_{4k+1}} F_\alpha(u(r))\,dr \leq (r_{4k+1}-r_{4k})F_\alpha(u(M_k))$$

$$\leq 2(r_{4k+1} - M_k)F_\alpha(u(M_k)) \leq 2\sqrt{2}(\frac{r_{4k+1}}{M_k})^{N-1}(u(M_k) - c_2)\sqrt{F_\alpha(u(M_k))}. \quad (3.40)$$

We know from (3.38) that for $k \in \mathbf{B}$, $\frac{r_{4k+1}}{M_k} \to 1$ as $k \to \infty$ and also from (3.3) and (3.8) it follows that the right hand side of (3.40) approaches zero as $k \to \infty$ with $k \in \mathbf{B}$. Hence, whether $k \in \mathbf{A}$ or $k \in \mathbf{B}$, we obtain from (3.30) and (3.40) that:

$$\limsup_{k\to\infty} \, [r_{4k+1}E_\alpha(r_{4k+1}) - r_{4k}E_\alpha(r_{4k})] \leq 0.$$

This completes the proof of lemma 3.6 for $N \geq 3/2$.

The proof for $1 < N < 3/2$ proceeds precisely as in lemma 3.4 and using identical arguments we can obtain inequality (3.26), i.e.

$$\int_{r_{4k}}^{r_{4k+1}} (u')^2 \, dr \leq 2\sqrt{2}[u(M_k) - c_2]\sqrt{F(d) - \alpha}.$$

Therefore, using (3.30) if $k \in \mathbf{A}$ and (3.40) if $k \in \mathbf{B}$ we obtain respectively:

$$r_{4k+1}E_\alpha(r_{4k+1}) - r_{4k}E_\alpha(r_{4k}) \leq 2\sqrt{2}(\frac{3}{2} - N)[u(M_k) - c_2]\sqrt{F(d) - \alpha} + \frac{2N^2}{r_{4k}}$$

or

$$r_{4k+1}E_\alpha(r_{4k+1}) - r_{4k}E_\alpha(r_{4k}) \leq 2\sqrt{2}(\frac{3}{2} - N)[u(M_k) - c_2]\sqrt{F(d) - \alpha}$$
$$+2\sqrt{2}(\frac{r_{4k+1}}{M_k})^{N-1}(u(M_k) - c_2)\sqrt{F_\alpha(u(M_k))}.$$

Thus from (3.3), (3.8), and (3.38) we see that the right hand sides of both of these tend to 0 as $k \to \infty$.

Thus for $1 < N < 3/2$ we also have:

$$\limsup_{k\to\infty} \, [r_{4k+1}E_\alpha(r_{4k+1}) - r_{4k}E_\alpha(r_{4k})] \leq 0.$$

This completes the proof of lemma 3.6 for $1 < N < 3/2$ and along with the earlier proof for $N \geq 3/2$ we see that this completes the proof of lemma 3.6. □

A very similar proof shows:

LEMMA 3.7. *Suppose $f(c_1) = 0$ and for some $\varepsilon_1 > 0$ that $f(t)$ is nondecreasing on $(c_1 - \varepsilon_1, c_1)$.*
Then

$$\limsup_{k\to\infty} \, [r_{4k+3}E_\alpha(r_{4k+3}) - r_{4k+2}E_\alpha(r_{4k+2})] \leq 0.$$

4. Proof of the main theorem.

PROOF. We assume now that $N > 1$ and that u is a solution of (1.2)-(1.3) such that $\lim_{r\to\infty} u(r) =$ *does not exist.* We know from (3.6) that $f(c_1) \leq 0 \leq f(c_2)$ and we want to show that one of these inequalities must in fact be an equality so we assume by the way of contradiction that $f(c_1) < 0 < f(c_2)$.

Then from lemma 3.3 we have:

$$r_{2k}E_\alpha(r_{2k}) - r_{2k-1}E_\alpha(r_{2k-1}) \le -c_0 \tag{4.1}$$

for some $c_0 > 0$.

Also we see from lemmas 3.4-3.5 that there exists an $n_0 \in \mathbb{N}$ so that for $k \ge n_0$ we have:

$$r_{2k+1}E_\alpha(r_{2k+1}) - r_{2k}E_\alpha(r_{2k}) \le \frac{c_0}{2}. \tag{4.2}$$

Adding (4.1)-(4.2) gives for $k \ge n_0$:

$$r_{2k+1}E_\alpha(r_{2k+1}) - r_{2k-1}E_\alpha(r_{2k-1}) \le -\frac{c_0}{2}. \tag{4.3}$$

Finally summing this from n_0 to n gives:

$$r_{2n+1}E_\alpha(r_{2n+1}) \le r_{2n_0-1}E_\alpha(r_{2n_0-1}) - \frac{c_0}{2}(n+1-n_0) \to -\infty \tag{4.4}$$

as $n \to \infty$ which contradicts the fact that $rE_\alpha(r) > 0$ for all $r > 0$.

Therefore we see that we must have either $f(c_1) = 0$ or $f(c_2) = 0$. This proves the first part of the main theorem.

We now assume that either $f(c_1) = 0$ or $f(c_2) = 0$ but that $f \not\equiv 0$ on $[c_1, c_2]$. If $f(c_1) = 0$ then we also assume that f is nondecreasing on $(c_1 - \varepsilon_1, c_1)$ for some $\varepsilon_1 > 0$ and if $f(c_2) = 0$ then we also assume that f is nondecreasing on $(c_2, c_2 + \varepsilon_2)$ for some $\varepsilon_2 > 0$. Then again (4.1) holds and by lemmas 3.6-3.7, (4.2) also holds and consequently (4.3) and (4.4) hold which again yields a contradiction.

Therefore we see that we must have $f \equiv 0$ on $[c_1, c_2]$ and so this proves the second part of the main theorem.

We finally need to show that if u is a bounded nonconvergent solution of (1.2)-(1.3) and $f \equiv 0$ on $[c_1, c_2]$ with f nondecreasing in a neighborhood of c_1 and c_2 then we must have $1 < N \le 2$. So again we assume by the way of contradiction that $N > 2$ and we let

$$J = \frac{1}{2}[ru' + (N-2)(u - (c_1+c_2)/2)]^2 + r^2F_1(u) \tag{4.5}$$

where $F_1(u) = \int_{(c_1+c_2)/2}^{u} f(t)\,dt$. Then using (1.2) we obtain:

$$\begin{aligned} J' &= [ru' + (N-2)(u - (c_1+c_2)/2)](-rf(u)) + 2rF_1(u) + r^2 f(u)u' \\ &= r[2F_1(u) - (N-2)(u - (c_1+c_2)/2))f(u)]. \end{aligned}$$

Since $f \equiv 0$ on $[c_1, c_2]$ and f is nondecreasing on $(c_2, c_2 + \varepsilon_2)$ we obtain:

$$F_1(u) = \int_{(c_1+c_2)/2}^{u} f(t)\,dt = \int_{c_2}^{u} f(t)\,dt \le f(u)(u - c_2). \tag{4.6}$$

Therefore,

$$2F_1(u) - (N-2)(u - (c_1+c_2)/2))f(u) \le f(u)[2(u-c_2) - (N-2)(u-(c_1+c_2)/2))] \le 0$$

for $u \in (c_2, c_2 + \varepsilon)$ for some $\varepsilon > 0$. Similarly one can show that

$$2F_1(u) - (N-2)(u - (c_1+c_2)/2))f(u) \le 0$$

on $(c_1-\varepsilon, c_1)$ for some perhaps different $\varepsilon > 0$. Choosing the smaller of these and using the fact that $f \equiv 0$ on $[c_1, c_2]$, it follows that:

$$2F_1(u) - (N-2)(u-(c_1+c_2)/2))f(u) \le 0 \text{ on } (c_1-\varepsilon, c_2+\varepsilon) \tag{4.7}$$

for some $\varepsilon > 0$. Now by assumption $\lim_{r\to\infty} u(r) =$ *does not exist* so $c_1 = \liminf_{r\to\infty} u(r) < \limsup_{r\to\infty} u(r) = c_2$. Thus there is an r_0 such that for $r \ge r_0$ we have $c_1 - \frac{\varepsilon}{2} < u(r) < c_2 + \frac{\varepsilon}{2}$ Therefore we see from (4.5) and (4.7) that:

$$J' \le 0 \quad \text{for} \quad r \ge r_0.$$

Thus there exists an $L \ge 0$ (since both terms in (4.5) are nonnegative) such that $\lim_{r\to\infty} J(r) = L$. Consider the interval (r_{4k}, r_{4k+1}) (where the r_k are similarly defined as in section 3). We claim that $ru' + (N-2)(u-(c_1+c_2)/2) > 0$ for $r \in [r_{4k}, r_{4k+1})$. This is certainly true for $r = r_{4k}$ since $u'(r_{4k}) > 0$ and $u(r_{4k}) = c_2$. Also $(ru' + (N-2)(u-(c_1+c_2)/2))' = -rf(u) \le 0$ on (r_{4k}, r_{4k+1}). Now if there is an $p_k \in (r_{4k}, r_{4k+1})$ for which $p_k u'(p_k) + (N-2)(u(p_k) - (c_1+c_2)/2) = 0$, then

$$J(p_k) = p_k^2 F_1(u(p_k)). \tag{4.8}$$

On the other hand, since the function $r^2F_1(u)$ is zero at r_{4k} and r_{4k+1} and $F_1(u(r)) > 0$ for $r \in (r_{4k}, r_{4k+1})$, the function $r^2F_1(u)$ must have a maximum at some point $q_k \in (r_{4k}, r_{4k+1})$. It follows then from calculus that:

$$2q_kF_1(u(q_k)) + q_k^2 f(u(q_k))u'(q_k) = 0.$$

Substituting this into (4.5) we see that:

$$J(q_k) = \frac{1}{2}[-\frac{2F_1(u(q_k)}{f(u(q_k))} + (N-2)(u(q_k)-(c_1+c_2)/2)]^2 + q_k^2F_1(u(q_k)). \tag{4.9}$$

Since $\lim_{r\to\infty} J(r) = L$ we see by subtracting (4.8) from (4.9) that we must have:

$$\lim_{k\to\infty} \frac{1}{2}[-\frac{2F_1(u(q_k)}{f(u(q_k))} + (N-2)(u(q_k)-(c_1+c_2)/2)]^2 + [q_k^2F_1(u(q_k)) - p_k^2F_1(u(p_k))] = 0. \tag{4.10}$$

Since $q_k^2F_1(u(q_k)) - p_k^2F_1(u(p_k)) \ge 0$ (by definition of q_k,) we see then that we must have

$$\lim_{k\to\infty} [-\frac{2F_1(u(q_k)}{f(u(q_k))} + (N-2)(u(q_k)-(c_1+c_2)/2)] = 0. \tag{4.11}$$

Using (4.6) and (3.3) we see that:

$$0 \le \frac{F_1(u(q_k)}{f(u(q_k))} \le u(q_k) - c_2 \le u(M_k) - c_2 \to 0 \text{ as } k \to \infty. \tag{4.12}$$

Now taking limits in (4.11) and using (4.12) we see that:

$$(N-2)(c_2-c_1)/2 = 0.$$

But $N > 2$ and $c_1 < c_2$ hence we obtain a contradiction.

Therefore, we must have:

$$ru' + (N-2)(u-(c_1+c_2)/2) > 0 \quad \text{on} \quad [r_{4k}, r_{4k+1}).$$

Similarly we can show that:

$$ru' + (N-2)(u - (c_1+c_2)/2) < 0 \quad \text{on} \quad [r_{4k+2}, r_{4k+3}).$$

Thus by the intermediate value theorem there exists an $r_0 \in [r_{4k+1}, r_{4k+2})$ such that:

$$r_0 u'(r_0) + (N-2)(u(r_0) - (c_1+c_2)/2) = 0. \tag{4.13}$$

Further, since $f \equiv 0$ on $[c_1, c_2]$, it follows that:

$$[ru' + (N-2)(u - \frac{c_1+c_2}{2})]' \equiv 0 \text{ on } [r_{4k+1}, r_{4k+2}] \tag{4.14}$$

and so (4.13) and (4.14) imply:

$$ru' + (N-2)(u - \frac{c_1+c_2}{2}) \equiv 0 \text{ on } [r_{4k+1}, r_{4k+2}].$$

Thus evaluating this at $r = r_{4k+2}$ gives:

$$r_{4k+2}u'(r_{4k+2}) + (N-2)(\frac{c_1-c_2}{2}) = 0.$$

On the other hand $r_{4k+2}u'(r_{4k+2}) + (N-2)(\frac{c_1-c_2}{2}) < 0$ since $u'(r_{4k+2}) < 0$, $c_1 < c_2$, and $N > 2$. Therefore, we obtain a contradiction and so we see that we must have $1 < N \leq 2$. This completes the proof of the main theorem. □

5. Examples of f's for which every solution converges.

In this section we give examples of some nonlinearities f such that *every* solution of (1.2)-(1.3) converges as $r \to \infty$. We first prove the following:

LEMMA 5.1. *If u is a nonconvergent solution of (1.2)-(1.3) and f is infinitely differentiable at c_1 and c_2 then either $f^{(k)}(c_1) = 0$ for every $k \in \mathbb{N}$ or $f^{(k)}(c_2) = 0$ for every $k \in \mathbb{N}$.*

PROOF. If $f \equiv 0$ on $[c_1, c_2]$ and f is infinitely differentiable at c_1 and c_2 then the result clearly holds and so we assume that $f \not\equiv 0$ on $[c_1, c_2]$. From the main theorem we know that either $f(c_1) = 0$ or $f(c_2) = 0$. So let us suppose that $f(c_2) = 0$ and that $f(c_1) \neq 0$. In addition suppose that $f^{(n_0)}(c_2) \neq 0$ but that $f^{(n)}(c_2) = 0$ for $0 \leq n < n_0$ for some $n_0 \in \mathbb{N}$. Since u is a bounded nonconvergent solution of (1.2)-(1.3) it follows that (3.5) holds, i.e. $f(u)$ attains some positive values to the right of c_2. It then follows from the mean value theorem that we must have:

$$f^{(n_0)}(c_2) > 0.$$

However, if this holds, then $f(c_1) < 0$ and $f(c_2) = 0$ with $f(u)$ nondecreasing in a neighborhood of c_2. But for this type of f the main theorem applies and it implies that $f \equiv 0$ on $[c_1, c_2]$ which is a contradiction. Thus, $f^{(n)}(c_2) = 0$ for every $n \in \mathbb{N}$.

Similarly, if $f(c_2) > 0$ and $f(c_1) = 0$ we can show that $f^{(n)}(c_1) = 0$ for every $n \in \mathbb{N}$.

Finally, if $f(c_1) = f(c_2) = 0$ and $f \not\equiv 0$ on $[c_1, c_2]$ a similar argument shows that f and all of its derivatives must vanish at either c_1 or c_2. This completes the proof of the lemma.

COROLLARY. *If $f(u)$ is a polynomial or is real analytic then every solution of (1.2)-(1.3) converges as $r \to \infty$.*

PROOF. If $f(u) \equiv 0$ then the result holds as can be proven directly by computation. If $f(u) \not\equiv 0$ and if there is a nonconvergent solution then from lemma 5.1, we see that f and all of its derivatives must be zero at either c_1 or c_2. But this is impossible for a nontrivial polynomial or nontrivial real analytic function. Thus for these types of f's, every solution of (1.2)-(1.3) must converge as $r \to \infty$.

NOTE: It would be interesting to investigate for what other types of equations a result similar to the main theorem holds. We suspect such a result for the p-Laplacian equation and would be interested in investigating the nonconverging solution phenomenom for other partial differential equations.

References

1. S. Maier, *Non-convergent radial solutions of semilinear elliptic equations*, Asymp. Analy. **8** (1994), 363–377.
2. G. Birkoff and G.C. Rota, *Ordinary Differential Equations*, Ginn and Company, 1962.

DEPARTMENT OF MATHEMATICS UNIVERSITY OF NORTH TEXAS DENTON TEXAS 76203
E-mail address: iaia@unt.edu

Contemporary Mathematics
Volume **357**, 2004

Traveling Wave Solutions to Nonlinear Evolution Equations

Zhaosheng Feng

ABSTRACT. A lot of physical phenomena in one- or higher-dimensional space can be described by nonlinear evolution equations. One of basic physical problems is to find traveling solitary wave solutions to nonlinear evolution equations, which play an important role in theoretical physics. Since many nonlinear differential equations can be reduced to ordinary differential equations such as the Lienard equation, to study those ordinary differential equations is of significance not only in mathematics itself, but also in physics. In this paper, a kind of exact solutions to the Lienard equation is obtained by using the method of variation parameters, and the applications of the result to the Burgers-KdV equation, the combined dissipative double-dispersive equation and the approximate sine-Gordon equation are presented.

1. Introduction

The last few decades have seen an enormous growth of the applicability of nonlinear models and of the development of related nonlinear concepts. This has been driven by modern computer power as well as by the discovery of new mathematical techniques, which include two contrasting themes: (i) the theory of dynamical systems, most popularly associated with the study of chaos, and (ii) the theory of integrable systems associated, among other things, with the study of solitons. However, not all systems arising from physical phenomena are integrable [1], for example, the Burgers-Korteweg-de Vries equation. Therefore, the direct methods together with new mathematical techniques to treat such kind of nonlinear systems appear to be

1991 *Mathematics Subject Classification.* Primary 34A05, 34C05; Secondary 35B05, 35Q53.

Key words and phrases. solitary wave, sine-Gordon equation, equilibrium point, variation parameters, orbit.

The author gratefully acknowledges partial support from the organizer (Dr. John M. Neuberger) of this conference.

more powerful and important. Since most of those nonlinear systems can be changed into the first- or higher-order nonlinear ordinary differential equations such as the Lienard equation through traveling wave transformation et al., to solve those nonlinear systems is somewhat equivalent to treating the corresponding first- or higher-order ordinary differential equations. Applications of nonlinear models range from atmospheric science to condensed matter physics and to biology, from the smallest scales of theoretical particle physics up to the largest scales of cosmic structure.

Consider the Lienard equation

$$u''(\xi) - ru'(\xi) - au^2(\xi) - bu(\xi) - d = 0, \tag{1}$$

where r, a and b are real constants, and satisfy $r \cdot a \cdot b \neq 0$. It has been pointed out that it is very difficult for us to find the explicit exact solution to eq. (1) by usual ways [2]. Nevertheless, studying eq. (1) is of physical significance. Many nonlinear differential equations, which arise from various physical phenomena, can be converted into (1) through traveling wave transformation, such as the Burgers-KdV equation [3-14]

$$u_t + \alpha u u_x + \beta u_{xx} + s u_{xxx} = 0, \tag{2}$$

where α, β and s are real constants with $\alpha \cdot \beta \cdot s \neq 0$; and the two-dimensional Burgers-KdV equation [15-19]

$$(u_t + \alpha u u_x + \beta u_{xx} + s u_{xxx})_x + \gamma u_{yy} = 0, \tag{3}$$

where α, β, s and γ are real constants; and the combined dissipative double-dispersive equation [20]

$$u_{tt} - \alpha_1 u_{xx} - \alpha_2 u_{xxt} - \alpha_3 (u)^2_{xx} - \alpha_4 u_{xxxx} + \alpha_5 u_{xxtt} = 0, \tag{4}$$

where α_i $(i = 1, 2, 3, 4, 5)$ are real constants; and the approximate sine-Gordon equation [21]

$$u_{tt} + r_1 u_t - r_2 \Delta u_{(n)} + u - \frac{1}{6}u^3 = d_3 u^3 + d_2 u^2 + d_1 u + d_0, \tag{5}$$

where r_1, r_2 and d_i $(i = 0, 1, 2, 3)$ are all real constants.

Eq. (3) is a two-dimensional generalization of the Burgers-Korteweg-de Vries equation (2), which is derived by Johnson as the governing nonlinear equation for the propagation of waves on an elastic tube filled with a viscous fluid [3]. Wijngaarden and Gao et al. used it as a nonlinear model in the flow of liquids containing gas bubbles [4] and turbulence [5,6]. Grad and Hu used a steady state version of (2) to describe a weak shock profile in plasmas [7]. The exact solutions to eq. (2) were obtained by various methods [8-14], but all these solutions are equivalent one another. Eq. (4) is used as the governing nonlinear equation for the propagation of the longitudinal strain kinks in an elastic rod due to its interaction with a viscoelastic external medium or cubic nonlinearity of the elastic materials [20]. Eq. (5) can be regarded as an approximate equation to the damped sine-Gordon equation as follows in the case where $|u|$ is very small [21]

$$u_{tt} + r_1 u_t - r_2 \Delta u_{(n)} + \sin u = f. \tag{6}$$

Since when f is a nonzero physical function, by the cubic spline interpolation, we can use a cubic polynomial $g = d_3 u^3 + d_2 u^2 + d_1 u + d_0$ approximating it. Eq. (6) describes the dynamical behavior of a continuous Josephson junction driven by a current source by taking into account of damping effect [22].

In our previous work [23], we proposed a new approach called the first integral method to study nonlinear differential equations. In our last paper [24] we applied it to obtain an exact solution to eq. (1) as follows

$$u(\xi) = \frac{3a}{2} \cdot \left[\frac{\pm\frac{2r}{5a} e^{\frac{r}{5}\xi}}{e^{\frac{r}{5}\xi} + c}\right]^2 - \frac{(k+1)b}{2a}, \tag{7}$$

where $k = \frac{6r^2}{25b}$ and c is an arbitrary integration constant. Motivated by the form (7), In the present paper, we study eq. (1) for the more general exact solution by utilizing the method of variation parameters. By means of this solution, traveling solitary wave solutions for eqs. (2)-(5) are presented, which extend the corresponding results in the previous literature.

The rest of the paper is organized as follows. In Section 2, applying the qualitative theory of ordinary differential equations, a qualitative analysis to a two-dimensional plane autonomous system is presented, which indicates that under some appropriate conditions, eq. (1) has the nontrivial bounded solution. In Section 3, a more general exact solution to eq. (1) is obtained by using the method of variation parameters, and traveling solitary wave solutions for eqs. (2)-(5) are expressed in terms of this solution. At the end of this section, the solutions of eqs. (2)-(5) obtained in the previous literature are compared with ours. In Section 4, we give a brief conclusion and pose an open question.

2. A Qualitative Analysis to the Plane Autonomous System (8)

Letting $x = u, y = u_\xi$, eq. (1) is equivalent to

$$\begin{cases} \dot{x} = y = P(x,y) \\ \dot{y} = ry + ax^2 + bx + d = Q(x,y). \end{cases} \tag{8}$$

Notice that (8) is a two-dimensional plane autonomous system, and $P(x,y)$, $Q(x,y)$ satisfy the conditions of the uniqueness and existence theorem ([25]). The integral orbits of autonomous system in the Poincare phase plane depict graphically the types of motions determined by eq. (1). Note that the orbits point to the right, to the left, or are vertical according as $\dot{x} > 0$ (upper half-plane), $\dot{x} < 0$ (lower half-plane), or $\dot{x} = 0$ (x-axis). This is because x is increasing, decreasing, or stationary in these three cases, respectively. For any solution $u = u(\xi)$ of eq. (1), if $u(\xi) \equiv u_0$ for $\xi \in (-\infty, \infty)$ (u_0 is a constant), the corresponding integral curve to $u(\xi) = u_0$ in the (ξ, u)-plane is a line which is parallel to ξ-axis, and the associated orbit with $u(\xi) = u_0$ in the Poincare phase plane is a point $u_0(x_0, y_0)$ which is usually called an equilibrium point (or critical point).

Physically, for a given plane autonomous system, an equilibrium point in the Poincare phase plane corresponds to a static state. If $u_0(x_0, y_0)$ is an equilibrium point of (8), then any orbit except $u_0(x_0, y_0)$ itself can not approach $u_0(x_0, y_0)$ within finite time. Vice versa, if an orbit of (8) approaches $u_0(x_0, y_0)$ as $\xi \to \infty$ (or $-\infty$), then $u_0(x_0, y_0)$ must be an equilibrium point of (8). If (8) has an isolated and closed orbit without any equilibrium point on it in the Poincare phase plane, then this closed orbit represents a periodic oscillation to eq. (1) in the (ξ, u)-plane. If the isolated and closed orbit emanates from an equilibrium point and also terminates at the same equilibrium point as $\xi \to \infty$ (or $-\infty$), then this closed orbit represents a bell solitary wave solution to eq. (1) in the (ξ, u)-plane (see Figure 1). Without

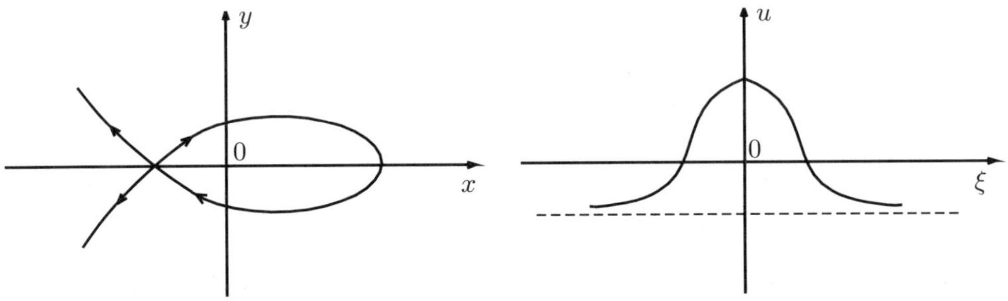

FIGURE 1. An isolated and closed orbit with an equilibrium point in the Poincare phase plane represents a bell solitary wave solution in the (ξ, u)-plane.

loss of generality, we suppose that

$$a < 0, \qquad r > 0, \qquad b^2 - 4ad > 0. \tag{9}$$

The arguments for other cases are closely similar. Under the assumption (9), (8) has two equilibrium points

$$P(-\frac{b}{2a} + \frac{\sqrt{b^2 - 4ad}}{2a}, 0), \qquad Q(-\frac{b}{2a} - \frac{\sqrt{b^2 - 4ad}}{2a}, 0).$$

The coefficient matrices of the linearizing systems with respect to P and Q are as follows, respectively,

$$\begin{aligned}\mathbf{M_P} &= \left.\begin{pmatrix} P'_x(x,y) & P'_y(x,y) \\ Q'_x(x,y) & Q'_y(x,y) \end{pmatrix}\right|_P \\ &= \begin{pmatrix} 0, & 1 \\ \sqrt{b^2-4ad}, & r \end{pmatrix},\end{aligned}$$

and

$$\begin{aligned}\mathbf{M_Q} &= \left.\begin{pmatrix} P'_x(x,y) & P'_y(x,y) \\ Q'_x(x,y) & Q'_y(x,y) \end{pmatrix}\right|_Q \\ &= \begin{pmatrix} 0, & 1 \\ -\sqrt{b^2-4ad}, & r \end{pmatrix}.\end{aligned}$$

Let λ_1, λ_2 designate two eigenvalues of $\mathbf{M_P}$, and μ_1, μ_2 designate two eigenvalues of $\mathbf{M_Q}$, then we have

$$\begin{aligned}\lambda_1 &= \frac{r}{2} + \frac{1}{2}\sqrt{r^2 + 4\sqrt{b^2-4ad}} \\ \lambda_2 &= \frac{r}{2} - \frac{1}{2}\sqrt{r^2 + 4\sqrt{b^2-4ad}},\end{aligned}$$

and

$$\begin{aligned}\mu_1 &= \frac{r}{2} + \frac{1}{2}\sqrt{r^2 - 4\sqrt{b^2-4ad}} \\ \mu_2 &= \frac{r}{2} - \frac{1}{2}\sqrt{r^2 - 4\sqrt{b^2-4ad}}.\end{aligned}$$

Note that when (9) holds, P is a saddle. Furthermore, if $r^2 > 4\sqrt{b^2-4ad}$, then Q is an unstable node; if $r^2 < 4\sqrt{b^2-4ad}$, then Q is an unstable focus.

Now, we consider the local behavior at the infinite equilibrium point if it exists. Through the Poincare transformation:

$$x = \frac{1}{z}, \qquad y = \frac{u}{z}, \qquad d\tau = \frac{dt}{z}, \qquad (z \neq 0),$$

(8) becomes

$$\left\{\begin{aligned} \frac{du}{d\tau} &= a + bz + dz^2 + ruz - u^2 z \\ \frac{dz}{d\tau} &= -uz^2. \end{aligned}\right. \tag{10}$$

There is no equilibrium point on u-axis because of $a \neq 0$. This implies that u-axis is an orbit of (10).

On the other hand, through the Poincare transformation:

$$x = \frac{w}{z}, \qquad y = \frac{1}{z}, \qquad d\tau = \frac{dt}{z}, \qquad (z \neq 0),$$

(8) becomes

$$\begin{cases} \frac{dw}{d\tau} = z - rwz - aw^3 - bw^2z - dwz^2 \\ \frac{dz}{d\tau} = -rz^2 - aw^2z - bwz^2 - dz^3. \end{cases} \tag{11}$$

The origin is the unique equilibrium point of (11). This implies that (8) has two infinite equilibrium points E and F in y-axis. Furthermore, by analyzing the linearization of system (11), we can see that E is a local source point, and F is a local sink.

Since $\frac{\partial P(x,y)}{\partial x} + \frac{\partial Q(x,y)}{\partial y} = r > 0$, using the Bendixson Theorem ([25]), (8) has no closed orbit in the Poincare phase plane. That is, eq. (1) neither has bell solitary wave solution, nor has periodic traveling wave solution. The global behavior of (8) is depicted in Figure 2. From Figure 2, we can con-

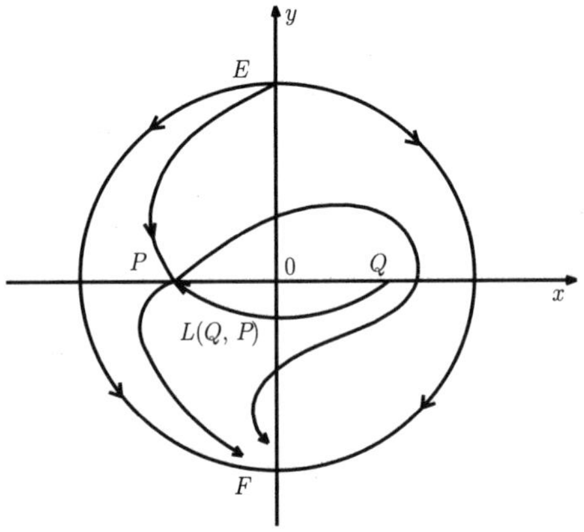

FIGURE 2. The global behavior to the plane autonomous system (8) when (9) holds.

clude that both x and y-coordinates of each point which lies on all orbits except P, Q and $L(Q,P)$ are unbounded. Notice that except the equilibrium points P, Q and the orbit $L(Q,P)$, all other orbits in the Poincare phase plane either emanate from the infinite equilibrium point or approach the infinite equilibrium point as $\xi \to +\infty$. This implies that the y-coordinates of the points which lie on the orbits except P, Q and $L(Q,P)$ must be unbounded. We now show that the x-coordinates of those points which lie on the same orbits must be unbounded either. By way of contradiction, assume that there exists a positive number δ such that $|x| < \delta$ as $y \to \infty$. By the Mean-value Theorem, $\frac{dy}{dx}$ is unbounded.

On the other hand, since the slope of the tangent line to those orbits can be expressed

$$\frac{dy}{dx} = r + \frac{ax^2 + bx + d}{y}. \tag{12}$$

(12) implies that $\frac{dy}{dx} \to r$ as $y \to \infty$. This yields a contradiction. Consequently, the proof of the above conclusion is complete.

Since the plane autonomous system (8) is equivalent to eq. (1), each nontrivial bounded traveling wave solution $u = u(\xi)$ of eq. (1) in the (ξ, u)-plane corresponds to an orbit of (8) in the Poincare phase plane, on which the x-coordinate of each point is bounded. By the above analysis, the unique orbit which satisfies this requirement under the assumption (9) is $L(Q, P)$. Namely, $u(\xi) \to x_P$ as $\xi \to +\infty$ and $u(\xi) \to x_Q$ as $\xi \to -\infty$, where x_P is the x-coordinate of P and x_Q is the x-coordinate of Q. Thus, we have

$$\begin{aligned} u(-\infty) &= \lim_{\xi \to -\infty} u(\xi) = -\tfrac{b}{2a} - \tfrac{\sqrt{b^2-4ad}}{2a}, \\ u(+\infty) &= \lim_{\xi \to +\infty} u(\xi) = -\tfrac{b}{2a} + \tfrac{\sqrt{b^2-4ad}}{2a}. \end{aligned}$$

This means that the quotient of a and b is determined by the sum of the limits of the traveling wave solution $u(\xi)$ as ξ goes to $\pm\infty$. That is,

$$\frac{b}{a} = -[u(-\infty) + u(+\infty)].$$

Furthermore, utilizing the qualitative theory of ordinary differential equations, we can conclude that when $r^2 > 4\sqrt{b^2 - 4ad}$, the traveling wave solution $u(\xi)$ of eq. (1) is strictly monotone decreasing with respect to ξ. To prove this, we only need to show that $u'(\xi) \neq 0$ for any $\xi \in \mathbb{R}$ because of the fact $u(+\infty) < u(-\infty)$. That is, except P and Q, $L(Q, P)$ does not intersect with x-axis.

Consider the autonomous system (8) in the Poincare phase plane. Letting $Q(x, y) = 0$, we obtain that the parabola

$$y = -\frac{1}{r}(ax^2 + bx + d),$$

which is the trajectory on which each orbit points to the left or right. Denote the vertex of the parabola by $V(-\frac{b}{2a}, \frac{b^2}{4ar})$. Construct two lines l_1 and l_2. l_1 is passing through V and parallel to the x-axis, and l_2 is passing through Q with the slope $K = \frac{2}{r}\sqrt{b^2 - 4ad}$, i.e.,

$$l_2: \quad y = \frac{2}{r}\sqrt{b^2 - 4ad}\left(x + \frac{b}{2a} + \frac{\sqrt{b^2 - 4ad}}{2a}\right). \tag{13}$$

Suppose that T is the intersection point of l_1 and l_2. Denote x-coordinate of T by x_T. Immediately we have the following

$$-\frac{b}{2a} < x_T < -\frac{b}{2a} - \frac{\sqrt{b^2 - 4ad}}{2a}. \tag{14}$$

Denote by Ω the domain bounded by line segments PQ, QT, TV and the curve $\overset{\frown}{VP}$. It suffices to prove that any integral orbit which starts from a point outside the domain Ω, can not enter the domain Ω as $\xi \to \pm\infty$. It is obvious that there is no integral orbit entering Ω through PQ and the curve $\overset{\frown}{VP}$, because all orbits at each point of x-axis between P and Q are orthogonally along the direction of positive y-axis and all orbits at each point of the curve $\overset{\frown}{VP}$ point to the left. Next, we only need to prove that each orbit at the points of the line segments QT and TV points outward. By (12) and (13), the slope of the tangent line to the integral orbit at the intersection point of the integral orbit itself and l_2 is

$$\begin{aligned} \frac{dy}{dx}\Big|_{(x,y)\in l_2} &= r + a\cdot \frac{(x+\frac{b}{2a}-\frac{\sqrt{b^2-4ad}}{2a})(x+\frac{b}{2a}+\frac{\sqrt{b^2-4ad}}{2a})}{y}\Big|_{(x,y)\in l_2} \\ &= r + \frac{ar\cdot(x+\frac{b}{2a}-\frac{\sqrt{b^2-4ad}}{2a})}{2\sqrt{b^2-4ad}}. \end{aligned}$$

Since $x_T \le x \le -\frac{b}{2a} - \frac{\sqrt{b^2-4ad}}{2a}$, using (14), then we have

$$\begin{aligned} \frac{dy}{dx}\Big|_{(x,y)\in QT} &\ge r - r\cdot \max\Big|\frac{a\cdot(x+\frac{b}{2a}-\frac{\sqrt{b^2-4ad}}{2a})}{2\sqrt{b^2-4ad}}\Big| \\ &> r - \frac{r}{2} = \frac{r}{2}. \qquad (15) \end{aligned}$$

By (15) and our previous assumption $r^2 > 4\sqrt{b^2-4ad}$, then we obtain

$$\frac{dy}{dx}\Big|_{(x,y)\in QT} > \frac{r}{2} > \frac{2}{r}\sqrt{b^2-4ad} = K.$$

This implies that except at Q, each orbit at the points of line segment QT

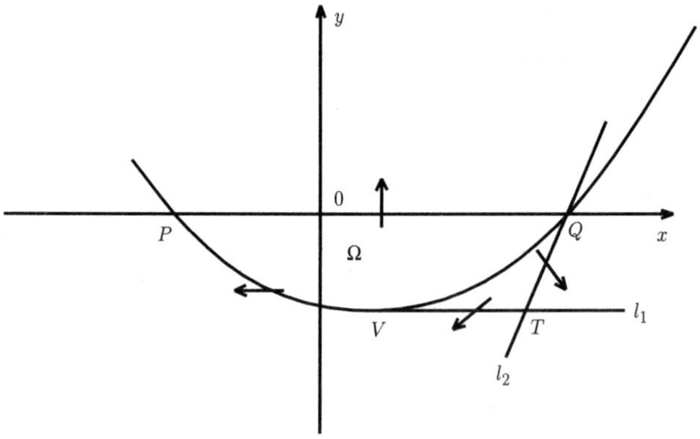

FIGURE 3. The domain Ω bounded by $PQTVP$.

points outward.

The argument for VT is closely similar to the above, so we omit it. Therefore, the orbit $L(Q,P)$ must lie inside the domain Ω. In other words, except the end points P and Q, the orbit $L(Q,P)$ can not intersect with

x-axis. Since $u(+\infty) < u(-\infty)$, we obtain that the traveling wave solution $u(\xi)$ to eq. (1) is strictly monotone decreasing with respect to ξ (see Figure 3).

3. Exact Solution to Eq. (1) and Its Applications

In our last paper [24], an exact solution in the form (7) to eq. (1) was obtained by using a new approach called the first integral method. Notice that the first part of (7) is the product of $e^{\frac{2r}{5}\xi}$ and $\frac{6r^2}{25a}\cdot(e^{\frac{r}{5}\xi}+c)^{-2}$. Motivated by this important fact, we are going to apply the method of variation parameters to study eq. (1) for a more general exact solution. Assume that eq. (1) has the solution in the following more general form

$$u(\xi) = \eta\cdot e^{k_0\xi}\cdot\rho(k_1\cdot e^{k_2\xi}+k_3)+k_4, \tag{16}$$

where η and k_i $(i=0,1,2,4)$ are constants which may depend on r,a,b and d, k_3 is an arbitrary constant, and ρ is a function to be determined. Then, we have

$$\left\{\begin{array}{lcl} u'(\xi) & = & \eta k_0\cdot e^{k_0\xi}\cdot\rho(k_1\cdot e^{k_2\xi}+k_3)+\eta k_1k_2\cdot e^{(k_0+k_2)\xi} \\ & & \cdot\rho'(k_1\cdot e^{k_2\xi}+k_3) \\ u''(\xi) & = & \eta k_0^2\cdot e^{k_0\xi}\cdot\rho(k_1\cdot e^{k_2\xi}+k_3)+\eta k_1k_2[k_2+2k_0]\cdot e^{(k_2+k_0)\xi} \\ & & \cdot\rho'(k_1\cdot e^{k_2\xi}+k_3)+\eta k_1^2k_2^2\cdot e^{(2k_2+k_0)\xi}\cdot\rho''(k_1\cdot e^{k_2\xi}+k_3). \end{array}\right. \tag{17}$$

Substituting (16) and (17) into eq. (1), we have

$$\begin{aligned} &\eta k_1^2k_2^2\cdot e^{(2k_2+k_0)\xi}\cdot\rho''+\eta k_1k_2\cdot[k_2+2k_0-r]\cdot e^{(k_2+k_0)\xi}\cdot\rho' \\ &-a\eta^2\cdot e^{2k_0\xi}\cdot\rho^2-(2ak_4+b+k_0r-k_0^2)\eta\cdot e^{k_0\xi}\cdot\rho \\ &-(ak_4^2+bk_4+d)=0. \end{aligned} \tag{18}$$

Setting $2k_2+k_0=2k_0$, we get $k_2=\frac{1}{2}k_0$. Let $k_0=\frac{2r}{5}$, then the term containing the first derivative ρ' disappears. Setting the coefficient of $e^{k_0\xi}\cdot\rho$ to zero, we then have $k_4=-\frac{b}{2a}-\frac{3r^2}{25a}$. Thus, when $d=-ak_4^2-bk_4=\frac{b^2}{4a}-\frac{9r^4}{625a}$, (18) becomes

$$\rho''=\frac{25a\eta}{r^2k_1^2}\rho^2. \tag{19}$$

This implies that while ρ is the Weierstrass elliptic function with a condition $g_2=0$ satisfying (19), then (1) has the exact solution as follows

$$u(\xi)=\eta\cdot e^{\frac{2r}{5}\xi}\cdot\rho\big(k_1e^{\frac{r}{5}\xi}+k_3\big)-\frac{b}{2a}-\frac{3r^2}{25a}. \tag{20}$$

Note that (20) is a new solution to eq. (1). In fact, it is the general solution to eq. (1) in the case where $d=\frac{b^2}{4a}-\frac{9r^4}{625a}$. It is easy to see that (7) is a particular case of (20) while choosing $k_1=1$ and $\eta=\frac{6r^2}{25a}$, since in this case $\rho(y) = y^{-2}$ is a particular solution of (19). Notice that (20) contains the Weierstrass elliptic function. There are many textbooks introducing and describing the Weierstrass elliptic functions in details such as references 26-28 et al.. We refer the reader to one of those textbooks or any handbook of mathematics to get more information about (20).

Now, let us study the traveling wave solutions for eqs. (2)-(5) by means of (20). Assume that eq. (2) has the solution in the form

$$u(x,t) = u(\xi), \quad \xi = x - vt, \quad (v \in \mathbb{R}). \tag{21}$$

Substituting (21) into eq. (2) and performing one integration then yield (1), where $r = -\frac{\beta}{s}$, $a = -\frac{\alpha}{2s}$, $b = \frac{v}{s}$ and d is integration constant. Thus, we can obtain that eq. (2) has the exact solution as follows immediately

$$u(x,t) = \eta \cdot e^{-\frac{2\beta}{5s}(x-vt)} \cdot \rho(k_1 e^{-\frac{\beta}{5s}(x-vt)} + k_3) + \frac{6\beta^2}{25\alpha s} + \frac{v}{\alpha}, \tag{22}$$

where k_3 is an arbitrary constant, and ρ is the Weierstrass elliptic function satisfying (19) with the coefficient in terms of η and k_1.

In particular, suppose that $k = \frac{6r^2}{25b}$ (i.e., $k = \frac{6\beta^2}{25vs}$), and set $k_1 = 1$ and $\eta = \frac{6r^2}{25a}$ in (22). Since $\rho(y) = y^{-2}$ is a particular solution of (19), we obtain that eq. (2) has a particular solution

$$u(x,t) = -\frac{12\beta^2}{25\alpha s} \cdot \frac{e^{-\frac{2\beta}{5s}(x-vt)}}{[e^{-\frac{\beta}{5s}(x-vt)} + k_3)]^2} + \frac{6(k+1)\beta^2}{25ks\alpha}, \tag{23}$$

where k_3 is an arbitrary constant. Note that in the case $k = 1$ and $k_3 = 1$ in (23), by using the equality $4A[\frac{e^{2t}}{1+e^{2t}}]^2 = -A\text{sech}^2 t + 2A\tanh t + 2A$, the traveling solitary wave solution to eq. (2) can be rewritten as

$$\begin{aligned} u(x,t) &= \frac{3\beta^2}{25\alpha s}\text{sech}^2\Big[\frac{1}{2}\big(-\frac{\beta}{5s}x + \frac{6\beta^3}{125s^2}t\big)\Big] \\ &\quad -\frac{6\beta^2}{25\alpha s}\tanh\Big[\frac{1}{2}\big(-\frac{\beta}{5s}x + \frac{6\beta^3}{125s^2}t\big)\Big] + \frac{6\beta^2}{25\alpha s}. \end{aligned} \tag{24}$$

In the case of $k = -1$ and $k_3 = 1$ in (23), the traveling solitary wave solution to eq. (2) can be rewritten as

$$\begin{aligned} u(x,t) &= \frac{3\beta^2}{25\alpha s}\text{sech}^2\Big[\frac{1}{2}\big(-\frac{\beta}{5s}x - \frac{6\beta^3}{125s^2}t\big)\Big] \\ &\quad -\frac{6\beta^2}{25\alpha s}\tanh\Big[\frac{1}{2}\big(-\frac{\beta}{5s}x - \frac{6\beta^3}{125s^2}t\big)\Big] - \frac{6\beta^2}{25\alpha s}. \end{aligned} \tag{25}$$

It is easy to see that the choices $\alpha \neq 0$, $\beta \neq 0$, and $s = 0$ lead eq. (2) to the Burgers' equation

$$u_t + \alpha u u_x + \beta u_{xx} = 0,$$

and the choices $\alpha \neq 0$, $\beta = 0$, and $s \neq 0$ lead eq. (2) to the KdV equation

$$u_t + \alpha u u_x + s u_{xxx} = 0.$$

It is well known that both the Burgers' equation and KdV equation are exactly solvable, and have traveling solitary wave solutions as follows, respectively,

$$u(x,t) = \frac{2k}{\alpha} + \frac{2\beta k}{\alpha}\tanh k(x - 2kt),$$

and

$$u(x,t) = \frac{12sk^2}{\alpha}\mathrm{sech}^2 k(x - 4sk^2t).$$

(24) and (25) can be regarded as a composition of a bell solitary wave and a kink solitary wave, which also confirm the qualitative analysis presented in Section 2. Notice that the traveling wave solution given by (24) has the limit zero as $\delta_1 = -\frac{\beta}{5s}x + \frac{6\beta^3}{125s^2}t \to -\infty$ and the limit $\frac{12\beta^2}{25\alpha s}$ as $\delta_1 \to +\infty$, and the traveling wave solution given by (25) has the limit zero as $\delta_2 = -\frac{\beta}{5s}x - \frac{6\beta^3}{125s^2}t \to -\infty$ and the limit $-\frac{12\beta^2}{25\alpha s}$ as $\delta_2 \to +\infty$. All orders of derivatives of u with respect to δ_i tend to zero as $|\delta_i| \to +\infty$ $(i = 1, 2)$. This type of traveling wave solution to the Burgers-KdV equation is the one studied by Bona et al. [29] who showed it is unique, but did not find its functional form. We remark that, as it stands, the traveling wave solution (24) and (25) can not be reduced to either the sech^2 type of traveling wave solution to the KdV equation in the limit $\beta \to 0$ or to the tanh type traveling wave solution to the Burgers' equation in the limit $s \to 0$. If such a reduction needs to be carried out, it is necessary to return to original Burgers-KdV equation and then to proceed to the limits $s \to 0$ or $\beta \to 0$, as a result of which the desired solutions will be obtained. The amplitudes, wave numbers, and frequencies of the traveling wave solutions given by (24) and (25) depend on the coefficients α, β, and s. Since the amplitudes are inversely proportional to α and s, the shock strengthens if either α or s becomes small. If the dissipative effect is weak, (24) and (25) give rise to a weak shock wave provided that α or s are $o(1)$ quantities (see figure 4).

Assume that eq. (3) has the exact solution in the form

$$u(x,y,t) = u(\xi), \quad \xi = hx + ly - wt, \tag{26}$$

where h, l, w are real constants to be determined. Substitution of (26) into eq. (3) yields

$$-whu_{\xi\xi} + \alpha h^2(uu_\xi)_\xi + \beta h^3 u_{\xi\xi\xi} + sh^4 u_{\xi\xi\xi\xi} + \gamma l^2 u_{\xi\xi} = 0.$$

Integrating the above equation twice with respect to ξ, then we have

$$sh^4 u_{\xi\xi} + \beta h^3 u_\xi + \frac{\alpha}{2}h^2u^2 + \gamma l^2 u - whu = R,$$

where R is the second integration constant and the first one is taken to zero.

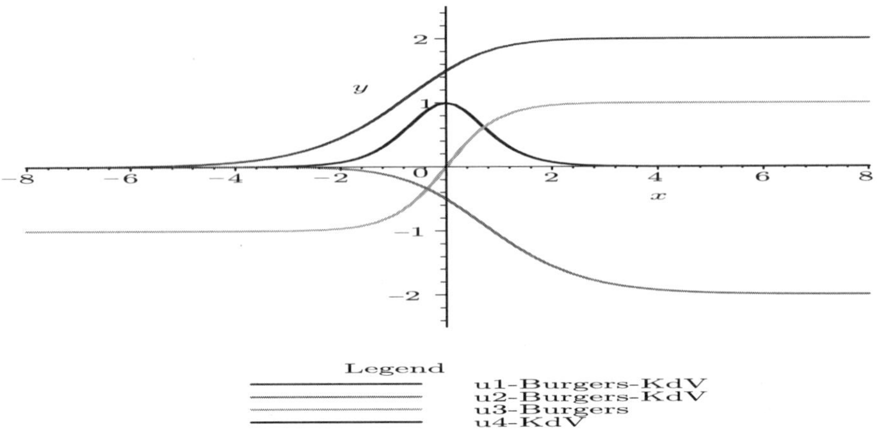

FIGURE 4. Traveling wave solutions for case $\alpha = 2$, $\beta = 5$, $s = 3$. u1-Burgers-KdV is given by (24), u2-Burgers-KdV is given by (25), u3-Burgers is the solution for Burgers' equation and u4-KdV is the solution for the KdV equation.

Following from (20) immediately, we obtain that eq. (3) has the exact solution

(27)

$$u(x,y,t) = \eta \cdot e^{-\frac{2\beta}{5sh}(hx+ly-wt)} \cdot \rho(k_1 e^{-\frac{\beta}{5sh}(hx+ly-wt)} + k_3) + \frac{6\beta^2}{25\alpha s} + \frac{wh-\gamma l^2}{\alpha h^2},$$

where k_3 is an arbitrary constant, and ρ is the Weierstrass elliptic function satisfying (19) with the coefficient in terms of the constants η and k_1.

Particularly, taking $k_1 = 1$ and $\eta = \frac{6r^2}{25a} = -\frac{12\beta^2}{25\alpha s}$ in (27), then we can obtain a bounded traveling wave solution to eq. (3) directly

$$u(x,y,t) = -\frac{12\beta^2}{25\alpha s} \cdot \left[\frac{e^{-\frac{\beta}{5sh}(hx+ly-wt)}}{e^{-\frac{\beta}{5sh}(hx+ly-wt)} + k_3}\right]^2 + \frac{wh-\gamma l^2}{\alpha h^2} + \frac{6\beta^2}{25\alpha s}, \tag{28}$$

where k_3 is an arbitrary constant.

Assume that eq. (4) has the traveling wave solution in the form (21), then eq. (4) becomes (1), where $r = \frac{\alpha_2 v}{\alpha_4 - v^2\alpha_5}$, $a = \frac{\alpha_3}{v^2\alpha_5 - \alpha_4}$, $b = \frac{\alpha_1 - v^2}{v^2\alpha_5 - \alpha_4}$, and d is integration constant. By making use of (20), we obtain that eq. (4) has the exact solution

$$\begin{aligned} u(x,t) &= \eta \cdot e^{\frac{2\alpha_2 v}{5(\alpha_4 - v^2\alpha_5)}(x-vt)} \cdot \rho\Big(k_1 e^{\frac{\alpha_2 v}{5(\alpha_4 - v^2\alpha_5)}(x-vt)} + k_3\Big) \\ &\quad - \frac{\alpha_1 - v^2}{2\alpha_3} + \frac{3\alpha_2^2 v^2}{25\alpha_3(\alpha_4 - v^2\alpha_5)}, \end{aligned} \tag{29}$$

where k_3 is an arbitrary constant, and ρ is the Weierstrass elliptic function satisfying (19) with the coefficient in terms of the constants η and k_1.

Next, consider eq. (5) in the case while $d_3 + \frac{1}{6} = 0$ by means of the method of variation parameters. The arguments for the case of $d_3 + \frac{1}{6} \neq 0$ has been presented in [21]. Assume that eq. (5) has the exact solution of the form

$$\begin{aligned} u(\xi) &= u(x_1, x_2, \cdots, x_n, t) \\ &= \beta \cdot e^{(\sum_{j=1}^n \alpha_j x_j + \gamma t + C_0)} \cdot \\ &\quad q[k_1 \cdot e^{k_2(\sum_{j=1}^n \alpha_j x_j + \gamma t + C_0)} + k_3] + k_4. \end{aligned}$$

where β, γ, k_1, k_2 and k_4 are constants which depend on r_1, r_2 and d_i $(i = 0, 1, 2)$, and $q(x)$ is a function to be determined. Not all real α_j equal zero, and C_0 and k_3 are arbitrary constants.

Using the closely similar arguments as the above, we obtain that while $(d_1 - 1)^2 - 4d_2d_0 > 0$, eq. (5) has the exact solution

$$\begin{aligned} u(x_1, x_2, \cdots, x_n, t) &= \beta \cdot e^{(\sum_{j=1}^n \alpha_j x_j + \gamma t + C_0)} \cdot q[k_1 \cdot e^{\frac{1}{2}(\sum_{j=1}^n \alpha_j x_j + \gamma t + C_0)} \\ &\quad + k_3] + \frac{1 - d_1}{2d_2} \pm \frac{\sqrt{(d_1 - 1)^2 - 4d_2d_0}}{2d_2}, \end{aligned} \tag{30}$$

where k_3 and C_0 are arbitrary constants, β and γ satisfy

$$\begin{cases} \gamma^2(k_2 + 2) + \gamma r_1 - r_2(k_2 + 2) \cdot (\sum_{j=1}^n \alpha_j^2) = 0 \\ \beta\gamma^2 + r_1\beta\gamma - r_2\beta \cdot (\sum_{j=1}^n \alpha_j^2) = 2d_2k_4\beta + d_1\beta - \beta, \end{cases} \tag{31}$$

and q is the Weierstrass elliptic function satisfying

$$\frac{k_1^2}{4} \cdot [\gamma^2 - r_2 \cdot (\sum_{j=1}^n \alpha_j^2)] \cdot q'' = d_2\beta \cdot q^2. \tag{32}$$

Particularly, using the fact that $q(y) = \frac{3k_1^2[\gamma^2 - r_2 \cdot (\sum_{j=1}^n \alpha_j^2)]}{2d_2\beta} \cdot y^{-2}$ is a particular solution of (32), eq. (5) has the exact solution

$$\begin{aligned} u(x_1, x_2, \cdots, x_n, t) &= \frac{3[\gamma^2 - r_2 \cdot (\sum_{j=1}^n \alpha_j^2)]}{2d_2} \cdot \frac{e^{(\sum_{j=1}^n \alpha_j x_j + \gamma t + C_0)}}{(e^{\frac{1}{2}(\sum_{j=1}^n \alpha_j x_j + \gamma t + C_0)} + k_3)^2} \\ &\quad + \frac{1 - d_1}{2d_2} \pm \frac{\sqrt{(d_1 - 1)^2 - 4d_2d_0}}{2d_2}, \end{aligned}$$

where k_3 and C_0 are arbitrary constants, and γ satisfies (31).

Let us end this section by comparing the solutions obtained in the previous references with ours. In [14,24], we studied eq. (2) for the traveling solitary wave solution and obtained (23) by applying the first integral method, which is apparently the particular case of (22). It is notable that the solutions obtained in [8-13] are particular cases of (23) in the case where $k = \pm 1$, i.e., either equivalent to (24) or (25). For example, Xiong used the analytic method to study the Burgers-KdV equation

$$u_t + uu_x - cu_{xx} + \beta u_{xxx} = 0. \tag{33}$$

With the assumption $6c^2 = 25\beta v$, an exact solution to eq. (33) is presented in [8] as

$$u(\xi) = 2v - \frac{3c^2}{25\beta}(1 + \tanh\frac{c\xi}{10\beta})^2. \tag{34}$$

By using the equality $\frac{Ae^t}{1+e^t} = \frac{A}{2}\tanh(\frac{t}{2}) + \frac{A}{2}$, it is easy to see that (34) is in agreement with (24) when $\alpha = 1$, $\beta = -c$ and $s = \beta$.

By means of a direct method and a series method, Jeffrey and Xu et al. investigated the Burgers-KdV equation in [9,10]

$$u_t + 2auu_x + 5bu_{xx} + cu_{xxx} = 0,$$

and found the solutions as follows provided that (2.9a,b) and (2.10a,b) [10, p.561] hold

$$u_1 = \frac{3b^2}{2ac}\left[\operatorname{sech}^2(\frac{\theta}{2}) + 2\tanh(\frac{\theta}{2}) + 2\right], \tag{35}$$

where $\theta = \frac{b}{c}x - \frac{6b^3}{c^2}t + d$, and

$$u_2 = \frac{3b^2}{2ac}\left[\operatorname{sech}^2(\frac{\theta}{2}) - 2\tanh(\frac{\theta}{2}) - 2\right], \tag{36}$$

where $\theta = -\frac{b}{c}x - \frac{6b^3}{c^2}t + d$. Since $\tanh(-x) = -\tanh(x)$, it is easily seen that (35) and (36) are identical with (24) and (25) respectively while $\alpha = 2a$, $\beta = 5b$, and $s = c$. By using partial use of a Painlevé analysis, the same results were presented by Halford and Vlieg-Hulstman in [11].

Wang applied the homogeneous balance method to (2) in [12] when $s = -s$, $\alpha = p$, $\beta = r$. The main result ((3.18) when taking $b = 0$) in [12] appears to coincide with (24) and (25). Notice that there are three errors on p. 286 in [12]: one is that the first term in (3.16) should be $-\frac{6r^3}{125s^2}$, as a result of which the denominator $125s$ in (3.18) should be replaced by $125s^2$.

The solution to eq. (3) derived by Ma [15, p.L19] is identical to (28) only in the case where k_3 is positive. The method used in [15] is not applicable to the case while k_3 is negative. By using the Hopf-Cole transformation and a computer algebra system, Li and Wang obtained a traveling wave solution to (3) in [16] simultaneously, which is equivalent to one obtained in [15]. The same result is obtained by Parkes too [17].

Recently, Fan et al. investigated the following two-dimensional Burgers-KdV equation by making use of an extended tanh-function method with symbolic computation

$$(u_t + uu_x - \alpha u_{xx} + \beta u_{xxx})_x + \gamma u_{yy} = 0, \tag{37}$$

and claimed that a new complex line soliton to eq. (37) is obtained as follows [18, p. 378]

$$u = -(d + c^2\gamma) \pm \frac{6\alpha^2}{25\beta} \tanh \xi + \frac{6\alpha^2}{25\beta} \mathrm{sech}^2 \xi + i\frac{6\alpha^2}{25\beta}(1 \mp \tanh \xi) \cdot \mathrm{sech}\xi, \tag{38}$$

where $\xi = \mp\frac{\alpha}{5\beta}(x + cy + dt)$.

Indeed, (38) is the particular case of (28) while $k_3 = i$. Denote that $\xi_1 = -\frac{\alpha}{5\beta}(x+cy+dt)$ and $\xi_2 = \frac{\alpha}{5\beta}(x+cy+dt)$. Making use of two equalities $\mathrm{sech}\xi = \frac{2}{e^\xi+e^{-\xi}}$ and $\tanh \xi = \frac{e^\xi-e^{-\xi}}{e^\xi+e^{-\xi}}$, we have

$$\begin{aligned} & -(d + c^2\gamma) + \frac{6\alpha^2}{25\beta} \tanh \xi_1 + \frac{6\alpha^2}{25\beta} \mathrm{sech}^2 \xi_1 \\ & +i\frac{6\alpha^2}{25\beta}(1 - \tanh \xi_1) \cdot \mathrm{sech}\xi_1 \\ = & -\frac{12\alpha^2}{25\beta} \cdot e^{2\xi_2} \cdot \frac{1}{(e^{\xi_2} + i)^2} + \frac{6\alpha^2}{25\beta} - (d + c^2\gamma). \end{aligned} \tag{39}$$

It is easy to see that (39) is identical to (28) while $k_3 = i$.

Similarly, making use of the equalities $\sinh y = \frac{1}{2}(e^y - e^{-y})$ and $\mathrm{csch} y = \frac{1}{\sinh}$, it is very easy to verify that the new solution claimed by Elwakil et al. [19, formula (31), p.183] is indeed the particular case of (28) while $k_3 = -1$.

The solutions (29) and (30) to eqs. (4) and (5), respectively, are not shown in [20,21]. One can see that (30) is significantly different from one we recently presented in [21].

4. Conclusion

In summary, in the present paper, first we give a qualitative analysis to a two-dimensional plane autonomous system (8) which is equivalent to the Lienard equation (1). Then, a more general exact solution to the Lienard equation is obtained in the case where $d = \frac{b^2}{4a} - \frac{9r^4}{625a}$ by using the method of variation parameters. Finally, by means of this solution, the traveling wave solutions to the Burgers-KdV equation equation, the combined dissipative double-dispersive equation and the approximate sine-Gordon equation are expressed explicitly, which generalize the previous results.

It is worthwhile to mention that we can also apply this technique to many other nonlinear differential equations. For instance, the generalized Klein-Gordon equation (40) and the generalized Schrödinger equation (41) [30-32]

$$u_{tt} - (u_{xx} + u_{yy}) + \alpha^2 u + g(uu^*)u = 0 \tag{40}$$

$$iu_t + u_{xx} - u_{yy} + \beta g(uu^*)u = 0, \tag{41}$$

where $g(v) = \gamma\sqrt{v} + \mu[\sqrt{v}]^{-1}$. Here we omit the detailed arguments for (40) and (41), since both of them can be reduced to eq. (1) by using the proper transformation [32].

In view of (20) obtained in the case of $d = \frac{b^2}{4a} - \frac{9r^4}{625a}$ and the result presented in [24] in the case of $d = 0$, here, we would like to pose an open question:

Can we apply the variational methods or any other mathematical techniques to obtain the general solution to eq. (1) for an arbitrary given d? It does not seem that the solutions for the cases $d \neq 0$ and $d \neq \frac{b^2}{4a} - \frac{9r^4}{625a}$ have been presented previously. If the answer is yes, the more general solutions to many nonlinear evolution equations can be obtained immediately by making use of them, which must play an important role in theoretical physics.

Acknowledgment: The author would like to thank Professor Goong Chen for his kind help.

References

[1] M. D. Kruskal, A, Ramani and B. Grammaticos B, *Partially Integrable Evolution Equations in Physics*, Kluwer Academaic Publication, Dordrecht, 1989.
[2] K. Y. Guan and G. Gao, The qualitative Theory of the mixed Korteweg-de Vries-Burgers equation, *Sci. Sinica* (Ser. A), **30**, (1987) 64-73.
[3] R. S. Johnson, A nonlinear equation incorporating damping and dispersion, *J. Fluid Mech.* **42**, (1970) 49-60.
[4] L. van Wijngaarden, On the motion of gas bubbles in a perfect fluid, *Ann. Rev. Fluid Mech.* **4**, (1972) 369-373.
[5] G. Gao, A theory of interaction between dissipation and dispersion of turbulence, *Sci. Sinica* (Ser. A), **28**, (1985) 616-627.
[6] S. D. Liu and S. K. Liu, KdV-Burgers equation modelling of turbulence, *Sci. Sinica* (Ser. A), **35**, (1992) 576-586.
[7] H. Grua and P. W. Hu, Unified shock profile in a plasma, *Phys. Fluids*, **10**, (1967) 2596-2602.
[8] S. L. Xiong, An analytic solution of Burgers-KdV equation, *Chinese Sci. Bull.* **34**, (1989) 1158-1162.
[9] A. Jeffrey and M.N.B. Mohamad, Exact solutions to the KdV-Burgers equation, *Wave Motion*, **14**, (1991) 369-375.
[10] A. Jeffrey and S. Xu, Exact solutions to the Korteweg-de Vries-Burgers equation, *Wave Motion*, **11**, (1989) 559-564.

[11] W. D. Halford and M. Vlieg-Hulstman, The Korteweg-de Vries-Burgers' equation: a reconstruction of exact solutions, *Wave Motion*, **14**, (1991) 267-271.

[12] M. L. Wang, Exact solutions for a compound KdV-Burgers equation, *Phys. Lett. A* **213**, (1996) 279-287.

[13] E. J. Parkes and B. R. Duffy, Travelling solitary wave solutions to a compound KdV-Burgers equation, *Phys. Lett. A* **229**, (1997) 217-220.

[14] Z. Feng, The first-integral method to the Burgers-Korteweg-de Vries Equation, *J. Phys. A: (Math. Gen.)* **35**, (2002) 343-349.

[15] W. Ma, An exact solution to two-dimensional Korteweg-de Vries-Burgers equation, *J. Phys. A* (Math. Gen.), **26**, (1993) L17-L20.

[16] Z. B. Li and M. L. Wang, Travelling wave solutions to the two-dimensional KdV-Burgers equation, *J. Phys. A* (Math. Gen.) **26**, (1993) 6027-6031.

[17] E. J. Parkes, Exact solutions to the two-dimensional Korteweg-de Vries-Burgers equation, *J. Phys. A* (Math. Gen.) **27**, (1994) L497-L501.

[18] E. Fan, J. Zhang, B.Y. Hon, A new complex line soliton for the two-dimensional KdV-Burgers equation, *Phys. Lett. A* **291**, (2001) 376-380.

[19] S. A. Elwakil, S. K. El-labany, M. A. Zahran, R. Sabry, Modified extended tanh-function method for solving nonlinear partial differential equations, *Phys. Lett. A* **299**, (2002) 179-188.

[20] A. V. Porubov and M. G. Velarde, Strain kinks in an elastic rod embedded in a viscoelastic medium, *Wave Motion*, **35**, (2002) 189-204.

[21] Z. Feng, Exact solution to an approximate sine-Gordon equation *Phys. Lett. A*, accepted, to appear in 2002.

[22] M. Levi, F. C. Hoppensyeadt and W. L. Miranker, Dynamics of the Josephson junction, *Quart. Appl. Math.* **7**, (1978) 167-198.

[23] Z. Feng, On explicit exact solutions to the compound Burgers-KdV equation *Phys. Lett. A* **293**, (2002) 57-66.

[24] Z. Feng, Exact Solutions to the Burgers-KdV equation, *Dynamics of Continuous, Discrete and Impulsive Systems*, accepted, to appear in 2003.

[25] T. Ding and C. Z. Li, *Ordinary Differential Equations* , Peking University Press, Peking, 1992.

[26] D. F. Lawden, *Elliptic Functions and Applications* , Springer-Verlag, New York, 1989.

[27] K. Chandrasekharan, *Elliptic Functions*, Springer-Verlag, New York, 1985.

[28] N. I. Akhiezer, *Elements of the Theory of Elliptic Functions*, Amer. Math. Soc. Providence, RI. 1990.

[29] J. L. Bona and M. E. Schonbek, Traveling wave solutions to Korteweg-de Vries-Burgers equation, *Proc. Roy. Soc. Edinburgh*, **101** A, (1985) 207-226.

[30] G. B. Whitham, Linear and Nonlinear Waves, Springer-Verlag, New York, 1974.

[31] W. G. Zhang, Analytic solutions to some nonlinear evolution equations, *J. Math. Res. Expo.* **12**, (1992) 421-426.

[32] Z. Feng, W. D. Ji and S. M. Fei, Analytic solutions of a generalized Klein-Gordon equation, *Math. Prac. Theory*, **27**, (1997) 222-226.

DEPARTMENT OF MATHEMATICS, TEXAS A&M UNIVERSITY, COLLEGE STATION, TX 77843

E-mail address: zsfeng@math.tamu.edu